LOVE AND PHYSICS

THE PEIERLSES

LOVE AND PHYSICS

THE PEIERLSES

MIKHAIL SHIFMAN

University of Minnesota, USA

NEW JERSEY · LONDON · SINGAPORE · BEIJING · SHANGHAI · HONG KONG · TAIPEI · CHENNAI · TOKYO

Published by

World Scientific Publishing Co. Pte. Ltd.
5 Toh Tuck Link, Singapore 596224
USA office: 27 Warren Street, Suite 401-402, Hackensack, NJ 07601
UK office: 57 Shelton Street, Covent Garden, London WC2H 9HE

Library of Congress Cataloging-in-Publication Data
Names: Shifman, Mikhail A.
Title: Love and physics : the Peierlses / [edited by] Mikhail Shifman (University of Minnesota, USA).
Other titles: Peierlses
Description: New Jersey : World Scientific, 2019. | Includes bibliographical references and index.
Identifiers: LCCN 2019003718| ISBN 9789813279902 (hardcover : alk. paper) |
 ISBN 9789811201387 (pbk. : alk. paper)
Subjects: LCSH: Peierls, Rudolf E. (Rudolf Ernst), 1907–1995. | Peierls, Eugenia, 1908–1986. |
 Physicists--Great Britain--Biography. | Nuclear physics--History--20th century.
Classification: LCC QC16.P375 L68 2019 | DDC 539.7092--dc23
LC record available at https://lccn.loc.gov/2019003718

British Library Cataloguing-in-Publication Data
A catalogue record for this book is available from the British Library.

Cover design by Polina Tylevich

Copyright © 2019 by World Scientific Publishing Co. Pte. Ltd.

All rights reserved. This book, or parts thereof, may not be reproduced in any form or by any means, electronic or mechanical, including photocopying, recording or any information storage and retrieval system now known or to be invented, without written permission from the publisher.

For photocopying of material in this volume, please pay a copying fee through the Copyright Clearance Center, Inc., 222 Rosewood Drive, Danvers, MA 01923, USA. In this case permission to photocopy is not required from the publisher.

For any available supplementary material, please visit
https://www.worldscientific.com/worldscibooks/10.1142/11266#t=suppl

Contents

1. Foreword — 1
   Rudolf Peierls's last interview *One Man and His Bomb*, conducted by Brian Cathcart, 1995 — 29
   References — 34

Part 1: The Beginning — 37

2. Love and Physics. The Beginning — 39
   References — 47

3. Jazz Band — 49
   References — 60

4. A Romance Taking Shape in Letters — 61
   References — 110

5. Flashback — 111
   References — 140

6. Marriage and Trepidation. March 1931 — 143
   References — 152

7. Flashback 2 — 155
   References — 171

Part 2: War and Peace 173

8. Glimpses 175

    1940: Selected Correspondence 176
    Los Alamos 185
    1961: Genia's Letter to *New Scientist* 199
    1969: Rudolf Peierls' AIP Interview 203
    Eugenia Peierls' Security Questionnaire, Los Alamos, 1944 205
    Memories of Genia 207
    1989–1993: Selected Correspondence 218
        References 222

9. Betrayal 227

        References 253

10. Tragedy of That Generation 255

        References 268

11. On the Other Side of the Iron Curtain 271

        References 302

12. Rudolf Peierls: The Lesson of the Fuchs Case 303

13. The Peierls FBI File 311

Part 3: Sir Rudolf Peierls. His Diary 323

14. Sir Rudolf Peierls by Sabine Lee 325

        References 353

15. Farewell 355

16. Rudolf Peierls' Diary 359

17. Peierls, Heisenberg and Farm Hall Transcripts 429

        References 438

18. Rudolf Peierls and Nuclear Responsibility 439

        References 449

Index 451

List of Photographs and Graphic Material

page 2:	The title page of Genia Kannegieser's student identification and record book, 1929
23:	The British Mission party
28:	The issue of *The London Gazette* in which conferment of Knighthood on Rudolf Peierls was announced
4:	Odessa Opera, 1930s
8:	The first page of a letter written by Rudolf to Genia in Russian
32:	Professor Sabine Lee, 2017
33:	The cover of Volume 2 of *Sir Rudolf Peierls* by Sabine Lee, 2009
40:	Genia Kannegieser, 1930
41:	Luzanovka beach near Odessa. Pauli talking with Frenkel and Tamm, (1930). Photograph by R. Peierls
42:	George Gamow, Abram Ioffe, Rudolf Peierls, Odessa, 1930
43:	Genia's and Rudolf's first journey together, 1930
44:	The Congress participants boarding *Gruzia*. Odessa, 1930
50:	Mokhovaya Street, 26
51:	Eugenia Kannegieser's student identification and record book, Leningrad, 1929
52:	George Gamow, Genia Kannegieser, and Lev Landau. Approximate date: before 1929
58:	Nina and Genia Kannegieser, Matvei Bronshtein, and others, 1931
59:	Lev Landau in 1929
89:	Lev Landau and Dmitri Ivanenko
114:	Isai Mandelshtam, Genia's stepfather. Date unknown
127:	Reply to my inquiry (2017) to the Federal Security Service (former KGB) regarding Isai Mandelshtam's arrest in 1951
136:	Postcard from Genia to her parents exiled to Ufa, 1936
138–140:	The Mandelshtam family tree
144:	On the day of Genia's and Rudolf's wedding, Leningrad, 1931
153:	Genia and Rudolf on skiing vacation, 1931

156: Nina and Genia Kannegieser, circa 1929
160: Pavel Zaltsman. Self-portrait, 1940s
162: Nina Kannegieser in the 1960s. Leningrad
165: Nina Kannegieser and Ronald Peierls, Brookhaven, NY, 1977
175: The Peierls family, 1961
186: On the way to Lamy, NM
187: Aerial view of Los Alamos road on mesa
188: Oppenheimer's parties
189: On one of the streets of Los Alamos
191: Somewhere in New Mexico
193: The Peierls in Los Alamos, 1944
194: The Peierls in Los Alamos, 2
195: Gaby Peierls in Los Alamos
198: Rudolf Peierls in 1946
213: Rudolf Peierls, Gerald Brown, and Victor Weisskopf at Peierls' retirement celebration, Oxford, 1979
223: Paul Dirac, Wolfgang Pauli, and Rudolf Peierls, London, 1953
224: William Penney, Otto Frisch, Rudolf Peierls, and John Cockcroft, 1946
224: Bryce Seligman (DeWitt), Mrs. Blackett, Rudolf Peierls, and Bernard Peters, circa 1947
225: Genia Peierls, circa 1956
226: Rudolf Peierls and C. N. Yang, 1969
229: L. Kochankov, the title page of *Atom Intelligence and KB-11*
244: Klaus Fuchs. From police file. Circa 1950
247: The cover page of the TNA Klaus Fuchs file KV-2/1263
260: Top secret Stalin telegram of July 3, 1937, which triggered the Great Terror
262: The memorial plaque at Butovo-Kommunarka firing range where mass executions took place in 1937–38
263: Children of the executed "enemies of the people"
264: Feuchtwanger and Stalin, Moscow, 1937
268: First page of a 1953 issue of *Les Lettres Françaises*
312: The front page of the Peierls FBI dossier
313: One of the pages of the FBI file
328: Leipzig University, 1929. In the front row are Rudolf Peierls and Werner Heisenberg, with Georg Placzek standing between them
335: The 1937 Copenhagen Conference organized by Niels Bohr. Rudolf Peierls is in the second row
336: Rudolf Peierls in Copenhagen in 1938
342: George Placzek, Jan Blaton, and Rudolf Peierls in Copenhagen in 1947
345: A cartoon in connection with the 50th birthday of Rudolf Peierls
351: Rudolf Peierls at home in Oxford, 1990

- 360: Rudolf Peierls Center for Theoretical Physics, Oxford, 2017
- 362: Rudi and Genia in Hong Kong, 1979
- 363: Rudolf Peierls and Elevter Andronikashvili
- 379: R. Peierls in Coimbra: Award of honorary doctorate, 1988
- 390: Rudolf Peierls' Coimbra outfit
- 395: A rally in Moscow in 1997 under the slogans "Yids and bourgeois out of Russia!"
- 431: Farm Hall, circa 1945

Chapter 1

Foreword

This book is not a monograph by a professional historian. Rather it presents a collection of essays — some of them are written by myself, others by relatives of Genia Peierls (née Eugenia Kannegiser) or her friends. Compared to Western scholars, my opinions will inevitably differ for a number of reasons, which, perhaps, can be viewed as advantages. Firstly, I spent forty years of my life in the communist country, the (currently non-existent) Soviet Union. Living there certainly impacted my perspectives, so that I an able to assess issues which are not quite familiar to westerners. Secondly, Russian is my native language, which allowed me to investigate the youth years of Genia Peierls and the tragic fate of her parents. Thirdly, I am a professional theoretical physicist, rather than a historian of science, which may be viewed simultaneously as a strength and a weakness. I cannot be as systematic as historians usually are. However, because this is not a monograph by a professional historian, I had the freedom to focus on topics which interest me most.

I came to this book from a rather surprising direction. From my early university years I heard the famous "Landau stories." Some of them refer to Landau's Leningrad years. Genia Kannegiser was a prominent figure in these stories. I was always curious about what happened to her after her departure from the USSR in 1931.

My PhD thesis adviser, Professor Boris Ioffe, late in the evenings at the Institute,[1] used to share with me some of his memories about his participation in the Soviet nuclear program. When he was a very young man he carried out some theoretical research on the hydrogen bomb.[2] Needless to say, he knew all the key players: Zeldovich, Sakharov, Ginzburg, Landau, etc. His assessments of various events did

---
[1] Institute of Theoretical and Experimental Physics in Moscow.
[2] See B.L. Ioffe, "A Top Secret Assignment," in *At the Frontier of Particle Physics*, Ed. M. Shifman, (World Scientific, 2001), Vol. 1, page 18.

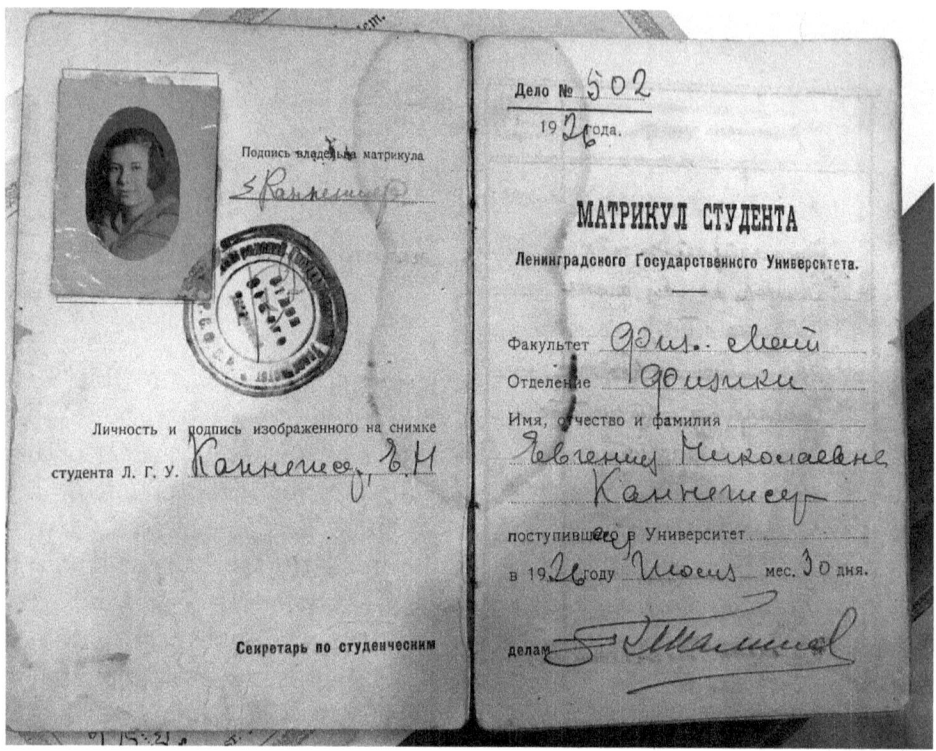

Figure 1.1 Genia Kannegiser's student identification and record book. Personal data are on pages 1 and 2, subsequent pages (e.g. page 51) present her progress in academic disciplines over the years. On the left side of the student book you see Genia's photo, her Russian signature and the university seal. The right-hand side reads:

File 502, 1926, MATRIKUL of the Leningrad University student. Faculty of Physics and Mathematics, Department of Physics. Evgeniya Nikolaevna Kannegiser entered the University in 1926, July 30.

not necessarily coincide with generally accepted views. This ignited my interest in the nuclear weapons programs in other countries, primarily Germany and the United States. As is well known, Rudolf Peierls and Otto Frisch (with their famous Memorandum [1]) were the driving force behind the inception of the Anglo-American program. The outcome of this program has shaped world history ever since. To a large extent it determined the lives of Genia and Rudolf Peierls.

*****

This book might have had more than one beginning, but I will start from a remarkable love story that lasted for over half a century. Love, friendship, and physics intertwined together.

The year of 1930 did not seem particularly alarming.

Of course, in hindsight symptoms of the approaching disasters were evident. The National Socialist Workers Party of Germany — Hitler's party — won 107 seats in Germany's parliament (18.3% of all the votes), making them Germany's second largest party after the Social Democrats (with 24.5% of all votes). In three years Hitler would become the Chancellor of the Third Reich. The crash of the New York Stock Exchange on September 29, 1929, was a harbinger (and a trigger) of the full blown Great Depression which swept across the world two to three years later. In Russia, the relatively liberal years of New Economic Policy[3] came to an end in 1928. In 1930, Stalin ordered accelerated forced collectivization: all peasant households were united in *kolkhozes* against the will of the owners. Millions of the wealthiest peasants (the so called *kulaks*) were sent in exile to Siberia. Just two years later, this would result in *Holodomor*, an artificial famine in Ukraine and other parts of the country in which millions died of starvation.

In physics, 1930 was a quiet year too. The breakthrough discoveries of fundamental quantum laws were earlier in the past while most of the breakthrough applications would shake the world a few years later. Arguably, the most important theoretical development of the year due to Heisenberg and Pauli [2] referred to the nascent quantum field theory. They showed that material particles could be understood as the quanta of various fields, in just the same way that the photon is the quantum of the electromagnetic field. Looking through the summer issues of *Physical Review* I came across a single paper which attracted my attention — that by Gregory Breit [3]. It was devoted to spin interaction of electrons. Nuclear Physics was in its infancy in 1930.

For the main characters of this book — Eugenia (Genia) Kannegiser and Rudolf Peierls — the summer of 1930 was fateful. In August 1930 they met for the first time in Odessa (currently, a Black Sea port in Ukraine), and spent about two weeks together which shaped their destinies.

---

[3]New Economic Policy (NEP), the economic policy in the USSR from 1921 to 1928, representing a temporary retreat from the doctrines of socialism. NEP included the return of most agriculture, retail trade, and small-scale light industry to private ownership while the state retained control of heavy industry, transport, banking, and foreign trade. Money was reintroduced into the economy in 1922 (it had been abolished under War Communism of 1917–1921).

As their daughters Gaby and Joanna told me, it is hard to imagine more contrasting people. Rudolf was quiet and reserved, while Genia was exuberant, straightforward and outgoing. This will become evident from their letters below. In a conversation I had with Gaby Gross (née Peierls) on August 17, 2017, she told me:

> You know, at a certain point in my life a thought crossed my mind that Genia married Rudolf just to get out of the USSR. But then I read the letters and understood that it was a pure and true love, of the type that rarely happens. But it happened.

The conference at which they met brought together almost all actively working Soviet scientists and a number of foreign guests, Wolfgang Pauli and Rudolf Peierls, among others. At that time 23-year old Peierls was Pauli's assistant. Genia had graduated from the Department of Mathematics and Physics of Leningrad University in 1930 and worked at Leningrad Geophysical Laboratory. She was 22 years old and went to the conference at her own expense out of interest. Genia Kannegiser's family had ties with Odessa from time immemorial, as we will see in Chapter 5.

Figure 1.2 Odessa Opera, 1930s.

The above statement that Genia graduated from Leningrad University should be understood in the context of Soviet realities of that time. In 1969, in an interview conducted by Charles Weiner for the American Institute of Physics Oral History project, Rudolf Peierls explained [4]:

> Genia had essentially got what we'd call a bachelor's degree, except she hadn't got a degree because of her background [from a *bourgeois* family]. Her training was that of a first degree in physics, and she then worked as what we'd call a research assistant, a technician. I mean she was taking readings at one time on some nuclear physics and at other periods in some geophysical work.

Apparently, in no time a spark connected the hearts of the young people. They spoke in English as this was the only language they had in common. They talked for two weeks or so, and then Rudolf returned to Zürich and Genia to Leningrad. They made arrangements to meet again in March 1931. That's how the romance encompassing exotic locations and love that overcame obstacles started. Rudolf Peierls' book [5] gives a sketch of this story:

> The boat took us to Batumi, and this was my first experience in a subtropical climate with unfamiliar vegetation, at its best in the beautiful botanical gardens, some way out of town, which we visited. But coming back, we had a little adventure. We were just in time to catch the train that was to take us back into town, but there was a long queue at the ticket office, and it was obvious that we would miss the train. I mentioned that at home we would just get on without a ticket and pay on the train, and our Russian friends said, "Let's try that!" But the conductor on the train was not amused. He called the armed guards that accompanied the train. Two soldiers stood over us with fixed bayonets, and in Batumi they marched us to the station master's office. After difficult negotiations it was ruled that the Soviet citizens in the party would have to pay a fine; the foreign visitors, who could not be expected to know the rules, were let off. Needless to say, the fine was shared.
>
> We went on by train to Tbilisi (Tiflis)[4] with a diminishing

---
[4]Currently, the capital of the Republic of Georgia.

group. It was an old town, beautifully situated among mountains. The Georgian people were not as wild as they looked, with the men's enormous handlebar mustaches and traditional costumes. One heard many tales about the wild temperament and the drinking habits of the Georgians, but our stay was too short to witness any of it.

An even smaller group, but still including my new friend Genia, continued in a hired car to Vladikavkaz beyond the mountains, where there is a railway to the north. Here the group dispersed; Genia and I decided to visit Kislovodsk, a resort in the high mountains. This involved another overnight journey by train. I was anxious to see how the locals travel and wanted to go by the "hard" class, where seats are not reserved. So Genia and I got into a crowded carriage full of wild-looking local types. Genia squeezed into a seat between them. There was no seat left for me, but I spotted some empty space on the wooden luggage rack. I climbed up and tied myself down with my belt so that I would not fall off while asleep. I slept so soundly that Genia had to shake me when we reached our destination. The stay in the mountains passed only too quickly, and when Genia and I parted I left with the feeling that something new and permanent had entered my life.

After Genia and Rudi parted, they started corresponding almost on a daily basis. Six months separated their first encounter in Odessa from their wedding in Leningrad on March 15, 1931. It was another six months or so before the exit visa from the USSR was issued to Genia. Miraculously, 67 letters exchanged between Genia and Rudi in that year survived. Some of them are collected in Chapter 4. After Rudi's return to Zurich in September 1930, he started studying Russian. It was remarkable that in three months, on November 30, he wrote his first letter to Genia in Russian! Moreover, when Peierls went to Leningrad in March 1931 he was asked to deliver a course of lectures on quantum theory of condensed matter in Russian, and he proved to be up to the task.

At the end of April it was time for Rudi to leave Leningrad and join Pauli in Zurich. He returned to Leningrad on August 15 in anticipation of the approval of Genia's petition to leave the USSR. In fact, it was not ready until weeks later. Six weeks of adventures in the Caucasus mountains and misadventures with the exit documents followed. Genia's visa materialized literally in the last minute. The Peierlses were finally able to leave for Zurich in late September.

They were incredibly lucky, Rudi and Genia. If they happened to meet two or three years later under the same circumstances, it is highly unlikely that the above events could have happened. From Stalin's ascent to absolute power till the collapse of the Soviet Union, mere contact between Soviet citizens and foreigners was strictly controlled, let alone emigration.[5] This control was established shortly after the *Bolshevik* coup d'état in 1917.

The rules, introduced on June 1, 1922, for traveling abroad required a special permit from the People's Commissariat of Foreign Affairs (NKID). This made the process highly restrictive, eventually transforming it into a complete ban by 1934–35.

In *Regulations on Entry and Departure from the USSR* published on June 5, 1925, all foreign countries were declared "hostile capitalist environment." In addition to previous constraints, one extra was added — permission from the GPU.[6]

Two examples of the forced isolation of Soviet science (imposed by Stalin) are widely known in the world physics community. They will be mentioned in Chapter 14.

In 1933, George Gamow was invited to give a talk at the Solvay Conference in Brussels. The exit visa was issued to him, but not to his wife Lyubov Vokhmintseva (whom he called Rho). It was the interference of Nikolai Bukharin, a high-ranking government official,[7] that helped to solve the issue. Bukharin helped Gamow get an appointment with Vyacheslav Molotov, a top Communist Party official. Eventually, the exit visa was issued to Lyubov Vokhmintseva too. They never returned to the USSR.

---

[5] It was somewhat relaxed by Mikhail Gorbachev in 1985. The USSR ceased to exist on December 26, 1991.

[6] Abbreviation for the Soviet Political Police in 1922–1923. This name changed many times without changing its essence — the organ of repression and implementation of the regime of brutal dictatorship. In 1917–1922 it was Cheka, in 1923–1934 OGPU, in 1934–1946 NKVD (with the exception of 1943 when the Soviet Political Police was renamed as NKGB), in 1946–1953 MGB, and, finally, in 1954–1991 KGB.

[7] Nikolai Bukharin was a prominent Bolshevik revolutionary, one of the co-authors of the first Soviet Constitution. He was secretly arrested in January 1937 and was expelled from the Communist Party for being a "Trotskyite." In March 1938 he was a defendant in the last public purge trial, falsely accused of counterrevolutionary activities and of espionage, found guilty, and executed. His wife was sent to Gulag.

Just a year later, in the summer of 1934, Pyotr Kapitza left Cambridge for the USSR for a vacation. When he was about to return

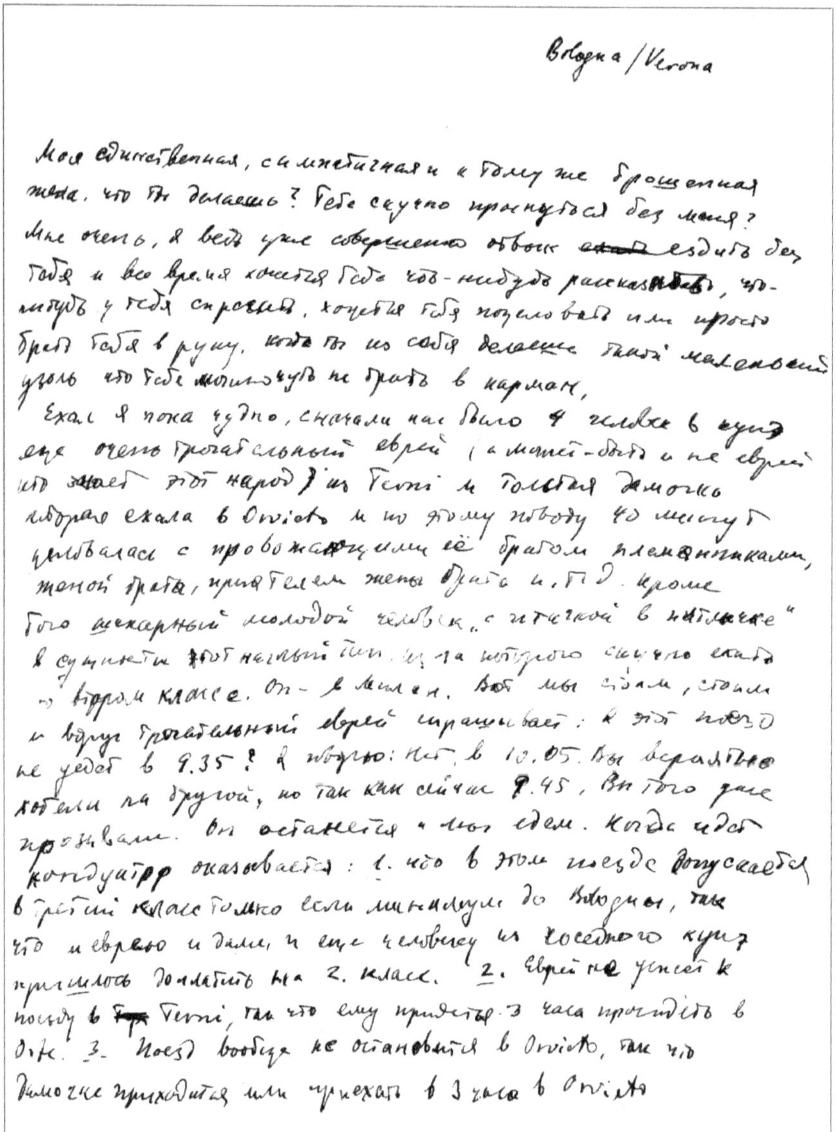

Bologna / Verona

Моя единственная, симпатичная и к тому же брошенная
жена. Что ты делаешь? Тебе скучно проснуться без меня?
Мне очень, я ведь уже совершенно отвык ездить без
тебя и всё время хочется тебе что-нибудь рассказать, что-
нибудь у тебя спросить, хочется тебя поцеловать или просто
брать тебя в руку, когда ты из себя делаешь такой маленький
узел, что тебе можно чуть не брать в карман.
Ехал я пока чудно, сначала нас было 4 человека в купэ:
еще очень трогательный еврей (а может быть и не еврей,
кто знает этот народ) из Terni и толстая дамочка,
которая ехала в Orvieto и по этому поводу 40 минут
целовалась с провожающими её братом, племянниками,
женой брата, приятелем жены брата и т.д. Кроме
того шикарный молодой человек „с птичкой в затылке"
в сущности этот наглый тип, из-за которого скучно ехать
во втором классе. Он — в Милан. Вот мы стоим, стоим
и вдруг трогательный еврей спрашивает: а этот поезд
не уедет в 9.35? Я говорю: Нет, в 10.05. Вы вероятно
хотели на другой, но так как сейчас 9.45, вы того даже
прозевали. Он остается и мы едем. Когда идет
кондуктор, оказывается: 1. что в этом поезде допускается
в третий класс только если минимум до Болоньи, так
что и еврею и даме, и еще человеку из соседнего купэ
пришлось доплатить на 2. класс. 2. Еврей не успеет к
поезду в Terni, так что ему придется 3 часа просидеть в
Orte. 3. Поезд вообще не остановится в Orvieto, так что
дамочке приходится или приехать в 3 часа в Orvieto

Figure 1.3 The first page of one of many letters written by Rudolf to Genia in Russian.

back to Cambridge his passport was revoked and he was not allowed to leave for England.[8]

Finally, the law of June 9, 1935, a logical continuation in the chain of tightening regulations, established the death penalty for illegal border crossings and basically closed it. The "Iron Curtain" had fallen. Relatives of defectors were treated as criminals and sent to Gulag.

Marriages between Soviet citizens and foreigners, which became exceptionally rare after 1934, were formally forbidden by law in 1947.

The early 1930s were exceptional. The first five-year plan was launched by Stalin in 1928. It was designed to industrialize the USSR in the shortest possible time, human cost notwithstanding, making the USSR self-sufficient in heavy industry. It required the involvement of Western companies and a large number of foreign engineers and scientists. In the West it was the time of the Great Depression and the Exodus of Jewish scientists and engineers from Germany and Austria [6]. Many of them came to the USSR to help rebuild the country after the devastating civil war of 1918–22. In the early 1930s foreigners were not just tolerated but, were in fact, welcome. Their appalling end came later, at the peak of the Great Terror.

I will give two examples. Hans Hellmann (1903–1938) is known as the founding father of quantum chemistry. On December 24, 1933, he was dismissed from the University of Hanover as "undesirable" because of his Jewish wife, and chose to emigrate to the Soviet Union. In Moscow, Hellmann assumed the leadership of Karpov Institute's Theoretical Group. On March 9, 1938, he was arrested by the NKVD and accused of spying for Germany. Hellmann was sentenced to death on May 28, 1938 and executed that same day by firing squad.

Fritz Noether (1884–1941), an outstanding German mathematician, Emmy Noether's brother, was not allowed to work in Germany under the Nazi racial laws. He emigrated to the Soviet Union, where he was appointed to a professorship at the University of Tomsk. In November 1937, he was arrested by the NKVD at his home in Tomsk. On October 23, 1938, Professor Noether was found guilty of spying for Germany and committing acts of sabotage. He was sentenced to 25 years of Gulag. However, on September 11, 1941, Fritz Noether was executed in Orel.

*****

---

[8]By the way, this unfortunate event for Kapitza turned out to be a blessing in disguise for Peierls. The money for Kapitza's salary remained unused, and Lord Rutherford managed to establish two new fellowships at the Mond Laboratory, one of which was offered to Rudolf Peierls.

The departure of Rudolf and Genia from Leningrad in September 1931 was the beginning of their "nomadic" existence in Western Europe which lasted five long years. Even now the life of young physicists is not easy. In the 1930s it was harder still. Finding a job was the most difficult problem they faced. Post-doctoral positions, common at present, were non-existent. After PhD, a young physicist had only two choices. A few professors with established reputations who could support assistants offered assistantships to the best. Another option was to apply to a private foundation (say, the Rockefeller Foundation) for a fellowship.

When the Peierlses arrived in Zurich, Rudolf still had a year with Pauli. He spent this year working with enthusiasm on the quantum theory of metals and looking for a new job. In his search he had to compete with a number of bright young men who began working on quantum physics at approximately the same time: the late 1920s — the second quantum generation. To name just a few, let me mention Hans Bethe, Otto Frisch, George Placzek, Wolfgang Panofsky, Felix Bloch, Robert Oppenheimer, Lev Landau, Edward Teller, and Victor Weisskopf. As we know from Max Born's letter to Arnold Sommerfeld, all assistant positions in Germany were filled at that time and no new ones were anticipated in the foreseeable future.

Fortunately, Rudolf won a one-year grant from the Rockefeller Foundation and decided to split it in two parts, between Rome and Cambridge, six months each. Rudolf, who was getting more and more interested in nuclear physics, chose Rome because of Enrico Fermi's group, which at that time was at the cutting edge of nuclear research, both in theory and experiment. Experiencing peculiarities of life under the Mussolini regime and discussing them with Genia probably played a role in the fateful decision Peierls made in 1933, of which I will explain below.

The Peierlses arrived in Rome in October of 1932. While in Rome, Rudolf received a letter from Wilhelm Lenz of Hamburg University offering him an assistantship at Otto Stern's institute, a position that Pauli had held before 1929. Lenz himself was not an outstanding physicist; he was quite famous, however, for his students. Suffice to say, Ernest Ising was one of them.

At first Peierls was very happy with this offer and even communicated to Lenz his tentative agreement. As the end of 1932 was approaching, however, he realized that the ascent of Hitler to power in

Germany was imminent. It was a matter of weeks rather than months. The past experiences of Rudi and Genia suggested to them that settling in Germany, even for a few years, would be too risky.

Thus, Rudolf Peierls turned down Lenz's offer despite a grave job market situation. In April 1933, they moved to Cambridge.

*****

On their way to England, the Peierlses made a stop in Berlin. Hitler was already *Führer und Reichskanzler*, "Aryans are the supreme race" was already the official slogan, and the Dachau concentration camp had commenced operation. Rudolf again tried to convince his father to leave Germany, to no avail.

After Chadwick's discovery of neutrons in 1932, Cambridge became a center of attraction for young nuclear theorists and experimentalists alike. Peierls delved into this new discipline, while still working on his previous research topics: the electron theory of metals and relativistic field theory. The presence of Dirac developing his "hole theory" gave him a boost. The Peierlses rented a tiny furnished house not far from the Cavendish Laboratory, Rudi's workplace-to-be for the next six months. In Cambridge, Rudolf became acquainted with Pyotr Kapitza whose creative imagination he admired. Kapitza had been at Cavendish since 1921 and was a Fellow of the Royal Society since 1929, although he was a Soviet citizen, and the Royal Society Fellowship was open only to the British. Nobody could solve this mystery. Peierls had many physics discussions with Victor Weisskopf who was in Cambridge also on a Rockefeller Fellowship. Genia was expecting a baby. Rudolf recollects [5]: "Genia's very obvious pregnancy and my lack of a job after the end of the summer sent ladies bursting into tears when looking at her."

*****

On August 20, 1933, in Cambridge, Genia gave birth to their daughter. They wanted her name to be easily pronounceable in any language since they had no clue as to where they would eventually settle. Thus, they decided on Gaby. The Rockefeller foundation required from its fellows a summary report of their achievements, so Rudolf sent them a list of physics questions he worked on and added a copy of Gaby's birth certificate in the envelope. Rumor has it that they were not amused.

The sensational news of Gaby's birth spread at the speed of light among the "second quantum generation" — a tightly knitted and relatively small physics community. In Bohr's institute in Copenhagen somebody pinned up an announcement on the bulletin board on the next day.

*****

Six weeks after Gaby's birth the Peierlses moved again, this time to Manchester where an assistant lectureship was advertised by Manchester University. Rudolf applied but was rejected. Lawrence Bragg, Langworthy Professor of Physics at the Victoria University of Manchester and a Nobel laureate in physics (the youngest ever), felt uneasy about this situation and arranged for a two-year grant from the Academic Assistance Council which supported refugees from funds raised by private donations. Although the grant was quite modest, it was better than nothing. Simultaneously they offered Hans Bethe a temporary position to replace someone on leave. The house the Peierlses rented in Manchester was six miles from the University. Hans Bethe settled with them. Not only did they become friends for life, Hans and Rudi became enthusiastic collaborators.

Genia used the book "Feeding and Care of Baby" by Sir Truby King for guidance in raising Gaby. This book was first published in 1913, and was reprinted many times through the 1920s, 1930s and 1940s. Apparently, something went wrong with this guidance. In a short while Genia realized that the instructions in this book did not provide enough milk for the baby. Applying common sense proved to be much more successful.

The Peierls were not used to the English methods for heating (or, rather, not heating) the houses. Rudolf notes [5]: "We had some radiant gas fires, which heated the immediate area around the fireplace, but left the rest of the house damp and icy."

Bethe and Rudi bought second-hand bicycles for their everyday commute of six miles to the University. Given rainy and foggy English weather, this was a good exercise for endurance.

Physics-wise, the time in Manchester was very fruitful in Peierls' career. He completed a few papers on the electron theory of metals, continued his work on Dirac's hole theory and joined the cohort of pioneers of the nascent nuclear physics. In February 1934, he published with Hans Bethe his first paper on nuclear physics. Shortly after, a number of other joint papers on this subject followed.

*****

In March of 1935, four years after Genia's and Rudi's departure to Switzerland, in the beginning of the Great Terror, all her family in Leningrad — her mother, stepfather and sister Nina — were deported to the eastern part of the USSR. The stepfather, Isai Benediktovich Mandelshtam spent a few years in exile, then was arrested and sentenced to Gulag, then another exile, another arrest, and finally the last exile from which neither he nor Genia's mother returned. They died at the end of Stalin's era and never saw their beloved Leningrad again. Genia's sister Nina was exiled too, but later managed to finish her course on biology, became an epidemiologist, and spent many years of her life in Kazakhstan, in the middle of nowhere, localizing epidemics of plague, hemorrhagic fever, and syphilis in rural areas. These diseases, officially exterminated in the Soviet Union, had repeatedly spread during the WWII and post-war period in Kazakhstan. She returned to Leningrad after Stalin's death. Their tragic stories are narrated in Chapters 5 and 7. The contents of these chapters is based on so far unpublished documents or documents that exist only in Russian in obscure sources. Combined together, they shed a new light on the Peierls FBI dossier, as discussed in Chapter 13. It took me quite an effort to assemble this new information.

Genia saw her parents for the last time in late summer of 1934. She had gone to Leningrad to show Gaby to her grandparents. Rudolf who accompanied Genia from Manchester to Leningrad went mountaineering in the Caucasus with Landau and a friend of his, engineer Mikhail Styrikovich.

*****

In the spring of 1935, the Peierlses decided to rent a better house. Just after they finished remodeling, Rudolf received an offer from the Mond laboratory in Cambridge (see footnote 8 on page 9). Although this was not a permanent appointment, Cambridge was *the* center of physics in England, and the salary was almost twice that in Manchester. Rudolf was happy to accept the offer and returned to Cambridge. On September 8, 1935, their son Ronald (Ronnie) was born in Manchester, and in October they moved into a well-built house on the outskirts of Cambridge. Rutherford, who replaced Kapitza at the helm of the Mond Laboratory, welcomed Peierls. He remembered him from his Rockefeller term and recognized Peierls immediately.

The two years that followed were, again, quite productive for Rudolf. In Cambridge, he published his seminal work on the phase transition in the two-dimensional Ising model, a paper which is still cited today. He had many discussions with David Shoenberg, a graduate student who worked on magnetic properties, in particular on the de Haas-van Alphen effect in bismuth. The latter was explained by Peierls during his Rockefeller term in Rome.

David Shoenberg was born in St. Petersburg, Russia, on January 4, 1911, into a Jewish family. His father Isaac Shoenberg was knighted in 1962. David's first language was Russian, but his father encouraged the use of English in their family. It was only after a year (1937–1938) in Moscow that David spoke and read Russian fluently; this proved a valuable asset to him and his colleagues after 1945. The Mond Laboratory was opened in February 1933, and Shoenberg was the first research student to work there.

After successful defense of his PhD thesis in 1935, David Shoenberg was invited by Kapitza to spend a year in Moscow at the Institute for Physical Problems. He arrived in Moscow in September 1937 and left in September 1938. Upon arrival from Moscow he brought to the West the sad news of Landau's arrest and incarceration on April 28, 1938. He was probably the last Western physicist to travel to the USSR until 1956.

In 1936, Rudolf Peierls published a paper on magnetic transition curves in superconductors which grew out of Shoenberg-Peierls discussions.

During the second year in Cambridge, Rudolf Peierls was asked to supervise three students. One of them was Charles Kittel, who later became a renowned solid-state physicist. This was a prelude to Rudolf's Birmingham labors. This was also the end of the nomadic life of the Peierlses.

In September of 1937 Rudolf Peierls was invited to a nuclear physics conference in Moscow. Genia planned to accompany him in the hope of seeing Nina and learning from her more about her parents. The Soviet organizers warned Rudolf, however, that her presence might compromise her relatives and friends. The Great Terror was in full swing, and people were arrested *en masse* for no reason. The participants of the 1937 conference were too preoccupied with what was going on in the USSR to focus on physics. Peierls notes (see page 147 in [5]): "I remember talking with Landau, not in the conference but on that occasion. I have a stronger memory of his worries about how the situation was deteriorating and how unhappy everybody was rather than about physics."[9]

Nevertheless, on his way to Moscow via Stockholm, Helsinki and Leningrad, Rudolf managed to meet Nina Kannegiser and had a brief conversation with her.

*****

In the spring of 1937 Mark Oliphant (who moved from Cambridge to assume the chair of physics at Birmingham University) asked Peierls whether he would be interested in a professorship at Birmingham. Peierls' charge was to establish a Chair of mathematical physics which did not exist yet. Needless to say, the answer was an enthusiastic yes.

Immediately after the interview with three applicants Peierls was informed that the decision was in his favor, and, inspired, he returned to Genia in Cambridge with the good news. The professorship in Birmingham was a permanent appointment! *Per se*, this was great. An extra bonus was the salary that went with it, 2.5 times higher than that in Cambridge.

The Peierlses arrived in Birmingham in October of 1937. At that time Rudolf was the only theoretical physicist at Birmingham University. Who could have anticipated that by the end of the 1940s, a center of theoretical physics — one of the best in Europe — would emerge in Birmingham? This was largely due to the effort of Rudolf Peierls.

The Peierlses bought an old car and learned to drive, Rudolf first, Genia second. Peierls notes in his book [5]: "It is said that the most severe test of the stability of a marriage is whether it can survive the husband teaching the wife to drive (or vice versa)." The Peierlses passed this test with flying colors.

At age 30, Peierls was fairly young for a professor, which once in a while caused confusion. In his book [5] he recollects:

> At a staff-student social, where one wore name labels, I danced with a girl student who looked at my name and asked, "are you a relation of the new professor?"

Peierls divided his time between teaching and administrative responsibilities on the one hand and active research in nuclear physics on the

---

[9]Indeed, Landau was arrested by the NKVD in April of 1938 and spent a year in prison. It was a miracle that he was released in April of 1939.

other. In 1936, Niels Bohr whose interests had gradually shifted from foundations of quantum mechanics to nuclear physics, suggested the so-called compound nucleus model, which served as an impetus for many further developments. Peierls played an increasingly active role in the research that followed.

Between 1937 and 1939 Peierls paid several visits to Bohr whose institute in Copenhagen became a major center of attraction for nuclear theorists. Rudolf Peierls, George Placzek, and Niels Bohr started collaborating on generic nuclear reactions in the continuum spectrum, with important results. Bohr, Peierls, and Placzek drafted and re-drafted their paper many times. Its short version was published in the July 29, 1939, issue of *Nature* in the Letter to Editor section, with the following footnote:

> The details of this and of the other arguments of this note will be published in the *Proceedings of the Copenhagen Academy*.

In this letter in *Nature*, Bohr, Peierls and Placzek focused on the photo-electric effect and photo-disintegration. Some of the more general considerations were briefly mentioned. The detailed paper never appeared. The authors did not manage to finalize a "perfect" text before the outbreak of the war. During the war they were preoccupied with bomb-related work. After the war, Bohr, Peierls, and Placzek discussed the issue and decided that, since their result was already common knowledge, there was no need for publishing the promised detailed version.

*****

On September 1, 1939, Germany attacked Poland. World War II broke out in Europe. In the spring of 1940 Germany invaded Denmark, Norway, Belgium, the Netherlands, and Luxembourg. On June 18, 1940, France surrendered. This left Great Britain alone, face-to-face with the enormous German military machine. The outbreak of war made the Peierlses "enemy aliens," with all ensuing consequences. The war-time restrictions on them were mild, however: basically the only limitation was that they were not allowed to own a car. Food and gas were rationed, but without their car, the gas rationing was not an immediate problem.

In February of 1940 the Peierlses received British passports which freed them from most, but not all restrictions.

The country was preparing itself for air raids. A blackout was in effect and people were urged to carry gas masks at all times. Rudolf was not allowed to join the Civil Defense team but he was accepted to the Auxiliary Fire Service. He acquired a uniform with helmet and axe and was summoned for duty every second night. Genia decided to start working as a nurse. The hospitals were being reinforced in anticipation of air-raid casualties. She was trained as a nurse during her student years at Leningrad University. However, her certificate was not recognized in England, so she took a course with the St. John Ambulance Association and then another one with the British Red Cross (graduating in late December of 1939). She worked as a nurse from 1939 until 1941 and left when she was told she was no longer needed. However, Genia did not want to stay at home and went on to work at a factory in Birmingham as a "marker out and setter off." "She was soon promoted as a forewoman in charge of an assembly shop, and then became a planning engineer. The manager director of the factory intended to make her his special assistant, when the departure of the Peierlses for the US in 1943 interrupted her business career."

Gaby's and Ronnie's school was moved to the countryside, to a country house near Shrewsbury, Attingham Park, and they were boarded there. This was a cautionary measure: the countryside was not supposed to experience as many *Luftwaffe* air raids as industrial cities, such as Birmingham. Shrewsbury is located about an hour's drive to the North-East of Birmingham. The gas ration for the Peierls family was sufficient to allow visits to the children every few weeks.

The situation drastically changed after the evacuation of the British Army from Dunkirk and the fall of France. The general mood became gloomy. The fear of a German invasion was palpable in the air. In the summer of 1940 it seemed that the invasion could occur any day. According to M. Verblovskaya (see page 130), in those days, Genia always carried poison with her, in case the Nazis landed in England.

By the time of the outbreak of war in 1939 Rudolf had lived outside of Germany for about a decade. Both Rudolf and Genia hated Hitler and despised the ideology that came with him. They loved England. Peierls felt obliged to serve the United Kingdom during a time of national emergency. He was not allowed to work on radars (this work was carried out at Birmingham University) because of his German origin. Little did he know that his scientific contribution would have a decisive impact on the UK and US war effort!

*****

In June of 1940 the University of Toronto in Canada generously invited the academic staff of the Universities of Oxford and Birmingham to send their children to Toronto (see the correspondence on pages 176–184). These were the darkest days of the war. The Peierlses accepted this invitation with gratitude. Rudolf writes [5] "We were not afraid of the raids [...] but we did fear a German invasion. If this happened, we, as former German and Russian, both Jewish, would be in danger, and it would be good at least to have the children out of the way. Genia was also eligible to go, but she did not want to: 'I could not bear reading in the papers about the fall of Europe; I wanted to be on the spot'."

*****

In December 1938, over Christmas vacation, physicists Lise Meitner and her nephew Otto Frisch, while trying to explain a puzzling finding made by Otto Hahn and Fritz Strassmann in Berlin,[10] realized that something previously thought impossible was actually happening; that a uranium nucleus had split in two parts.

In the summer of 1939, Otto Frisch came from Copenhagen to Birmingham, to discuss a possible appointment. Frisch understood that the German occupation of Denmark was imminent. And, indeed, on April 9, 1940, the Third Reich invaded Denmark, which meant that Frisch could not return to Copenhagen. He stayed with Birmingham University on a temporary teaching position. At this time Peierls was working on a simplified model of the uranium fission. Peierls hesitated on whether or not to publish his model, being afraid that it could have military implications. So, he decided to consult Frisch. "I see no reasons whatsoever against publication, said Frisch, hadn't Bohr shown that an atomic bomb was not a realistic proposition?"

Shortly after, in February or March of 1940, Frisch dropped by Rudolf's office and asked, "Suppose someone gave you a quantity of pure 235 isotope of uranium — what would happen?" The answer to this simple question gave Peierls and Frisch their place in history.

*****

---

[10]In fact, Lise Meitner was a participant in this discovery. However, shortly before the end of the experiment she had to flee Germany — she was Jewish — and was unfairly crossed out by Hahn from the authors list on the corresponding paper. Lise Meitner was an Austrian citizen. After *Anschluss Österreichs* on March 12, 1938, she became a non-person in Germany.

Since the number of secondary neutrons in U^{235} fission had been measured — albeit approximately — Frisch and Peierls had all the data necessary for their estimate of the critical mass. They inserted the data into the Peierls formula for the critical mass, and found that for U^{235} it was just about 0.5 kg. Previous estimates referring to natural uranium had been in the ballpark of tons! Thus, the answer summarized in the Frisch-Peierls Memorandum, which was ready before the end of March, 1940, ultimately became *the* atomic bomb.

Memorandum: On the Construction of a "Super-bomb" based on a Nuclear Chain Reaction in Uranium

The possible construction of "super-bombs" based on a nuclear chain reaction in uranium has been discussed a great deal and arguments have been brought forward which seemed to exclude this possibility. We wish here to point out and discuss a possibility which seems to have been overlooked in these earlier discussions. [...]

Frisch and Peierls divided their memorandum into two parts. The first was an outline of the implications of their calculations. It included a proposal that the best defense against such a weapon would be to develop one before Germany did so. They anticipated the policies of deterrence which would shape Cold War geopolitics later on. The second part provided a more detailed explanation of the underlying physics. The beginning of the second part is quoted on page 315.

Otto Frisch and Rudolf Peierls passed the Memorandum to M. Oliphant who passed it on to Henry Tizard in his capacity as the chairman of the Committee for the Scientific Survey of Air Warfare (CSSAW). After discussions between John Cockcroft, Mark Oliphant and George Thomson, CSSAW created the MAUD Committee to investigate the issue. As enemy aliens, Peierls and Frisch were initially excluded from its deliberations, but they were later added to its technical subcommittee.

The secret MAUD Committee report issued in June of 1941, concluded that an atomic bomb was feasible. This set in motion a chain of events in the UK, US and Canada which culminated in 1945, forcing Japan to capitulate and thus ending the World War II, the bloodiest war ever in the history of mankind.

*****

Part II of the book presents glimpses of the Peierlses' lives during the war years, in particular, in Los Alamos where Rudolf, as a member of the British Mission, played a crucial role in the Manhattan Project. Not only was he the Head of the T-1 group (implosion dynamics) in the Theoretical Division, but after Sir Chadwick's departure, Peierls became also the Head of the British Mission. The scientific side of his contribution is well covered in the literature. Therefore, I will focus on two aspects: everyday life in Los Alamos and Klaus Fuchs' betrayal. Fuchs was what I call a conscientious spy. He worked not for money or other material benefits but rather for an Idea. The driving force behind his actions was a fanatic belief in Communism as a Radiant Future of all Humankind. For Fuchs, Stalin's Soviet Union was an embodiment of a perfect society, a dream of Paradise on Earth. This obsession penetrated his mind and heart to the extent that he shut his eyes to the realities of the Great Terror, of mass arrests and executions in the USSR. Mad zigzags of Stalin's politics, his pact with Hitler, and his bloody dictatorship did not bother Fuchs.

Today's scholars argue that the "Sovietization" of Eastern Europe, the West Berlin blockade, and the Korean War would not have happened if the atomic bomb technologies were not stolen by Fuchs and passed to Moscow. My thesis adviser, B.L. Ioffe, had the same point of view.

Of course, Fuchs was not alone in this psychosis, which struck a significant (if not dominant) part of left-leaning Western *intelligentsia* in the 1930s–1950s. The moral support they provided Stalin — the tragedy of that generation — helped him to maintain his grip on power for 30 years, although this was not a decisive factor, of course. I try to explore this phenomenon from a modern perspective using documents "from the other side of the Iron Curtain" which became available after the collapse of the USSR. I present my point of view in Chapters 9, 10, and 11.

After Klaus Fuchs was exposed and arrested in 1950, the shadow of suspicion inevitably fell on the Peierlses. Rudolf had recruited Fuchs to work on the British nuclear program and brought him to Los Alamos as a member of the British Mission. As for Genia, rumors circulated in Los Alamos that she had been an officer in the Red Army and a member of the Communist Party. The unsubstantiated suspicion was a strong blow to both since they trusted Fuchs unconditionally, a blow that shook them. After Fuchs' incarceration Rudolf visited him in prison and the next day, on February 4, 1950, Genia wrote a letter to Fuchs (page 242) which, I believe, characterizes her better than a thousand words. Everybody should read it, as well as Fuchs' mind-boggling answer (page 244). Who knows, perhaps, professional psychologists would like to comment on it.

Before the Peierlses were cleared, the FBI and the British Secret Service investigated them for years. In Chapter 13, I discuss the FBI file. It seems strange to me that they could not establish even such essential facts as the date of Rudi's and Genia's marriage and the date of their departure from the USSR. Importantly, they could not rule out Genia's membership in communist parties (three options were considered: Soviet, German, and British). Whether or not she was a communist agent was a central point on which most of FBI's suspicions hung.

*****

In early 1942, Peierls with a few colleagues visited the United States on a reconnaissance mission regarding the prospects of joining the efforts of both countries. This was his first encounter with the New World. He flew to Canada in one of the bombers that crossed the Atlantic to return air crews who had flown American planes to the UK. The atomic bomb depended on the physics of fast neutrons and required the separation of uranium isotopes. This latter task of an enormous conceptual and technical complexity had preoccupied Peierls for a year or so. Today, the separation is performed by virtue of a large number of centrifuges. However, in the discussions on the UK side it was decided that the method of diffusion was more practical in wartime conditions. After visiting a number of major US universities and laboratories Peierls came to the conclusion that "nobody on the American side was giving much thought to the actual weapon." In Chicago Peierls had a conversation with Arthur Compton. It was agreed that priority should be given to the fast neutron work and separation of the uranium isotopes. Compton was a key figure in the Manhattan Project. In 1942, he became head of the Metallurgical Laboratory, with responsibility for producing nuclear reactors to convert uranium into plutonium, finding ways to separate the plutonium from the uranium and to design an atomic bomb. Peierls and Compton agreed that Robert Oppenheimer[11] would be the right person to lead this work.

---
[11]Rudolf Peierls knew Robert Oppenheimer from his Leipzig and Zurich days.

Peierls writes [5], "I do not recall whether he suggested Oppenheimer and asked for my opinion, or whether he asked me if I could suggest a name, but the conclusion was clear. Soon after, Oppenheimer was asked to form a committee to study the fast neutron and weapon design problems."

In addition to his British Mission activities Peierls used "windows" in his travel itinerary to see his relatives and old friends. More importantly, he visited Gaby and Ronnie who had been evacuated to Toronto. Genia was happy to have received a first-hand report on the wellbeing of their children.

*****

By the end of 1942, the American scientists decided that the United States no longer needed British help. However, in August of 1943 a secret summit was held in Quebec City, where Churchill and Roosevelt signed an agreement which resulted in a resumption of the UK-US cooperation. The Americans' progress and expenditures amazed the British. In 1943, the United States had already spent more than $1 billion ($11 billion today), while the United Kingdom had spent about £0.5 million (£22 million today). It was decided to immediately relocate the British Mission including Niels Bohr, Rudolf Peierls, Otto Frisch, Klaus Fuchs, and others to the United States. Rudolf and Genia took the *Andes*, a cruise liner that had been adapted as a troop transport, and arrived in Newport News, Virginia, in mid-December of 1943.

Rudolf was supposed to work with the Kellex Corporation who were charged with design and construction of the isotope separation plant. A room had been reserved for the Peierlses in the Hotel Tuft in the Times Square area. In his book [5] Rudolf notes,

> It was a depressing place: a noisy lobby full of milling crowds, narrow corridors in which one met rather unattractive characters. [...] Altogether, the neighborhood of Times Square was unattractive. The stranger who has seen pictures of modern skyscrapers and elegant mansions and expects to find this style to be typical of the city soon discovers that only a small part corresponds to this image — in 1943 chiefly Fifth Avenue, including Washington Square, Central Park South, and Riverside Drive.

In mid-January of 1944, Genia found a nice apartment on Riverside Drive, a ten-minute walk from Columbia University, and Gaby and Ronnie were collected from Toronto. The Peierlses found that the public schools in Manhattan were terrible, with huge classes presided over by helpless old ladies, and circulating stories about knifings and even shootings. They made up their minds to send Gaby and Ronnie to a private school until the end of the school year. By that time it was decided that Rudolf Peierls was needed in Los Alamos.

The Peierlses arrived in Los Alamos in late May or early June of 1944 (see pages 190–197 and also Genia's Security Questionnaire on page 205 dated July 11, 1944). In order to detonate an atomic bomb, one needs enough uranium or plutonium to ensure that neutrons released by fission will strike another nucleus, thus producing a chain reaction. The fissionable material in a bomb must be organized in spatially separated subcritical pieces, which have to be assembled to reach the critical mass. Plutonium spontaneous decays are much more intense than those in uranium. Assembling a critical mass of plutonium from smaller pieces requires a faster and more powerful process than a "gun process" used in the uranium bomb. Here comes *implosion* meaning that the plutonium pieces had to be surrounded by a sphere of conventional explosives. Upon their detonation the subcritical pieces are imploded compressing the mass to criticality. The implosion device posed great problems because generating a convergent spherical detonation wave is difficult. At first it was not even clear whether such a wave would be stable. The Nagasaki bomb called "Fat Man" was made of plutonium.

At 5:30 a.m. on July 16, 1945, Los Alamos scientists detonated a plutonium bomb at a test site located on the U.S. Air Force base at Alamogordo, New Mexico. Oppenheimer chose the name "Trinity" for the test site, inspired by the poetry of John Donne. The main reason for the test was to verify the reliability of implosion.

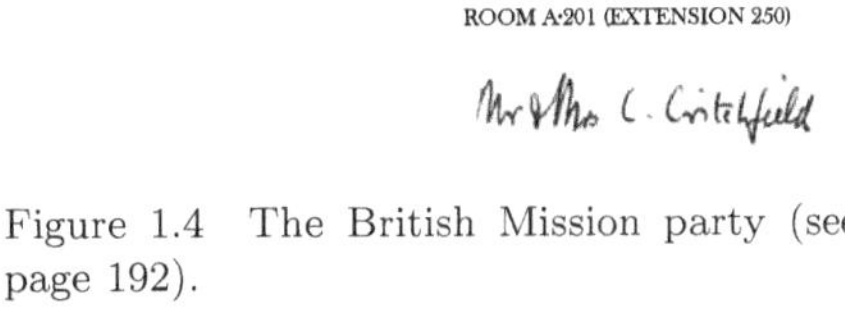

Peierls who spearheaded the implosion research was there to witness the test. He waited with colleagues on a hill 20 miles from "point zero" through the long night. As the final phase of the countdown began,

THE BRITISH MISSION

INVITES YOU TO A PARTY IN CELEBRATION OF THE ATOMIC ERA

THE BIRTH OF THE ATOMIC ERA

FULLER LODGE

SATURDAY, 22ⁿᴰ SEPTEMBER, 1945

DANCING, ENTERTAINMENT
PRECEDED BY SUPPER AT 8 P.M.

R.S.V.P TO MRS. W. F MOON
ROOM A201 (EXTENSION 250)

Mr & Mrs C. Critchfield

Figure 1.4   The British Mission party (see page 192).

he lay on the ground and then, a little before dawn, through a piece of thick dark glass, he watched what happened in the first ever atomic explosion. He was among the first few dozen scientists to see the now familiar and appalling shape of a nuclear mushroom.

*****

In December of 1946 the Peierlses returned to the UK after spending two years in the US (18 months in Los Alamos). Shortly before their departure from the US, Rudolf was offered a chair at Cambridge University, the most prestigious university in the UK. From Liverpool, Genia and the children went to stay with Rudolf's brother near London while Rudolf set out for Cambridge and Birmingham. He promised it would take him no more than seven days to make his decision. To solve his hesitations he drew inspiration from Robinson Crusoe: Peierls took a large sheet of paper, divided it in two parts drawing a line in the middle, and wrote on the left all reasons to stay in Birmingham and on the right all reasons to go to Cambridge. The list on the left was longer than that on the right. So, he returned to Birmingham. I think the decisive factor for Peierls was his desire to create a strong and dynamic group of young bright theorists, the best in the UK and, arguable in Western Europe. To this end his position at Cambridge University would have been more restrictive.

Peierls succeeded where many others failed because he had a clear vision and a determined devotion to his task. He had seemingly infinite enthusiasm for teaching and building up a viable group; and this paid off. To name just a few students or research fellows in the 1950s whom Peierls led to world acclaim, I should mention Freeman Dyson, Gerry Brown, Richard Dalitz, Stanley Mandelstam, John Bell, James Langer, Samuel (later Sir Samuel) Edwards, Nina Byers, and Brian Flowers.

In the post-war years Peierls invested much effort in social questions, in particular (but not exclusively), nuclear arms race. He was deeply concerned with the consequences of possible use of nuclear weapons. To my mind, he was a proponent of the only reasonable strategy in this issue — technical rather than political. In his book [5] Peierls notes: "Once the phenomenon of nuclear fission was discovered it could not be undiscovered." Thus, the issue he addressed was keeping the amount of nuclear weapons minimal but sufficient for nuclear deterrence.[12] As we see today, the idea of deterrence has worked well so far. There

---

[12]Nuclear deterrence is sometimes referred to as "Mutual Assured Destruction."

were no global wars after the end of the WWII. A more severe problem emerged, however, namely, nuclear proliferation. Rogue and intrinsically unstable countries with nuclear capabilities such as Pakistan or North Korea (or, at least, some elements in these countries) are not necessarily deterred by the threat of obliteration.

Assessing the minimal deterrence level of nuclear arms was a large part of Peierls activities as a member of the Atomic Scientists Association (ASA) and later the Pugwash Movement (see Chapter 18). The second important part was working out methods of control and verification. It seems to me that at the initial stages (i.e. in the 1940s and 1950s) Peierls was overly naive, believing that an agreement could have been achieved between the West and Stalin's Soviet Union. This point of view was shared by some of the Western scientists, as seen from the letters by Placzek and Bethe which I present on pages 446 and 448. Later Peierls arrived at more or less the same conclusion, "On the face of it the idea that any arms control agreement should leave the two sides [US and USSR] in approximate balance seems very reasonable, but in reality it creates prohibitive difficulties. [...] Verification to be complete would require a very pervasive inspection of each country, which Soviet Union, with their traditional suspicion of foreign intervention and spies, would never accept. The development of the reconnaissance satellites [could solve] the problem." [5]

Now we live in the age of a network of quite sophisticated reconnaissance satellites, and hence, verification is no longer an issue (or, at least, not such a crucial issue as it was in the past). Did it solve the problem of eliminating the possibility of nuclear apocalypse? Alas... the answer is negative.

As another example of Peierls' social activism I would like to mention his 1948 memorandum entitled "Our Relations with German Scientists."[13] In this memorandum, among other things, Peierls expressed his dissatisfaction with the way denazification was carried out in Germany after the end of the WWII. For years this was a taboo question. Currently, academic papers are available substantiating unfairness of the denazification process (see e.g. [10] and references therein).[14]

*****

Journalists used to ask Peierls the following questions:

---

[13]Sabine Lee, Volume 2, page 115.
[14]A sad story in Göttingen which happened with Fritz Houtermans, Peierls' colleague whom he knew personally from his visits in Russia, is described in [11].

– Did you not know when you worked to help develop the atomic bomb what horrors it would lead to?
– You were motivated initially by the fear of Hitler getting there first. Why did you continue with the work after the defeat of Germany?
– How can anyone live with such recollections?

In doing so, they make a gross mistake which has become quite common at present; they take events of the past out of their historical context pretending that only they — journalists — represent high moral grounds ignored back then. Peierls answered them with dignity. I will briefly summarize his answers and then mention my own argument which is usually forgotten about. Here is a quotation from [5]:

> Nobody could look at the reports and the pictures about Hiroshima and Nagasaki with anything but horror, and nobody would feel any pride at having had a hand in bringing this about. But this was war, and in war, death, suffering, and destruction are unavoidable. The number of casualties in the atomic-bomb raids were no greater than in a big fire raid on Tokyo. It is not the scale of destruction that gave war a new dimension with the introduction of the atom bomb; what was new was the ease with which the weapon can be used, with a single plane creating the kind of destruction that could previously have been accomplished only by a massive military operation. We knew the destructive powers of the bomb, and its radiation effects, and Frisch and I pointed this out in our first memorandum. We knew the ease with which it could be used, and therefore the terrible responsibility it would impose on the political and military leaders who would have to decide whether and when to use it. [...]
>
> Given all the wisdom of hindsight, what should we have done? Should we have refrained from working on the atom bomb from the beginning, or stopped work after the defeat of Germany? The first would have meant an intolerable risk; at the later stage there was still a bloody and cruel war going on, which could have been (and was) shortened by the new weapon. Besides, once [...] the possibility of an atom bomb was understood, it was inevitable that it would be developed sooner or later by someone.

I would like to continue the above argument from the point at which Rudolf Peierls stopped. Let us suppose for a moment that after the capitulation of Germany in May of 1945, President Truman had ordered complete cessation of further work on nuclear arms and disassembly of the already existing equipment. One should not forget that at that time the work on the atomic project in the USSR was in full swing.[15] Under no circumstances would Stalin have stopped it — he was obsessed with the idea of world domination. In four or five years, having the nuclear monopoly, Stalin would not hesitate a minute to start blackmailing the world or — quite probably — use atomic bombs both in Europe and Asia. Then by 1960, the Union of Soviet Socialist Republics would probably extend to Lisbon and London in the West and Tokyo in the East. The Marxist dream of global communism would finally come true. Believe me, this future would not be radiant.

*****

In late May of 1968, a phone rang in the Oxford house of the Peierlses. The call was from Buckingham Palace. A man on the other side of the wire introduced himself, and then Rudolf heard "Professor Peierls, I am authorized to inform you that in the next issue of *The London Gazette*[16] Her Majesty Elizabeth II will announce the following: 'The QUEEN has been graciously pleased on the occasion of Her Majesty's Birthday, to signify her intention of conferring the Honor of Knighthood   Knight Bachelor to Rudolf Ernst PEIERLS, Commander of the Most Excellent Order of the British Empire (CBE), Wykeham Professor of Theoretical Physics, University of Oxford'."

Thus, in recognition of Rudi's contributions to physics, he became Sir Rudolf and Genia, Lady Peierls. Paraphrasing a verse from page 236 of [5] I can write,
> Endowed by the British Lion
> With (official) birthday powers,
> Her Majesty made you Sir Rudolf,
> So Genia becomes Lady Peierls.

*****

---

[15] See Chapter 11.
[16] Supplement №44600, pages 6299–6300, May 31–June 8, 1968,
https://www.thegazette.co.uk/London/issue/44600/supplement/6300
The London Gazette is an official Crown newspaper. The Queen's birthday is April 21, 1926.

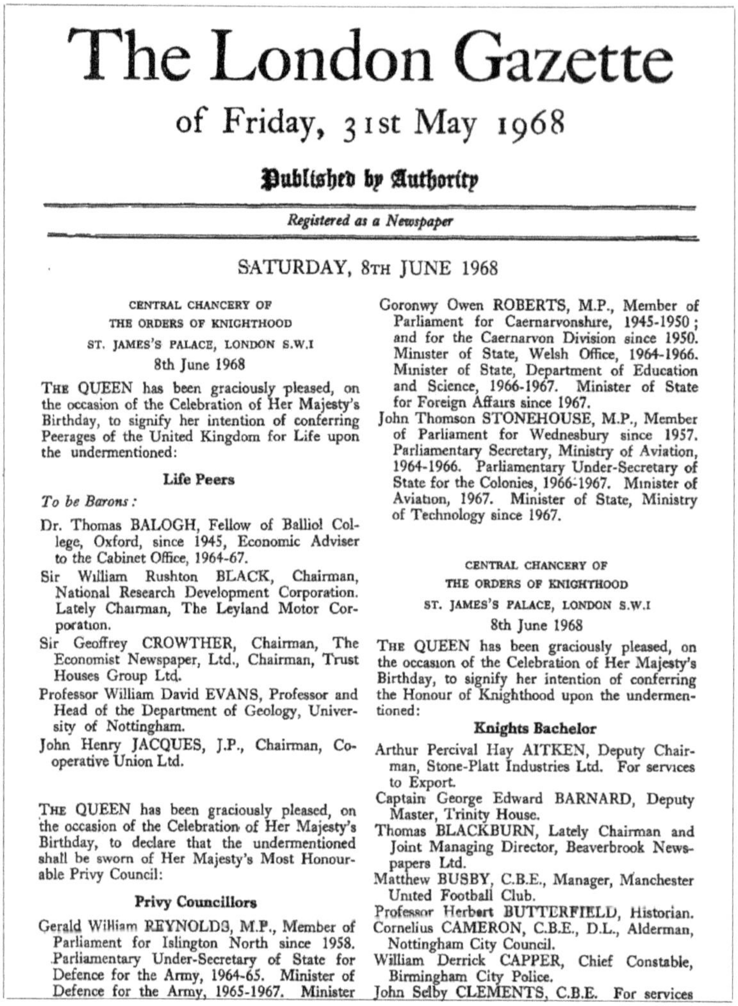

The London Gazette

of Friday, 31st May 1968

Published by Authority

Registered as a Newspaper

SATURDAY, 8TH JUNE 1968

CENTRAL CHANCERY OF
THE ORDERS OF KNIGHTHOOD
ST. JAMES'S PALACE, LONDON S.W.1
8th June 1968

THE QUEEN has been graciously pleased, on the occasion of the Celebration of Her Majesty's Birthday, to signify her intention of conferring Peerages of the United Kingdom for Life upon the undermentioned:

Life Peers

To be Barons:

Dr. Thomas BALOGH, Fellow of Balliol College, Oxford, since 1945, Economic Adviser to the Cabinet Office, 1964-67.

Sir William Rushton BLACK, Chairman, National Research Development Corporation. Lately Chairman, The Leyland Motor Corporation.

Sir Geoffrey CROWTHER, Chairman, The Economist Newspaper, Ltd., Chairman, Trust Houses Group Ltd.

Professor William David EVANS, Professor and Head of the Department of Geology, University of Nottingham.

John Henry JACQUES, J.P., Chairman, Co-operative Union Ltd.

THE QUEEN has been graciously pleased, on the occasion of the Celebration of Her Majesty's Birthday, to declare that the undermentioned shall be sworn of Her Majesty's Most Honourable Privy Council:

Privy Councillors

Gerald William REYNOLDS, M.P., Member of Parliament for Islington North since 1958. Parliamentary Under-Secretary of State for Defence for the Army, 1964-65. Minister of Defence for the Army, 1965-1967. Minister

Goronwy Owen ROBERTS, M.P., Member of Parliament for Caernarvonshire, 1945-1950; and for the Caernarvon Division since 1950. Minister of State, Welsh Office, 1964-1966. Minister of State, Department of Education and Science, 1966-1967. Minister of State for Foreign Affairs since 1967.

John Thomson STONEHOUSE, M.P., Member of Parliament for Wednesbury since 1957. Parliamentary Secretary, Ministry of Aviation, 1964-1966. Parliamentary Under-Secretary of State for the Colonies, 1966-1967. Minister of Aviation, 1967. Minister of State, Ministry of Technology since 1967.

CENTRAL CHANCERY OF
THE ORDERS OF KNIGHTHOOD
ST. JAMES'S PALACE, LONDON S.W.1
8th June 1968

THE QUEEN has been graciously pleased, on the occasion of the Celebration of Her Majesty's Birthday, to signify her intention of conferring the Honour of Knighthood upon the undermentioned:

Knights Bachelor

Arthur Percival Hay AITKEN, Deputy Chairman, Stone-Platt Industries Ltd. For services to Export.

Captain George Edward BARNARD, Deputy Master, Trinity House.

Thomas BLACKBURN, Lately Chairman and Joint Managing Director, Beaverbrook Newspapers Ltd.

Matthew BUSBY, C.B.E., Manager, Manchester United Football Club.

Professor Herbert BUTTERFIELD, Historian.

Cornelius CAMERON, C.B.E., D.L., Alderman, Nottingham City Council.

William Derrick CAPPER, Chief Constable, Birmingham City Police.

John Selby CLEMENTS, C.B.E. For services

Figure 1.5 The issue of *The London Gazette* in which conferment of Knighthood on Rudolf Peierls was announced.

The main characters of this book — Genia and Rudolf Peierls — went through trials and tribulations in the first half of the 20th century. "At my shoulders the wolf-hound age hurls itself...," wrote Osip Mandelshtam,[17] a great Russian poet who perished in a *Gulag* camp. And yet, their life in which love and physics were tightly intertwined was happy. It was full of adventures. They not only witnessed but in fact directly participated in some of the fateful events of the century.

[17] *To the thunderous glory of ages yet to come...*, by Osip Mandelshtam, March 1931, translated by D. Smirnov-Sadovsky. Osip Mandelshtam (1891–1938) was a distant relative of Genia Kannegiser, see page 165.

They had four children and innumerable friends. They knew many important people all over the world. Scientific achievements of Rudolf (in which, I dare say, there was Genia's contribution) were numerous: for example, the hole theory of the electrical conductivity in semiconductors, Brillouin zones (before Brillouin!), the Boltzmann equations for phonons and the *Umklapp* process, Bethe-Peierls photo-disintegration, to name just a few. His obituary in *Physics Today* [7] states that "his many papers on electrons in metals have now passed so deeply into the literature that it is hard to identify his contribution to conductivity in magnetic fields and to the concept of a hole in the theory of electrons in solids." Then it continues to describe Peierls as "a major player in the drama of the eruption of nuclear physics into world affairs..."

Concluding my introduction I would like to quote a few paragraphs from what was probably the last interview of Rudolf Peierls conducted by Brian Cathcart in 1995 [12].

> [...] Now aged 88, [Rudolf Peierls] lives in a retirement home near Oxford. You turn off the road into what was once a private country estate. Up the hill, among the trees, is the home, with a smart new wing overlooking a lawn and the car park below.
>
> Ask for Sir Rudolf and the young nurse shows you the way. "Rudi," she says after knocking, "your visitor." His room is L-shaped, compact, with a bed, a bay window, a desk and plenty of papers. He is sitting at the window with a magnifying glass and a notepad covered in spidery algebra.
>
> He is a small man with thick, whitening hair and a face that is slightly twisted, perhaps by his years of pipe-smoking or the struggle with poor eyesight. Though he settled in Britain more than 60 years ago, his accent and occasionally his grammar are unmistakably German. [...]
>
> Sir Rudolf Peierls is humane, gentle and thoughtful. He is precise in conversation, choosing his words with fastidious care, slow to generalise and even slower to criticize anyone personally. He is generous and scrupulous. Although he has no religion, and although the paradox may seem improbable, this little old man possesses a certain saintliness.

Bird of Passage, Peierls' autobiography which was published in 1985, contains just six lines on Hiroshima and Nagasaki, referring to the elation with which Los Alamos greeted this news that meant the war must soon end, and expressing sorrow at the suffering the bombs must have caused. [...]

All of this Peierls spells out slowly, thoughtfully, looking down at the floor or into the middle distance as he searches for the right words. His language is studiously unemotional; only once does he lapse into something like rhetoric, when he remarks: "As regards the casualties, I don't know whether I would rather be blown up by an atom bomb or perish in a fire raid."

The window bay is warm and bathed in bright sunlight, and we sip coffee as we talk. He came to this home after the death a few years ago of his wife Genia, a Russian-born physicist whom he met at a conference in Odessa in 1930. His eyesight is now so poor that he asks me to add milk to my coffee myself since he does not trust himself to pour. This explains the magnifying glass by his hand and the scanning machine on his desk, which enlarges type for him. His eyes are not his only problem; he also has regular dialysis treatment in Oxford. As a result, he can no longer keep pace with the academic literature in his field but — he taps his page of calculations — "I am looking at some old problems which have loose ends to them." [...]

Today, from his cosy room in Oxfordshire, Peierls takes little part in public affairs. Instead, progressively, he has become a historical resource. The past dozen years have seen the release of a flood of official documents on the origins of the nuclear age in the US, Britain, the Soviet Union and Germany, and in consequence the study of the field has become more meaningful and more fashionable.

Peierls still has that clarity of vision and combines with it an impressive memory for people and events. If you misquote someone, he spots it; if a date is wrong, he will usually put you right. Since he is among the two or three most senior survivors of wartime Los Alamos, his views are constantly sought; his name appears widely in acknowledgements and footnotes; he is still asked to review in the *New York Review of Books*[18] and elsewhere; and he corresponds with historians everywhere. And always with the same care, the same anxiety not to be wrong. Once, after a conversation, he wrote me a hasty letter of correction relating to some minor point in our discussion: "I have since found after discussion with some colleagues..." With good reason, Peierls wants the record straight, for his has been an astonishing life, straddling that line across history.

*****

Now, I would like to discuss particular technical issues regarding this book. First, in the above introduction I barely mentioned Chapters 10, 11, 16, and 17. They are very important to me. In Chapter 16 excerpts from Peierls diary (sometimes he calls it "weekery") are presented. I have selected those which are of general interest. In Chapter 17 the question of Heisenberg's role in the German nuclear program is discussed. Peierls addressed this issue in at least two *New York Review of Books* publications, as is seen in Chapter 16. Two points of view regarding Heisenberg's role exist in the literature. New data (e.g. an unpublished letter from Bohr to Heisenberg) which relatively recently (after Rudolf's death) became available in public domain, have completely changed the discourse. For me personally there is no doubt: Peierls' point of view was correct while his opponents were wrong.

Second, a few words about the names are in order. The surname "Kannegiser" was not uncommon in the Jewish community of the Russian Empire. It originated from the German word "Kannegießer," which means caster, smelter, tinsmith. The first carriers of this name lived in Moravia. As a result of migration through Poland to the east of Europe members of this family appeared in Courland and Semigallia (German *Herzogtum Kurland und Semgallen*), a Duchy that existed in the western part of modern Latvia from 1561 to 1795. Under the third partition of Poland in 1795 Courland was annexed to Russia where it became Courland Province. In Russian Kannegießer was transcribed in a number of ways, typically Каннегисер or Каннегиссер. In English Rudolf Peierls used mostly Kannegiser, see [5], although at least

---
[18]See Chapter 16.

in one of his letters he referred to Kannegiesser, see page 320. I will consistently use the first version. Americans often pronounce Genia's first name as "Djenia." It originates from Russian Женя, a diminutive from Евгения, Eugenia. The first "G" in Genia should be pronounced as French G, as in *Genève*.

Those readers who would like to familiarize themselves with a systematic scientific biography of Rudolf Peierls are referred to an excellent publication by Sabine Lee [8]. A Professor of History at Birmingham University, she carried out a work of enormous, truly gigantic proportions — I would say, a heroic deed — to systematize and publish two volumes of Peierls' *Selected Correspondence*, which amounted to almost 1000 letters on 2000 pages. Each chapter in her publication is supplemented with detailed commentaries and thoughtful biographical essays. I

Figure 1.6   Professor Sabine Lee, 2017.

reproduce one of them in this Collection (Chapter 14). In addition, she published a volume of Bethe-Peierls Correspondence [9]. Professor Sabine Lee conducted numerous interviews with Joanna Hookway, the daughter of Rudolf and Genia Peierls who resides in the United Kingdom. Each and every letter she published was thoroughly analyzed by her and discussed with the family. There is no historian in the world who would know more about Rudolf Peierls' career, his life and family.

I realize that this book is far from perfect. I had postponed its publication for two years in the hope that eventually I would be able to improve its contents, presentation and style, closing the gaps, etc. Alas... It became clear that either it will be published in the form it exists now, or never. I hope that it still has some merit, especially in the parts which were not covered by previous publications.

*****

Figure 1.7 The cover of Sabine Lee's volume.

Acknowledgments

In working on this book, I received invaluable assistance from Gaby Gross and Joanna Hookway (the daughters of Rudolf and Genia Peierls) with whom I consulted on various questions which arose *en route*. They also provided me with some documents and photographs from their private archives. I am extremely grateful to Sabine Lee for conversations that we had at the University of Birmingham in 2017 and correspondence that followed. Professor Lee made available to me her archive used in preparing the fundamental volumes [8] (see, for instance, page 8). Thank you, Gaby, Joanna, and Sabine.

Correspondence with Natalia Alexander, Masha Verblovskaya's daughter, is acknowledged. She kindly provided me with articles and unpublished recollections from her mother's archive. I am grateful to Katya Arnold, João da Providência, and Issachar Unna for writing articles specifically for this book. A number of documents were kindly provided by the following archives: American Institute of Physics, Bodleian Archive (Oxford University), Bradbury Science Museum — Los Alamos National Laboratory (Alan Carr), The National Archives (TNA), Kew, UK, Houtermans Family Archive (Annika Fjelstad, Minneapolis, MN), Cécile DeWitt-Morette Archive, Austin, TX (Chris DeWitt), and Sakharov Center Library (Moscow). I am grateful to the above archives for permission to reproduce a few photographs and letters in my publication. I would like to thank Alina Chertilina for her permission to publish in English a part of her grandmother's story (Chapter 11). I would like to say thank you to Gennady Gorelik who kindly made available to me Genia's private letter of March 9, 1984.

I am grateful to Jeremy Christopher and Alicia Canfield for their kind assistance.

A number of chapters or fragments were translated from Russian. My sincere thanks go to Wladimir von Schlippe, Alice West, James Manteith, and Larry Bogoslaw who performed this difficult task.

Finally, I am deeply indebted to Polina Tylevich for graphic design including the book cover. As usual, it is excellent.

References

[1] "The Frisch-Peierls Memorandum of 1940," in *Selected Scientific Papers of Sir Rudolf Peierls*, Eds. R.H. Dalitz and R. E. Peierls, (Imperial College Press, London, 1997), pp. 277–282.

[2] W. Heisenberg and W. Pauli, Z. Phys., **56**, 1 (1929); *ibid.*, **59**, 168 (1930).

[3] G. Breit, *The Fine Structure of HE as a Test of the Spin Interactions of Two Electrons*, Phys. Rev., **36**, 383 (1930).

[4] Charles Weiner, Session I, II, and III, August 11, 12, and 13, 1969, https://www.aip.org/history-programs/niels-bohr-library/oral-histories/4816-1

[5] R. Peierls, *Bird of Passage*, (Princeton University Press, 1985).

[6] Gordon Fraser, *The Quantum Exodus: Jewish Fugitives, the Atomic Bomb, and the Holocaust*, (Oxford University Press, 2012).

[7] S. Edwards, "Rudolf E. Peierls," *Physics Today*, **49** (2), 71–74 (1996).

[8] Sabine Lee, *Sir Rudolf Peierls: Selected Private and Scientific Correspondence*, (World Scientific, 2007 and 2009), Volumes 1 and 2. In what follows to be referred as *Sabine Lee*.
[9] Sabine Lee, *Bethe-Peierls Correspondence*, (World Scientific, 2007).
[10] G. Rammer, "Allied Control of Physics and the Collegial Self-Denazification of Physicists," in *Physics and Politics*, Eds. H. Trischler and M. Walker, (Franz Steiner Verlag, Stuttgart, 2010).
[11] *Physics in a Mad World*, Ed. M. Shifman (World Scientific, Singapore, 2016).
[12] Brian Cathcart, *One Man and His Bomb*,
https://www.independent.co.uk/arts-entertainment/one-man-and-his-bomb-1590614.html

PART 1

THE BEGINNING

Chapter 2

Love and Physics. The Beginning

In August 19–24, 1930, the All-Union Congress of Physicists was held in Odessa. In fact, it was a rebranded Seventh Congress of Russian Physicists, organized by the Russian Association of Physicists. In attendance were over 800 delegates, with two hundred talks covering all branches of physics. Foreign guests were also invited [1; 2], from the venerable — Arnold Sommerfeld, Walther Bothe, Wolfgang Pauli, Richard von Mises, Carl Ramsauer and Franz Simon,[1] to the young — Rudolf Peierls and Fritz Houtermans.[2] Peierls was Pauli's assistant.

For the city of Odessa this Congress was a great event. Plenary sessions were held in the building of the City Council, and the opening was broadcast by the local radio...

*****

In a large, elegantly decorated hall, above the Presidium's table hangs red fabric, lettered with a proclamation in many languages: "Physicists of the world, unite! In the name of a bright future for all mankind!"

Pauli and his assistant Rudolf Peierls arrived in Odessa on August 18th, and Pauli and Tamm[3] were to give talks on the 20th. Peierls found the conference "lively and interesting." There he met Yakov Frenkel[4] and Igor Tamm, "one of the most charming personalities in physics. He had an agile mind, and an equally agile body, and the first impression he gave was of never standing still."

---

[1] Franz Simon (1893–1956) studied physics in Munich (1912–14) and Berlin (from 1919) and completed his PhD under Nernst in 1921. He continued research in Berlin. He was appointed to the chair of Physical Chemistry in Breslau in 1931. In 1933 he accepted an invitation by F. A. Lindemann to work at the Clarendon Laboratory in Oxford where he stayed until his death.

[2] For Houtermans' adventures and misadventures in the USSR see [1; 2].

[3] Igor Tamm (1895–1971) was a Soviet physicist who received the 1958 Nobel Prize in Physics for the interpretation of the Cherenkov effect.

[4] Yakov Frenkel (1894–1952) was a Soviet physicist renowned for his work in condensed matter theory.

Rudolf Peierls was taking pictures of everything he saw around. These snapshots became part of history.

*****

The city authorities took good care of the participants of the Congress, providing them with the best hotels. Delegates could also travel free on trams. All sorts of entertaiment were organized: tickets to theaters, cinemas, excursions, etc. However, the most popular entertainment in the free hours between the morning and evening sessions was the famous Odessa beach in Luzanovka. Rudolf Peierls, who visited the Soviet Union in June 1985, delivering a talk at Leningrad Physical-Technical Institute showed several 55-year-old pictures.

Figure 2.1   Genia, circa 1930.

Judging from the expressions on their faces, Pauli, Frenkel, Tamm and Simon, captured in bathing suits, continued scientific debate even on the beach. The Congress organizers arranged a boat trip on the steamer "Gruzia" (Georgia) to Batumi, Soviet Georgia, for the participants. In his recollections Peierls wrote: "After the conference, all participants were taken by boat across the Black Sea,... and I was sharing a cabin with Pauli."

Many young people gathered in Odessa. Among them were George Gamow, Evgeniya (Eugenia, Genia) Kannegiser and some other members of the *Jazz Band* which was in fact, neither jazz nor band — rather, a club of aspiring, inquisitive, and cheerful physicists or future physicists and their friends, in their early 20s, of which I will write later. This club also became part of history.

After the Congress Genia wrote a poem which remained in the memories of the Jazz Band:

> Don't seize stars from the heavens,
> Don't go searching far and wide –
> After all, physics conventions
> Are a marketplace for brides!

Among her Leningrad friends, Genia was known as a prolific poet!

Figure 2.2 Luzanovka beach near Odessa. Pauli (center) talking with Yakov Frenkel and Igor Tamm, 1930. Photograph by R. Peierls.

*****

I have already mentioned that the Peierls were an unusual couple. The circumstances of their life in those troubled times were such that, in a sense, they had their honeymoon way before the wedding rather than after. The chronology and geography of their brief honeymoon journey can be exactly established. Peierls and Pauli arrived in Odessa on Monday, August 18, 1930. The Congress opening was the next day, on Tuesday, August 19. Presumably, that was the day when Genia and Rudi met for the first time. The Congress lasted till Sunday, August 24. It is likely that "Gruzia" sailed for Batumi on Monday, August 25. She made a brief stop in Crimea, and reached Batumi on 27th. We know for sure that Rudolf Peierls and Wolfgang Pauli were present as witnesses at the wedding ceremony of Charlotte Riefenstahl and Fritz Houtermans on August 28, at Batumi City Hall. Then, as Rudi recollects (see page 5), they went to Vladikavkaz for a couple of days from where they took a train to Kislovodsk. Genia and Rudi parted in Kislovodsk on September 3 or 4, 1930. Genia continued her vacation in

Figure 2.3 George Gamow, Abram Ioffe, Rudolf Peierls, and others conversing outside of the Odessa All-Union Physics Conference. Courtesy of Emilio Segré Visual Archives, Frenkel collection.

Gagry[5] (see her letter to Rudolf of September 13, on page 45), while Rudolf took a train to Kharkov where he presumably spent just one day visiting UPTI.[6] He arrived in Leningrad not later than Sunday, September 7, and paid a visit to Genia's family, after which he wrote a letter to Genia and mailed it to Gagry. This letter was received by Genia on September 12.

---

[5] Gagry is a resort city which currently belongs to Abkhazia, a rebel province of Georgia.
[6] Ukrainian Physical Technical Institute, currently known as Kharkov Physical Technical Institute.

I compiled a map of the Peierls' first journey together...

Figure 2.4   From Odessa to Kislovodsk and beyond.

*****

From Leningrad Rudolf Peierls returned to Zürich, Switzerland. Genia returned to Leningrad. For a year they wrote letters to each other. They got married during the second visit of Rudolf to Leningrad (at that time it was still possible). Half a year later Genia finally obtained an exit visa and left the USSR for good. She joined her husband in Zürich. Eventually they settled in England.

The wedding ceremony was held on March 15, 1931, in Leningrad. It should be added that Rudolf's father was not enthusiastic about it, to put it mildly. In one of his letters (see page 149) he hints that Rudolf could have easily found an appropriate girl in Germany, from a reputable family.[7]

---

[7] Despite the fact that Heinrich Peierls, Rudi's father, had disapproved of his marriage, they were later able to establish decent relations between themselves. As Sabine Lee puts it, "they could talk to each other and hear what the other side has to say." Before 1938, Heinrich Peierls also provided some financial support to the young family, through wire transfers first to Switzerland and then to England, within the limits imposed by the Nazi government of Germany.

Figure 2.5 The Congress participants boarding *Gruzia*. Odessa, 1930. From the Houtermans Family Archive. Courtesy of Annika Fjelstad.

Yakov Frenkel who spent a year (1930–31) in the US wrote to his wife on April 1, 1931 [3]:

> I was overwhelmed by the news of the marriage of Rudolf Peierls and Genia Kannegiser. Now I understand. So that's why he so willingly agreed to come to Russia and studied Russian so hard!

According to Rudolf Peierls, it was Frenkel who introduced him to Genia in Odessa, and it was Frenkel who organized Peierls' trips to Kharkov, Moscow and Leningrad and invited Peierls to give lectures at Leningrad Physical Technical Institute in the spring of 1931.

***

Genia Kannegiser to Rudolf Peierls[8]

Gagry
September 13, 1930

Rudi,

Dearest, it is the second edition, the first I wrote this morning on the seaside and on cliff — a lot of foam and all was covered with a big wave, it was very jolly, but it was so long, long letter.

Darling, I am afraid that this will be also very long, and I beg your pardon for your hard labour. It will be so difficult for you to read and to understand something, but Rudi I cannot do anything else. I have no post paper, no dictionary and even no ink and pen and now it is the evening. I do not know what o'clock it is, because I live without watch. But I think that something like eleven, and the moon is yet under mountain and I have so bad lamp. So, darling, once more I beg your pardon. I received yesterday your letter, and I thank you for it very seriously, very very seriously, because I believe now that I was not for you only "adventure" — one of many other "girls" that you know that you will know. Certainly, I thought that before also, because if I have not, I should never give you my hands and my lips, they are too "clean" for "adventure," but I was not sure and that way very sad, some minutes. Now all is all right. I am afraid that in such long discussions you did not understand my meaning of "serious." I will try to explain it once more. "Serious" is not so awfully how you think it my darling.

Serious — it is when the person is "dear" for you, when you love not for your own pleasure or happiness, not for yourself, but for other, when you love not because you have not something other to do, but but because you cannot to love not, when your love is not an adventure, not *pour passe le temps* but an happiness, a great and deep thing, a "serious" thing and not only a play. But serious is only first step to marriage. Always — is a fiction, but sometime, now — is "always" — when you want to say "moment — stop." Certainly, one cannot know what will be tomorrow, but you can want such and such future, and your want to love me always is the greatest thing I can ask you. Darling

---

[8]Sabine Lee, *Sir Rudolf Peierls: Selected Private And Scientific Correspondence*, (World Scientific, Singapore, 2007); Volume 1, page 112. In what follows referred to as Sabine Lee.

(No attempts have been made to alter the oddities of phraseology which can be traced back to Russian grammar. The writings on a number of letters have faded, and some passages are illegible. -SL).

you write about my meaning of love — marriage and for you it is not so? Certainly, marriage is a logical development of love. If I love a man it is necessary for me to be together with him, to know all about his life and work to show him all that interests me, to see him when I want and so on and so on. It would be for me quite insupportable to eat when he is hungry to have something that he want and has not etc., etc. and this is marriage — when I cannot live without a man I marry him. I know you only two weeks, but I love you and it is so awfully that you are not here.

I have such a lot to tell you, I want to know about you a thousand things. Six months is very long, but perhaps with the letters it would be better will you, shall we write to each other very often (twice a week or something in this kind) and about all our life? Work, sport, books theatre, cinema, dances, new people, etc. You know now my people, my town, my family, and my home. I know about you less, certainly, but a little I know, so it would be not so difficult to write and we shall live a little together. So, darling, answer this letter, and write all that happens with you in Berlin, especially how goes the thing with Simon's institute. I shall relate you about Gagry in next letter, because that's too long. You know the letter from Leningrad with dictionary would be better I believe. It is awfully late now, and I must go up to morrow very early and go arrange my ticket to Leningrad. Good night, my dearest boy, I wanted to write you a lot of "philosophical" things more, but it is so tedious — the philosophy. Yes?[9] And now is so beautiful: the sea under the moon and big big stars fall down into the sea. Good night!

Continuation. September 14, 1930

You know that I cannot stop my letter, if it remains a bit of clean paper. I think that this side is quite well for my pencil and "beautiful" handwriting. So I go on and you must suffer an half an hour more. You are not angry? Write to me your preferable format and which I will try not to write so time some letters well? Now I am after my dinner which I get not without difficulties, such a crowd in restaurant! But it was splendid and I am in very "lyrical" state.

Darling, here is splendid this morning I swim very fare, so fare that I cannot see the people on the beach. Great?! Certainly I was not alone,

---

[9] In Russian, if one wants to convert a statement into a question, one just adds "yes?" at the end of the statement. This is similar to English "isn't it?" or "doesn't it?".

because I am [illegible] a little with me were two very good swimmer (two engineer, one of Moscow and other of Kharkov). But they were not necessary at all, and I swim, swim, swim in blue blue sea under the sun, and I think that I can go till Constantinopol.

Fabelhaft!![10] But I was tired in the return! My heart is not well at all. You know in the mountain near my home lives bears, jackals and wolves, here is a cave (I forgot the word, you know this thing in which we were sitting in Kislovodsk when it rains so hard). In this cave in the 6th century lives a saint-holy. Now I was there, very interesting. In five minutes I will go to see underground river(!) and tomorrow in he mountains to see the waterfall! It will be very interesting, even [kilometers] from my home.

Rudi, you can be jealous (are you jealous? I am not at all.) I am quite charmed with one Turkish merchant named [illegible]. He sells the fruit on the Basar of Gagry. He is as heavy as two Pauli, and so splendidly lazy! I think that he is charmed also, he gives me best peaches and laugh when I come in the shop. So if I will be in some harem, don't be afraid. Oh, the last leaf. Rudi, good bye! Till the March! Write the answer in Leningrad. I will be there. Have you received my first letter in Leningrad? Darling, I love you very very strong and am very happy. I am an optimist. It would be yet better that it was. Good bye.

Genia

References

[1] *Physics in a Mad World*, Ed. M. Shifman, (World Scientific, Singapore, 2016).
[2] M. Shifman, *Standing Together in Troubled Times*, (World Scientific, Singapore, 2017).
[3] Victor Frenkel, *Yakov Ilyich Frenkel*, (Moscow, Nauka, 1966), in Russian.

---

[10] Fabulous, magnificent (German).

Chapter 3

Jazz Band

In 1927 Lev Landau graduated from Leningrad State University (LU) and was admitted to the Graduate School of Leningrad Physical Technical Institute (LPTI). At LU he found himself in the company of other creative young physicists: George Gamow, Dmitri Ivanenko, and somewhat later, Matvei Bronshtein. Landau was the youngest, at that time he was only 19. They and Genia Kannegiser, a physics student at LU since 1926, formed the core of what became known as the "Jazz Band," a group glorious in the history of Russian theoretical physics. Gamow, Ivanenko, Landau and Genia were the heart of the "Band." At different times Jazz Band attracted other physics students who were involved in some of its activities, for instance, Irina Sokolskaya, who had a talent for drawing, and Andrei Anselm who disappeared from the "Band" in 1928 before its decline. Here is a paragraph from his 1972 memoir [1]:

> "Attractive for us was the house on Mokhovaya — a house with a flat roof in which the family of Isai Benediktovich Mandelshtam lived, his stepdaughters were Nina and Genia. In summer and spring we had a lot of fun on this flat roof. Isai Benediktovich Mandelshtam was an electrical engineer, but was a man with humanitarian likings. He did excellent translations from French and German: Balzac, Kellermann, and others. He was a charming man. We all owe him a lot, including Landau.
>
> Landau in my memory is almost a boy with awfully spiritual face, indecently young. I remember only one case (in my whole life!) when he asked something about physics, or did not understand something (seemingly he did not understand something from Lorentz, but asked Gamow). Gamow was no less gifted than Landau, but in the

technical sense — grasp, understand, do — Landau was probably the first in the world. Genia Kannegiser remembers that once she had to make a report but could not understand the appropriate material. She ran to Dau and told him: "I do not understand anything, in my opinion, this is nonsense." Dau had a quick look and answered: "Wait, this is Gibbs. Gibbs Statistics." In half an hour he explained Gibbs statistics in its entirety to Genia. This was around 1928.

Back then our studies were organized in a peculiar way. Exams were handed over essentially all the year round, in any order and sequence. Nobody cared about the right order, you could first take some disciplines from the graduate year, and then from the freshmen year. Just approach the professor and make arrangements with him.

Figure 3.1 The house on Mokhovaya 26 in St. Petersburg, in which Isai Mandelshtam lived with his family. It was built in 1913 by the architect S. Ginger. The house still exists, and even the name of the street is still the same.

The key members of the Jazz Band had their special nicknames. Gamow was Johnny, Ivanenko, Dimus, and Bronshtein the Abbot.

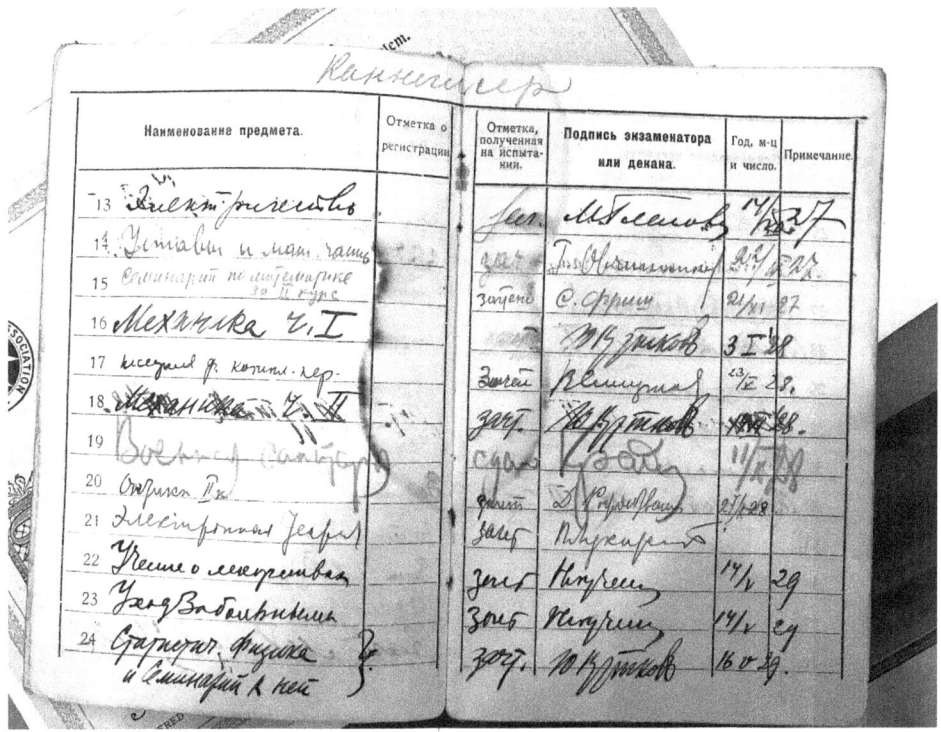

Figure 3.2 Eugenia Kannegiser's student identification and record book illustrating a chaotic examination order (all of them were graded on the pass/fail basis). For instance, "Electricity" was passed on November 14, 1927, while "Mechanics I" on May 3, 1928. It also shows that Eugenia Kannegiser passed examinations on "Sanitary Measures during Wartime" (October 11, 1928), "Theory of Medications" (May 14, 1929), and "Nursing" (May 14, 1929).

Landau was Dau. Once, Ivanenko told Landau that his family name could have been written in French as *l'âne* Dau, the first part meaning a donkey or a fool. Since then Landau demanded to be called Dau, rather than by his full name. Johnny, Dimus and Dau were also called the "three Musketeers." Ill-wishers (of whom there were plenty) by adding just one letter to the name of the group made it sound terrible in Russian. They would pronounce it as "Jazz-Band**a**" which in Russian means Jazz Gang.

It was Genia who brought Bronshtein into the Jazz Band. In her letter to G. Gorelik (March 9, 1984) she writes [2]:

> I'll do my best to describe everything I remember of Matvei Bronshtein. I first met him in early spring of 1927.

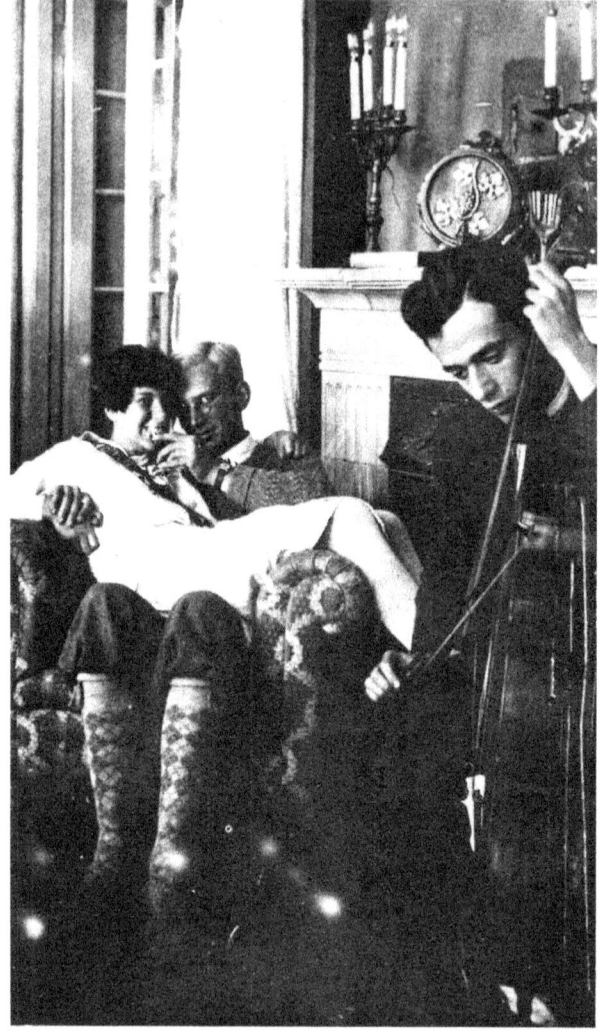

Figure 3.3 In a private room at a restaurant: George Gamow posing as the "guest," Genia Kannegiser as the "hostess" and Lev Landau as the "hired musician." From [3]. Approximate date: 1926–27. (In 1928 the so-called New Economic Policy (NEP) was folded, and all private enterprises, including restaurants, were shut down.)

There were puddles everywhere, sparrows were chirping in a warm wind. On emerging from a laboratory on the Vasilievsky Island I quoted a line from Gumilev[1] to a young man who happened to pass by. He was not tall, wore large glasses and had a head of fine nicely cropped

---

[1] Nikolai Gumilyov (1886–1921) was an influential Russian poet of the beginning of the 20th century, a traveler, and military officer. He was arrested and executed by the Cheka, in 1921.

dark hair. Unexpectedly, he responded with a longer quotation from the same poem. I was delighted; we walked side by side to the University quoting our favorite poems. To my amazement Matvei recited *The Blue Star* by Gumilev which I had no chance of reading.

At the University I rushed to Dimus and Johnny to tell to them about my new friend who knew all our favorite poems by heart and could recite *The Blue Star*.

That was how Matvei joined the Jazz-band. We were putting out the *Physikalische Dummheiten* and read it at university seminars. In general, we sharpened our sense of humor on our teachers and at their expense. I should say that by that time Johnny, Dimus and Dau had left all others far behind where physics was concerned. They explained to us all the new and amazing advances in quantum mechanics. Being a capable mathematician the Abbot (Matvei Bronstein) was able to catch up with them quickly.

I can visualize Matvei with his specs slipping down his nose. He was an exceptionally "civilized" and considerate person (a rare quality in a still very young man); he not only read a lot but also had the habit of thinking a lot. He accepted no compromises when his friends "misbehaved."

I cannot say who gave him his nickname, which suited him perfectly: he was benign in his skepticism, appreciated humor and was endowed with universal "understanding." He was exceptionally gifted.

As seen from the above letter, from time to time, the Jazz Band released typewritten issues of a jocular journal *Physikalische Dummheiten* (Physical Absurdities or Physical Gibberish) which was circulated among students of LU until the end of 1928. In some articles the Jazz Band members cheekily sneered at their teachers. Sometimes, at themselves. Was it friendly joking or sarcasm? The line remains blurred...

They also had a seminar of their own at which they heatedly discussed events in physics and culture:[2] ballet and poetry, Freudism and

---

[2] I guess they were smart enough to avoid politics, a dangerous topic even in the relatively "liberal" years of the late 1920s. See page 315.

the relations of the sexes were subjected to scathing and dissecting theoretical analysis [2].

Around 1926 Genia summarized the theoretical situation in physics in poetic form on the pages of *Physikalische Dummheiten*:

> Though Heisenberg's theories weren't a triumph,
> And Born's hard-earned laurels seemed withered a bit,
> Yet Pauli's principle, Bose's statistics
> Have long won your hearts, and your minds, and your wit.

> The Nature is still enigmatic and hidden,
> You still do not know all the secrets of light,
> The nuclear laws still remain undiscovered,
> And you are now trying to conquer the blight.

> When reading your cleverest papers in *Zeitschrift*,
> With all our problems becoming more vague,
> The only delight is the thought that Bothe[3]
> Will give you all, guys, a proper spank.

The Jazz Band thrived on jokes of all sorts. They conducted theatrical performances either in Irina Sokolskaya's apartment or, more often, in the apartment of Isai Mandelshtam, Nina's and Genia's stepfather. Costumes were made by Irina — her sketches were creative and funny — while Genia was a recognized queen of poetry. She wrote "scripts" for the shows. The shows were fiery, and everything was topsy-turvy... The sisters were nicknamed Shoutgirls, or Nina Shout and Genia Shout because they spoke (or rather shouted) loudly as was typical for many temperamental Jewish women. Genia was witty, had irresistible charm and could certainly outdo any beauty although some would say that she was not beautiful in the classical sense of the word. She produced poetry on any occasion, and gosh, these poems were sharp. For example, Dimus had once looked for a room to rent, and after a long search he found one, but for some reason was forced to cancel this arrangement. Genia responded[4] immediately [4]:

---

[3]Walther Bothe and Hans Geiger, *Zeits. für Phys.*, **32**, 639 (1925). This experiment ruled out the energy non-conservation hypothesis (in Compton scattering) which at a certain point was put forward by Bohr. The quotation above is a fragment. For the poem in its entirety see [2], page 21.

[4]This and two subsequent verses below are translated from Russian by James Manteith.

> Windows, three, and area average,
> Bathroom and a telephone,
> With an entrance foyer separate,
> In the academic zone.
>
> Though the offer sounds congenial,
> It'll have to be turned down:
> "I'd be too far off from Genia,
> And no trams there run to Dau."

Dau was also in search of a room to rent, and had made an offer to a famous actress. His offer, however, was refused. Genia's critical reaction was fast [4]:

> Hunched and squinting disconcertingly,
> Pale and skinny as a corpse,
> When Samoilova-Mishurina
> Needs a fiery man to board.

And here is another example [5]:

> Purple fire filled the sky above,
> Dmitri Dmitrich saddled up a skate,
> Every word a line from Gumilev,
> Akhmatova by the verse and phrase.
>
> Lengthy promenades along the alleys
> In a Summer Garden fantasy.
> Dimus propagates the work of Rayleigh.
> Genia falls asleep on her feet.

Gone with the stormy winds of the 20th century are almost all of Genia's jocular verses so valued by her friends in Leningrad. Only a few more (in addition to those quoted above) were preserved for us in English translation by George Gamow, who published them in his book [3] from which I quote:

> ...Our usual meeting place was the Borgman Library which grew from a large collection of books willed to the Physics Institute by the late Professor Borgman. The library occupying a couple of rooms lined with bookshelves was open to professors and graduate students and served as the forum for the discussion of the problems of modern

physics and other matters. Here is an English version[5] of a verse written by Genia about that cozy place:

> How snug the Bórgman athenaeum![6]
> For more than five-and-twenty years
> Within this cheerful mausoleum
> Our theorists have met their peers.
>
> Here, famed for scientific talent,
> Pillar of learning's *why* and *what*,
> Professor Bursián the gallant
> Lolls in his clothes of foreign cut.
>
> And here, as the exam is looming,
> Vladímir Alexándrych Fock,
> Mustache in shape from final grooming,
> Composes questions round the clock.
>
> Here Ivanénko listens, drowsing,
> Sucker in mouth, to shimmy-beat.
> And Gámow, munching while he's browsing,
> Eats all the choc'lates he can eat.
>
> To tuneful songs, Landáu the clever
> Who'll gladly argue anywhere,
> At any time, with whomsoever,
> Holds a discussion with a chair.

My arrival at Swinemünde[7] was scheduled for early morning, and awaiting the landing I wrote to Genia dating the postcard with a time unusual for me, 6-00 a.m. Next week the answer arrived:

> Greetings from you at six a.m.?
> They didn't seem the least surprising.
> Too overcome for poetizing,
> I simply shout "Hurrah!" *pro tem.*[8]
>
> "Hurrah! Hurrah!" For now the fates

---

[5] Translated by G. Gamow. The Russian originals are lost.

[6] In the names, the accented syllable is stressed. They are pronounced accordingly, for instance, Bur-si-án; Vla-dí-mir Al-ex-án-drych; I-vanén-ko; Gám-ov (broad *a*, of course, and *v* sound at end); Lan-dáu. -GG

[7] Swinemünde is a city and seaport on the Baltic Sea and Szczecin Lagoon. Currently Świnoujście, located in the extreme north-west of Poland.

[8] *Pro tem* (Latin) means for the present time but not permanently.

In your direction are inclining.
Planes! Cabs! Revues! (the silver lining!)
Bananas! *Fräuleins!* Chocolates!

You didn't change your milk-white slacks?[9]
You're hopeless, Geo!... Dau raved and ranted
Because his passport wasn't granted,
And cast about for rope or ax.
But now he yields to his defeat
(And boils with inner irritation!)
He'll sail the Vólga on vacation,
Play tennis till the August heat.

Dimus seeks fame the Alpine way,
Hoards hobnails in his new bandanna.[10]
Poltáva-bound, he and Oksána
Took off — God bless them! — yesterday.

Mercury arcs illuminate
My photographic coffin. In it
I'm snapping spectra every minute.
I like my job; it's really great.

It rains all summer — rain, rain, rain!
You lucky boy, your lot I covet;
Still, for Odessa (how I love it!)
I'll soon be leaving on the train.

From Gretchens what can you expect?
And so, hands off the *Fräuleins* (misses)!
To Fock and you I genuflect
And send — by mail, of course — some kisses

I hope this letter makes you glad.
Farewell, dear Geo, *my bon ami!*
Nina says: "Write!" and so does
Z
Fourteenth of June ... in Leningrad.

The truck was deserted and no help was in sight. I sat

[9] G. Gamow explains: "Because of the shortage of clothes in Russia at that time the only decent-looking pair of trousers I had to wear on the trip were white tennis pants."
[10] A large handkerchief, typically having a colorful pattern, worn tied around the head or neck.

there remembering the verses written by Genia concerning our skiing on the hills (if any) near Leningrad:

> Gee-Gee[11] fell and lost his ski.
> Dimus, Dau, Ksana, Ki![12]
> Oh, you lazy lads and lasses,
> Find for me in snow my glasses.

In 1928 Landau quarreled with Ivanenko, and their relations started deteriorating. In 1929 Ivanenko moved to Kharkov, and after that, the Jazz Band gradually disintegrated. However, friendly relations between Landau, Bronshtein, and Genia and Nina Kannegiser continued. Rudolf Peierls was sort of accepted in this circle too; soon he became an insider. This was facilitated by the fact that he could speak some Russian. He even got from Landau his own nickname — *Páinka*, which means "a good girl or boy" in Russian. In 1931 Landau and Peierls worked together, a collaboration which eventually led to a joint publication [6].

Figure 3.4 From left to right: Nina Kannegiser, Victor Ambartsumian, unknown, Genia Kannegiser, Matvei Bronshtein, 1931.

---

[11] Gamow's nickname at the time.
[12] Oksana Korzukhina and Kira Tinuleinen, the two medical coeds in our skiing and other adventures. -GG

Figure 3.5 In 1929, Landau was 21 years old. His friends recollect that at that time he was very shy and his ideas about life were extremely naive. Once, Dau accidentally found out that two of his friends — he and she — met secretly from others. The couple in love, according to Dau, violated the basic rules of friendship, and he distanced himself from them [4].

*****

All recollections of Genia's student years one can find in the Russian literature depict the pastime of the Jazz Band members as a chain of merry adventures, pranks and funny entertainments. However, this was only one aspect of their lives. One should not forget about another aspect which is familiar to everyone who lived in the USSR. In 1950 in a famous letter to Klaus Fuchs (page 242) Genia recollects:

> [...] My Russian childhood and youth taught me not to trust anybody, and to expect anyone and everyone to be a communist agent. Twenty years of freedom in England softened me somewhat and I learned to like and trust

people, or at any rate some of them. But early attitudes are deeper [...].

Landau and Gamow became celebrated in the history of physics. Matvei Bronshtein, a pioneer of quantum gravity, who also authored works in astrophysics, semiconductors, quantum electrodynamics and cosmology, was forgotten for many years. The fates of many "Jazz-Banders" were dramatic (or should I say, tragic?), as were the fates of millions who lived in the Soviet Union at this tragic time. Matvei Bronshtein was arrested by the NKVD on August 6, 1937, at the peak of the Great Terror, and was executed by firing squad on February 18, 1938. Dmitri Ivanenko was arrested on March 4, 1936, and sentenced to three years in exile. Lev Landau was arrested on April 28, 1938, and nearly died in the NKVD prison in Moscow. Miraculously, he was released in April of 1939 — as a result of a direct appeal from Pyotr Kapitza to Stalin. In 1935 Nina Kannegiser was sent (with her parents) in exile to Ufa.

References

[1] V.M. Berezanskaya, *A Man beyond Stereotype*, (LENAND/URSS, 2016), page 247, in Russian.
[2] Gennady Gorelik and Victor Frenkel, *Matvei Petrovich Bronstein and Soviet Theoretical Physics in the Thirties*, (Springer, 1994).
[3] George Gamow, *My World Line*, (The Viking Press, 1972).
[4] http://www.famhist.ru/famhist/landau/00002746.htm
[5] G. Sardanashvili, *I am a Scientist. Notes of a Theoretical Physicist*, (LKI, 2011, Second Edition), page 137, in Russian.
[6] L. Landau and R. Peierls, *Z. Phys.*, **69**, 56 (1931).

Chapter 4

A Romance Taking Shape in Letters

The lives of Genia Kannegiser and Rudolf Peierls spanned almost all of the 20th century. This was a troubled century. It saw great social cataclysms, the bloodiest wars in human history, the unprecedented evil of the murdering of millions of innocent people, a prolonged confrontation between two ideologies: fascism and communism on the one hand, and liberalism on the other. At the same time, the 20th century was the time of revolutionary progress in science. The advent of quantum physics and breakthrough discoveries in genetics changed the very basics of human society and human relations. Modern technologies can pave the way to general well-being, probably, for the first time ever in human history.

By the will of fate, Rudi and Genia found themselves at the epicenter of events that eventually shaped the modern world. As usual, at each given moment of time one cannot evaluate the scale and impact of changes, the more so that they come gradually and reflect the joint work of thousands, if not millions, of people. The scale becomes apparent only in hindsight, when the mosaic of personal fates and efforts fuse into a general Grand Picture. We should never forget, however, that behind each element of this picture there are real people with everyday lives and plans for the future which may or may not materialize, with their love and aspirations, victories, failures, and difficulties.

Genia's and Rudolf's life together lasted for 55 years, until Genia's death in 1986. Together, they had four children, went through trials and tribulations of the war time, dealt with personal tragedies, and made many good friends all over the world, yet, experiencing also a betrayal. Their life was eventful, exciting and happy — never boring.

Every life is made of thousands of fortuities and chances which, unfortunately, are often missed. I think that Genia Kannegiser and Rudolf Peierls were blessed: they proved time and time again that serendipity was on their side starting from the very beginning — the first encounter in Odessa... then Genia's timely emigration from the USSR (a few years later it would be impossible), Rudolf's positions in universities outside Germany, participation in the Manhattan project, and ascent to the academic Olympus.

In the following chapters, I will tell you much of what I was able to learn about milestones of their life journey. But for now, let's turn to the very beginning. Let old letters speak for themselves.

Rudolf Peierls' mother tongue was German. Genia did not speak German, however, so in the beginning their common language was English, with insertion of a few Russian words here and there. Genia's English was very peculiar, with many funny grammatical mistakes. In some instances she used Russian linguistic constructions, translating them in English literally, for instance, instead of "isn't it?" or "doesn't it?" at the end of a sentence she would write "Yes?" or "No?". Sabine Lee kept Genia's style intact, which, to my mind, was an excellent decision.

After his return from Odessa, Rudolf Peierls started learning Russian. His teacher was Fania Moskovskaya, a young girl whose parents left Russia (presumably) in 1917 and moved to Zurich where, 10 years later, she became a student of the Department of Law at Zurich University. Their first lesson was on October 8, 1930.

Rudolf made remarkably rapid progress in Russian, and after a few months most of their correspondence was in Russian. Genia wrote the first letter to Rudolf in mid-September 1930. By November of that same year parts of their letters were written in Russian. Apparently, Rudolf Peierls was extremely talented not only in physics, but also in linguistics. On November 30th Rudolf switched to Russian. In due place I will show a couple of examples. They exchanged letters if not daily, a couple of times a week, unless there were some unusual circumstances which prevented them from writing. These letters are about everything that might interest a girl and a young man; they are tender and moving. Surprisingly, they realized rather quickly that their backgrounds and aspirations were not that different as one might think given the differences in the environments in which they were raised. They read the same books, liked the same music, and had similar moral values. Their attraction was deeply mutual.

In the autumn of 1930 Rudolf Peierls was Pauli's assistant at ETH,[1]

---

[1] *Eidgenössische Technische Hochschule*, or Swiss Federal Institute of Technology.

Zurich. This was his second year of PhD work on solid-state physics under Pauli's supervision. Pauli used to say that although he invented modern solid-state physics, he disliked it. The topic he suggested to Peierls was on the heat conduction in insulating crystals (the heat is conducted by lattice vibrations, or phonons). Genia had just graduated from Leningrad University and found her first job, at the Leningrad Geophysical Laboratory.

***

Rudolf Peierls, Bird of Passage, 1985[2]

A good deal of our time was spent at Luzanovka, the beach in Odessa. There I met, among others, a girl from Leningrad, Eugenia (Genia) Nikolaevna Kannegiser, who had recently graduated in physics. She seemed to know everybody, was known to everybody, and was more cheerful than everybody. After the conference, all participants were taken by boat across the Black Sea. This included Genia, with whom I talked a good deal during the voyage. She did not speak German, and I knew no Russian, so our only common language was English, which Genia could speak reasonably but not fluently. If I was asked why I did not seek the company of the many other young people, who included many accomplished linguists, I could say only that the thought had never occurred to me. As a guest, I was traveling in first class, while Genia, as a junior member of the conference, had a berth in third class, which was so uninviting that she preferred to sleep on deck under a lifeboat. I felt uncomfortable about this, and would have liked to change places with the lady; but as I was sharing a cabin with Pauli, there was nothing I could do!

---

[2]Rudolf Peierls, *Bird of Passage: Recollections of a Physicist*, (Princeton University Press, 1985).

Rudolf Peierls to Genia Kannegiser[3]

Berlin
September 24, [19]30

I was so glad to get your letter, dear, and of course it was not too long, and if you write still longer letters the only change will be that I can only read them three times instead of 6 times. Also it is "fishing for compliments" when you say it is difficult to read your letters. I really understand every word in spite of the bad paper, pencil, language, philosophy and so on. I am afraid it will not be so easy with my letters, but when I get back to Zurich I will have my little friend again, my little Corona, and then letter-writing is much quicker and you can read them easier. (If you are being jealous, I must tell you that "Corona" is a portable typewriter.) As for jealousy, I am not jealous at all, and I should like to [think] I will never be, but so far one must be careful: "One must never say 'never'." Theoretically, jealousy is very silly, because you never l[o]se anybody by the influence of somebody else, but if somebody else has such an important influence, it only shows that you have lost that person before.

Now I am already in the middle of philosophy which, as you say, is very tedious. Still, I am afraid I cannot help bringing a lot of philosophy into this letter, but first I will speak about more agre[e]able things.

My last days in Leningrad were very interesting again. Did I tell you how much I liked *Fremitagc*?[4] I had one more evening in the theater with Mrs. Frenkel,[5] *Olga Gregorevna*, and the Simons,[6] saw the Bhuddist temple with *Bronshtein* and some things more. Then with some difficulties I got my steamer tickets, and I had a very nice time on the sea. I was glad that the sea was rather rough, because then most people could not eat their dinner and we got double portions. On the steamer, I met a very nice man, a professor of social hygiene from *Leningrad* who will come to Berlin in one of these days.

In the train from Stettin to Berlin, I suddenly saw some relatives who could tell me that my sister had born another child, a son this

---

[3] Sabine Lee, Volume 1, page 115.
[4] Eremitage (or Hermitage) is a museum of Fine Arts in Saint Petersburg (Leningrad in 1930) founded by Catherine the Great in 1754. Here and below words in italics were written in Russian.
[5] Sara Isaakovna, wife of Yakov Il'ich Frenkel.
[6] Franz Simon and his wife Charlotte. Sir Francis Simon, (1893–1956), was a German and later British physical chemist and physicist. In the summer of 1930, together with his wife, he visited the Soviet Union, namely, Odessa, Moscow and Leningrad.

time,[7] and I was very anxious to get back to Berlin and see her. The child is really charming and I have never seen such an intelligent-looking child of 4 days.

Now my other niece is not quite well, so she is not allowed to stay in the same house with the little one, and had to go to my parents and they put me to my brother's flat. I like that very much because I had never had much time for my brother and my sister-in-law and now I can see them very often. Also I have quite a flirt with my brother's step-daughter (aged 6 years) so there is another reason for you to be jealous.

I like to talk with my sister-in-law, Nina, because she can understand so many things and I think she is the only person of my family whom you would not call "European." She is also the only person of my family who has a slight idea that somewhere in *Leningrad* there is some girl...

Besides her, also Barbara Zarniko[8] knows a bit — that is that girl of Simon's institute I told you of and one of my best friends. She saw the picture I took of you and liked you so much — the picture really gives one little image of all the things I love in you — that I could not help telling her that she was not alone with her opinion. Generally I do not like to speak to other people about such things, but this time I could not help it. Are you angry about it? You are not. *Yes?* I had a splendid time here, of course, and had to see a lot of people, all of them were interesting to hear about my travel and I had to talk the same story three times a day.

I met all my friends again, but I cannot tell you about all of them, because then this letter would become a big book. But besides this, I did not do very much here and I am afraid it will be very hard to start working again, when I get to Zurich. On my way back I shall stop in Heidelberg to see my best friend who works in a hospital there[9] and in Freiburg to see an old uncle of mine,[10] a concession which I make to my parents.

---

[7]Günther Krebs. See Sabine Lee, the letter of Heinrich Peierls to Rudolf Peierls, October 10, 1930.
[8]Later Barbara Zarniko married Martin Ruhemann and became Barbara Ruhemann, an ardent communist. For more details on the Ruhemann family see *Physics in a Mad World*, Ed. M. Shifman, (World Scientific, 2016).
[9]Hans Thorner.
[10]Felix Weigert and his wife, see Sabine Lee, the letter of Heinrich Peierls to Rudolf Peierls, July 7, 1930.

September 25, [1930]

Darling, if marriage is such as you describe, of course love means marriage. But then marriage does not mean family as you once told me. "Family" is a responsibility which I do not yet feel able to undergo. Marriage in the sense of your last letter, and I was glad to read it — means a relation between two persons, but family means so much more (and less in certain respects), I do not want family, perhaps I shall want it later, perhaps never.

There is another word which astonishes me in your letter: "to be too clean for adventures". Do you really think adventures immoral? I can understand if you do not want an adventure in a special case. I could perhaps understand if you say you don't want adventures at all. But why generally say they are immoral? Of course this has nothing to do with our relation, because I do not want an adventure with you and even if I would it would be immoral to have what Landau calls an unsymmetrical situation. I mean that two persons mean a different thing and simply let things go without trying to get a solution. I realize that I speak dreadfully philosophical, i.e. ununderstandable, I only want to say: if you should want love and I want an adventure, it would be quite immoral. But adventures in general can be quite clean, why not? I told you that in many respects, I do not yet know my way and this is one of the points I will have to find out. Do I want adventures, and if so is it possible to have them without l[o]sing more important things? *Genia*, I am afraid you do not understand a single word of this letter. But if I would try to explain more, it would become more and more philosophical and you would understand less and less. So I better stop it.

You say I know much of you, but I do not. Which way do you think to go? Do you do your work now because you feel it impossible not to do it, have you serious relation to it, or is it just because you want some occupation? If you could do what you like, what would you do?

I had to ask you many more things, but if one only writes it, everything sounds so abstract and so silly.

Darling, I shall be so glad if you write to me very often, but you must not, through these letters l[o]se the persons you live with in *Leningrad*. It is sure now, that I come next March, if nothing extraordinary happens in the meantime. So, *good bye*. In my hands I have yours as in that car on the *Gruzinskaya* (do you remember?) and my lips ... well I better tell nobody what they are doing!

Your Rudi

Give my regards to *Nina*[11] and tell her how unhappy I was I could not say good-bye to her, but poor *Bronshtein* telephoned her one hour without getting her.

Your next letter to Zurich, Physikalisches Inst. d.E.T.H., Gloriastraße 35.

○○

Genia Kannegiser to Rudolf Peierls[12]

Leningrad
September 26, 1930

My very very dear!
I don't know why, but this three days I so awfully want to see you to speak with you, that it is rather hard to write a letter the answer on which I shall receive only in two weeks, and it is the minimum. Ou, *Rudi*

I received your "angry" letter five minutes before I start of moto car from Gagry to Sochi and I have read it in this "fabelhaft" way. Do you remember *Baidary* in *Crimea* and the way from *Sochi*[13] to *Ritsa*?[14] Gagry is 10^5 better: the high mountains covered with curled forests, roses, palm trees and high high pins right under your feet — the sea — blue and green and white foam of tide. *Great!*

I was sitting near the moto driver (as is the best place) and certainly began to speak with him a little and to ask him to go quicker; if you know what a scandal began between the people in the moto car! Ladies cried of horror and told me that I must immediately cease to speak because the way is dangerous and so I was obliged to be silent all the way! Poor girl! Yes? I was awfully tired because I packed all the night my luggage; think only I brought with me 30 klgr(!) of fruits and sweet things! And all this except the wine grapes was packed in my bag and my trunk (do you remember them, they are not so big at all!) But they were a marvel of packing — the peaches and pears were in mine dresses and after five days of way (I remained 2 days in *Moscow*) were in such a beautiful state how immediately from the garden. But the state of my dress wanted to be better.

---
[11] Nina Kannegiser, Genia's sister who was a year younger than her.
[12] Sabine Lee, Volume 1, page 119.
[13] A resort town 70 km north of Gagry, see the map on page 42.
[14] Lake Ritsa is a lake in the Caucasus Mountains, in the north-western part of Abkhazia, Georgia.

I made from my cushion a volume of 5 cm^3! Great. But if you know how it was heavy to carry all this things! In *Sochi* and in *Moscow* I had the porters, but in *Gagry* it was very difficult indeed to manage with these 4 things two of which were 15 klgr each. But *Nina* and my mother[15] were in right enthusiasm when they saw all this treasures.

I was two days in *Moscow* and walked all the time in the streets and was in Tretyakov and in the Museum of Western Art, Masereel[16] is very good. Do you remember the man with the little dog near him? And his self-portait — *fabelhaft*.

I awfully like Renoir, but why all his women resemble themselves so much, but they are really charming with blue eyes and such beautiful lips. Yes?

I was very sorry because on the place of Picasso and new French and German painters, (do you know Kisling?[17] We have there only one picture of him, a girl in rose dress, but I like it awfully) they were an exhibition of Rabindranath Tagore.[18] I don't like the dilletants and he is one in the painture, but this little museum is charming and I prefer it to all other *Moscow* museums. This time I was in a museum of old painters, something of the kind of *Eremitage* but very little one, and there I saw a "Madonna" of early Italian primitifs, it is one of the best picture I saw in all my life, and I want to sent you a post carte with it but they were all sold. You know long, long, hands and light hair. When you look on her very beautiful thing I am always a little sad, and you?

Now I am at home and to morrow will go in the institute. This night we go at the theater with a lot of people. I think that the company is of 8 persons! I am so sorry, that I was not in Leningrad when you were here, but I am so glad that you were charmed with it. I think Lening[rad] is one of the beautifulest town of the world. Such a lot of water and gardens, and such unity in style. I will show you a lot of things in March.

Darling, I cannot find my dictionary and the mistakes are awful; I beg your pardon for them and for the tedious letter also I want to say you such a lot and it is so difficult! And it is so sad when perhaps it is not interesting for you at all. Between your last letter and this are

---

[15] Maria Abramovna Levin, see page 112.
[16] Frans Masereel (1889–1972), Belgian-born woodcut artist who widely exhibited in communist countries.
[17] Moïse Kisling was a French painter of Jewish Polish descent best known for his highly stylized, unsettling portraits.
[18] Rabindranath Tagore (1861–1941), Calcutta-born writer and poet, who also painted.

two weeks and it is all the time from Odessa to *Kislovodsk*. I hope that you send me the photograph from Berlin; my mixt letter with Bubi[19] was very cool my dear sir, but it was mixt and you understand the conditions. Yes? I sent you my big photo. It is not very nice, but the others are far and I want to post the letter immediately by air post. Sent me your big one. Well?

Genia

P.S. Rudi, I forgot to tell you: Dimus Ivan[enko][20] told me that Sack[21] was awfully shocked that you remained in *Kislovodsk*. Is it true?

Write me often, because if you will not I should think that I am not interesting for you at all and it is very disagreeable to think.

P.P.S. *Rudi*, you know that one month from six is passed. Hooray!!! 1/6 is on! Oh, one sensation! Sara Isaakovna told me that Houtermans will remain in Leningrad. It will be very jolly in March. Yes?

∽

Commentary

Friedrich Georg (Fritz) Houtermans (1903–1966) was a nuclear physicist well-known in the 1930s. He was the first to suggest that the source of energy release in stars was associated with thermonuclear reactions. Fritz Houtermans completed his PhD at Göttingen under James Franck in 1927 and worked as Gustav Herz's assistant in Berlin in 1928–1933, when his affiliation with the German Communist Party forced him to leave Germany. He did not stay in Leningrad long-term in 1930, as Genia speculated in this letter, but in February of 1935 he took up a position at UPTI in Kharkov. On December 1, 1937, Fritz Houtermans was arrested by the NKVD. His wife Charlotte and two small children were left in a desperate situation. They had no means of survival. At first Charlotte did not even have her passport, which was with Fritz at the moment of his arrest and was confiscated by the NKVD. Only by a miracle Charlotte and her children had escaped

---

[19] Sergei Mandelshtam, a son of the well-known Soviet physicist Leonid Isaakovich Mandelshtam. Sergei himself was a physicist and later specialized in optics. He was known among his friends as Bubi. A short letter from Genia to Rudolf (September 23, 1930) mentioned above had been a joint letter with Sergei Mandelshtam.

[20] Dmitri D. Ivanenko (1904–1994), Russian physicist and close friend of Landau's in the late 1920s and early 1930s.

[21] Heinrich Sack, physicist, married to chemist Charlotte Sack.

from the USSR to Riga, Latvia, which was an independent country until August 1940. On May 2, 1940, Fritz Houtermans was extradited to Germany and arrested by the Gestapo at the Soviet-Polish border.

Incredible adventures and misadventures of the Houtermans family caught between two evils — German National Socialism and Soviet Communism — are described in detail in three recent books [1; 2; 3].

*****

Rudolf Peierls to Genia Kannegiser[22]

Berlin
September 29, 1930

My dear Genia,

It is my last day in Berlin, and I must quickly write you a bit of a letter. In the last days I thought very much about the possibilities of your coming to Zurich and living with me, but the result is rather bad. First of all I decided after some calculations, that it is absolutely impossible for us both to live on my salary which amounts to 400 Swiss francs (150[rb]) a month.[23] This is very much for one person, but not far enough for two. Of course it could be possible that you have a profession also, and anyhow, you would not like sitting around the whole day and doing nothing, but earning money with scientific work is much more difficult in Europe than in Russia now, and the probability of getting this position in the same town I am living in is very near to zero. And even if it were greater, it would be a great risk as you had to come here without knowing how to live. The worst thing is that I have no hope to improve my position. If for instance, this offer with Simon's Institut converges,[24] and it seems to converge, as far as I can see, although it is still too early to decide, as it cannot be before next winter, even if this converges, I shall not get more money than now. Nobody knows how long it will take till I get a better position, but don't forget that I am terribly young.

I tell you all this, dear, not to make you sad, but that you do not lay too many hopes in this marrying possibility.

---

[22]Sabine Lee, Volume 1, page 122.
[23]Around $1100 in USD-2018.
[24]See Genia's letter of September 13 on page 46. In 1931 Franz Simon was appointed to the chair of physical chemistry at the Technical University of Breslau where he began to assemble his low-temperature equipment. With Nazis in power he was forced to emigrate to England in 1933.

I am ashamed to confess that I was not a single time in the theatre during these two weeks in Berlin, but we have no good earnest plays now, and I am tired of comedies, because they have so plenty of them here or in Zurich and always the same kind of thing. So I spent all evenings with friends or with my family. I am a bit dubious about the marriage of my brother and Nina now. They are very happy now, but she has too much time, she has a servant, and as she is very clever, housekeeping only takes a very small part of her time. My brother is out the whole day and he is very tired when he gets home. Of course, she ought to find an occupation, if not a position, but she cannot make her mind. I am very doubtful what will become of it. I am curious what will happen in Zürich. You remember that evening on *Grusiya* when I told you I would only stay a short time and then go to Zürich where there is some girl. You know I think differently about this, and the girl became rather unimportant to me, so unimportant that I did not even tell you the details. Still, she does exist, and I have to do something about it when coming back. She is a very nice girl much younger than I am, violinist, a very good violinist, as I know, and awfully well educated, so that she cannot do anything without asking her parents.[25] My relation to her was very young and on the way of developing, and so we had not spoken very much about our relation. But now, when I came back from Leningrad, I wrote her a letter, and told her that in the meantime something had happened which could not disturb or affect my relation to her, but showed me what feelings and how intense feelings were possible besides this relation. Now perhaps everything will be all right, and this will stay as it was; a very good friendship — surely not more for many reasons — but as she is very "European" it is also possible that she will not want to see me again. In this case, I would be sorry — but only sorry. We will see what happens. On September 6, Gamow got the following telegram: "Sensationelle Flucht Jenny Peierls nach Gagry falsch oder trivial Dimus Kiebel,"[26] and Gamow sent it to me with the remark: "What does that mean?" I explained as far as I understood it myself and laughed very much about it. Perhaps you better do not tell Ivanenko that I told you this.

---

[25] Lilly Fenigstein (1911–1999), Swiss violinist who later married the English-born Herbert Herz. See www.gbnf.com/genealog2/herz/html/d0006/I574.HTM. -SL
[26] Sensational Escape Jenny Peierls to Gagry false or trivial Dimus [Dmitri Ivanenko] Kiebel. (Ilya Efroimovich Kibel (1904–1970) was a notable Soviet mathematician and meteorologist. In the 1930s he was known to his friends as Kib and belonged to a circle of "astronomers" closely related with Genia, a group which also included Ambartsumyan, Gamow and Bronshtein).

Please read the last lines of Gamow's "Fine structure of alpha rays," *Nature*, September 13, 1930.[27]
Very sincerely yours

Rudi

Answer soon!

∽

Rudolf Peierls to Genia Kannegiser[28]

Zürich
October 2, [19]30

My dear and distant Genia,

It is really awful with the long time between a letter and an answer because I cannot remember why you call my letter from Leningrad an "angry" one. But now I really ought to be angry with you. You apologize for everything, for writing long letters, for writing tedious letters, for writing letters with mistakes, for writing letters with Bubi Mandelshtam[29] and so on, and I only wonder why don't you apologize for being born! Do you really not know how glad I am about these long and "tedious" letters which, of course, instead of being tedious are the more amusing the longer they are. Don't you know that I like these letters with all mistakes, that I like them more the more mistakes they contain, because I can hear you speak the more distinctly. Or did you expect I would not understand why the style of a "mixed" letter must be different? In the contrary I was very glad about this letter because there were some very little and very fine traces of true meaning in it, which, of course, nobody else could understand. Sometimes one can be nearer to each other than ever, if one is not alone and cannot speak frankly. But of course you know all this and you only wanted me to tell you once more.

As for the time it takes for you to get an answer, the minimum is much less than two weeks: you will even get this letter earlier although yours had to wait here for me and although I could not answer it yesterday, as I was busy the whole day, and although I have not yet the time-table of the airplanes.

---

[27] G. Gamow, "Fine Structure of Alpha-Rays," *Nature*, September 13, 397 (1930).
[28] Sabine Lee, Volume 1, page 124.
[29] See letter of September 26, 1930, on page 67.

The distance from Berlin to Zürich seemed very short to me now. The two hours in Heidelberg were very nice,[30] but very short, because this friend, Thorner — I told you about him, do you remember, as my friend in München, he is now a doctor working in internal medicine and neurology — I had not seen since last Christmas. Last winter it happened that I thought with whom could I go skiing to the mountains. So I sent a post-card to this Thorner. When I returned from the letterbox there was a post-card from him: "What do you think about making a ski-tour together?" Landau came with us, then, and I think they liked each other, though they are very different types and quarreled at every occasion. Since then I had not seen this Thorner, and we wrote letters very seldom, so we had a lot of things to tell. We talked during every second during these two hours, till my train left. Then I went to Freiburg and payed a visit to an old uncle of my father. He and his wife are really charming people, although it is very sad, as he is now entirely blind, and begins to become deaf also, although his mind is still quite fresh and young. (He is aged 85.)

Then I came here, on a splendid evening with stars, mountains and so on, one can see the mountains from my street! The whole day I had to do a lot of things, speak with Pauli and discuss with him the letters which arrived in the meantime, and so on. Our relation is very strange; usually he says: "Here is a letter of Mr. So and so who has calculated something, but I cannot yet see through it. I am sure there must be a mistake." Then usually I can show him the mathematical proof that he is right, and that there is actually a mistake. So Pauli always says that although I am not very interesting physically, I am very comfortable and a very good help for his laziness. "The ideal assistant."

But I was terribly tired last night, because I am no more used to do something reasonable from morning to evening!

Bloch[31] is here now. He had a bad accident in the mountains. He was climbing with a friend and suddenly lost a step and fell down. He was on a rope, of course, so his friend could keep him from falling down entirely, but he had a broken leg, and had to wait there two days in a bad position till they could come and carry him down with many difficulties. I telephoned with Lilly Fenigstein (the girl I wrote to you about in my last letter)[32] and it seems as if she is not offended at all,

---

[30]See letter of September 24, on page 64.
[31]Felix Bloch (1905–1983) was a Swiss-American physicist who was awarded the 1952 Nobel Prize for "development of new ways and methods for nuclear magnetic precision measurements."
[32]See letter of September 29, 1930, on page 70.

which would mean that I was less important to her than I feared. I am glad about it; we will probably be good friends.

I think it nearly impossible that Sack was shocked,[33] perhaps he was a little bit angry that I let him go to Moscow alone. Dimus seems to take everything 100 times more important than it is. Sacks photos came out well.

Also, Mandelshtams photos are very nice, in spite of the "negress." About your photo I was very glad, but send me a better one occasionally, I am sorry I have no large one of me, I shall try and get one and send it to you.

But it is terribly late now and I must go to the institute. Before, I will look for somebody who can give me Russian lessons. How do you enjoy working again, and how do you enjoy being in Leningrad again with so many people and with all the pleasures of a big city?

Regards to everybody and my love to you, darling!

Your Rudi

I am anxious to get your letter! I do not send this letter recorded delivery because then it goes quicker. I hope it will not be lost.

∞

Genia Kannegiser to Rudolf Peierls[34]

Leningrad
October 3, 1930

I am awfully pride that you understand my letters, darling, but you are really very clever boy because they are many people who cannot understand even my Russian letters and write so angry answers that I am quite ashamed when I read them, and I certainly understand your every word. So all is quite right.

I want to say you so many things, that I don't know when I shall finish this letter. Now it is evening, I am a little tired because I worked all day very hard, but at home is very nice and my cat is sitting near me and looks on our toys with very sweety eyes and sings, sings, sings.

You asked me if I like my work, or it is only "some occupation." I love the physics very much and I work now with "clean physics." I like to think.

---

[33] See letter of September 26, 1930, on page 67.
[34] Sabine Lee, Volume 1, page 126.

October 4, 1930

Rudi, on this word I fall asleep and when I opened my eyes it was late and Nina slept so well that I went right to bed and dreamed all night about you. (What a material for Freud and his prophet Bronshtein!) Dear, I go on, so I like to think and sometimes I have some ideas but I am not of high opinion about me. It is true that I understand all very easily, perhaps too easily, but I am too "woman," too "surfacing" not "deep" (you understand?) and so lazy, so awfully lazy! But certainly I prefer this work to any other and it will be for me very tedious without it. I write to you in my *laboratoria* all the people is going away I am alone (it is 9 o'clock in the evening) and awfully angry — my Kohlhörster[35] (you know such electroscope for penetrating rays) became quite mad, to day I measured the first time after *vacances,* and after his traveling on [illegible] with my patron and in the same condition he show now 21.0 instead of 11.5 before summer. I cannot understand what is the matter with him and dance around 11 hours without any result. I am afraid that my patron dirtied him and some radioactive products I cannot find other reason and if it is so it is awful!

I don't know what will be my way, now I am on the quite path of "laboratoria girl" and I enjoy it very much, but Rudi I am an adventurist in the high sense of the word and too young also, so it is a great probability that I shall have many adventures and perhaps many professions. You know, dear, it is so tedious and even so sad to be sure in your future. Future must be unknown and bright. Yes? When you know your future you are big slave in present and it is immoral. Rudi, it is too philosophical but you understand not only the words but their deep sense and it is so easy to speak with you. You know I write not only the funny verses but and lyrical [illegible] also, I am sad that you cannot understand them, I don't read them to anyone (there were some exceptions of the rule) but sometimes I like them, your opinion interest me very much, but it is nothing to do. I think now a little about a novel, but I am too lazy to write it and it is too difficult.

If I can do what I want, perhaps I will travel and write books about towns, peoples and stars. Sometimes I want to be a writer. And there are such moments when I want to be an ingeneer, not an ingeneer of fabrique near the big city, but somewhere in the forest or in the mountains with wilde people and wilde nature around. In such conditions

---

[35] Werner Kohlhörster (1887–1946) was a German physicist, a pioneer of cosmic rays research who found in 1929 (with Walther Bothe) that cosmic rays are not photons, but consist of charged particles. In physicists' jargon, Kolhörster means an elecrostatic device to measure cosmic rays.

you can very well think and see and write. About "adventures" — you distinguish between "love" and "adventure," so it is easy to explain my point of view on this thing. Is an "adventure" immoral? What is moral? I understand only self-moral, if remembering some thing you do b-r-r-r (if you are uncomfortable in the middle) it is immoral, so can exist moral adventure, such after which you are quite well, but it is "unhonest" it is "profanation" of love, of words, of kisses, it is fox-trott under 7 symphonie of Beethoven (you understand). And the love for the money — it is "dirty" also in this sense — "profanation" of love and of woman. Ou how late, I must see for the last time my Kohlhörster. Till to morrow, dear.

October 6, 1930

When I returned at home the 4th, I found your letter from Berlin[36] on my dining table and yesterday your Luftpost from Zürich[37] awakened me in the morning. It was so nice, just as you entered in the room. Darling I was glad of both your letters, but really I had not want compliments with "long, tedious, mixed etc. letter" but they are awful in comparison with that which I have in my head, I am so sad I cannot write you Russian yet, I am a little afraid of it because I remembered the little tale of Maupassant: a frenchman fell in love with an English girl and married her she was charming and spoke an awful French and they were very very happy until she learned French quite well. Then he saw that she was stupid and vulgar and fell into deep spleen and was unhappy all his life. It is not our "case." No?

My dear, dear boy — I had not many hopes in such nice and simple end of our story — as life with you in Zurich, but I am an optimist and, it is a long time to wait that it can happened a lot of things in the style of moral novels you found 100,000 from on the street, or the death of American uncle bring you a house in New Orleans and so on. Yes?

No, "seriously," darling, when a man want something very strong, he always get it. Rudi, I want to see you! It will be splendid in March.

Now is very beautiful here — "the gold autumn," and the town with blue rivers and gild and red gardens is charming. I walk sometimes very long time through the streets and you know it is such a pleasure to look around when the sky is high and blue and dry leaves under your feet. How I love Leningrad if you know how I love it! Why you are not

---

[36] See letter of September 29, 1930, on page 70.
[37] See letter of October 2, 1930, on page 72.

here now, we should go in Detskoye Selo[38] and in Pavlovsk.[39] The park of Pavlovsk is one of the best in the whole world and in autumn there is really *fabelhaft*.

But in early spring (March and April) is also very beautiful another style certainly but you shall see it.

I related you about Moscow in last letter but I forgot to tell you about my Moscow friend. Do you remember this evening in Kislovodsk, our first tête-à-tête when I tell you my story (it is true that you were not very interested in it). And I relate you that last summer when I was in Crimea I meet there one "not ingeneering ingeneer." He is forty two, traveled very much, likes verses and pictures and understood very easily many things. I walked with him and we spoke many hours and afterwards we wrote letters to one another sometimes. I was very pride that such aged man (42) is so interested in me. Now in Moscow we walked and spoke again and you know I thought it before and now I am sure that he loves me awfully, why I think so I do not know because we never spoke a single word about love, but there is something in eyes and laugh that make sure that there is a big and sad love — the "lost love."

I don't love him at all, he is only good friend, very intelligent and more aged, but I am very serious in such things and I don't like "to play" with love, but it is not my fault. I never want that he loves me. So I am sorry with it, but for other side. Love is such splendid thing, not perhaps it is better a not divided love that not at all. I don't know. Rudi, dear, I forgot to thank you for photo. They are very nice, the best I think are Pauli in the train and a view of green with pin and palm tree and my portret.

Don't you find that all our story is awfully romantic? Such your plane of love as from the Caucasus to the Black Sea! When I remember all I am quite astonished how it can happen when we were alone only two last days.

Do you remember Sack with two trunks in our wagon? And the first evening in the *Grusiya* when came quite afraid Bronshtein thinking that we fell into the water! It was only two weeks but we saw such a lot of things and it passed so many between us that it seems to me much longer time.

---

[38]The town was founded in the 18th century as Tsarskoye Selo (Royal Village); it was renamed Detskoye Village (Children's village) in 1918. In 1937 the name was changed again into Pushkin to commemorate the centenary of Pushkin's death. -SL

[39]A town 25 km south of Leningrad, famous for its grand palace and parks. -SL

Good night my darling, I love you so strong

Yours, Genia

I write this letter 3 days, but yesterday I cannot write a word it was my free day and all the day came all sorte of friends. I was not very charming with them at the end. Nina sent you her regards and beg you to sent her a face of Pauli if you have it.

∽

Rudolf Peierls to Genia Kannegiser[40]

Zürich
October 8, [19]30

Genia,

Dearest, when will I receive your letter? I have waited so long!
My life here begins to become more "stationary" without any exciting events. I am in the institute the whole day (which does not mean, of course, that I am working all the time, only sometimes). I am with Bloch, who has not yet quite recovered from his accident. It must have been a dreadful thing, lying 8 hours alone on a small place on a rock, where he could not move without falling down, and then he had to stay another two nights in this place (3000 meters high) and only within the next two days they could bring him down with the help of about 12 guides. But he was very brave all the time, although it must not be a pleasure to be carried down the rock with a broken leg.

Yesterday I was careless enough to show him the pictures of Caucasus and he was quite upset and if there is an opportunity, he will surely climb one of the summits there.

This week I started my Russian lessons, and I very much enjoy them. My teacher is a student of Laws, a very clever girl. If you like, you may sometimes write me in *Russian* (на русском), but in the first time, I will have to read it together with my teacher — you understand! This is all my life here now; not very much. I make many efforts to work, but this is very hard (трудно), if you are so used to laziness and if there is nobody to control you. Most of my time is spent in preparing a lecture

---
[40] Sabine Lee, Volume 1, page 130.

on wave mechanics, which I shall give in Pauli's place this winter. This is very interesting (интересно).

What about Houtermans? In a post card to Pauli it seemed they would go back to Berlin? Did you hear the details of their marrying story?[41] In Germany there is a very good talking film now, "Die Affare Dreyfuss." If it comes to Leningrad (Ленинград) you must go there! I had never thought they could make such a good earnest but not sentimental film in Germany.

I spoke with Pauli, and I can stay away from Zurich from first of March to first of May, so I shall have at least seven weeks in Leningrad. But until then there are still nearly 5 months. Genia, do you know how often (Женя, ты знаеш как часто) I think of you, talk with you, dream of you! Aren't we silly, you and I? There are so many boys and girls in the world, and just this one in Leningrad and this one in Zurich... But who can help it? I love you very much.

But I promised to Pauli I would read some papers before he comes and he will be here "early morning" today (i.e. $13\frac{1}{2}$ o'clock). Good bye, my dear!

Your Rudi

The included pictures are the best ones I could find.

---

[41] See commentary on page 69 and M. Shifman, *Standing Together in Troubled Times*, (World Scientific, Singapore, 2017).

Rudolf Peierls to Genia Kannegiser[42]

[Zürich]
October 9, [19]30

My dear child,

If only you knew how happy I was when I got your letter. I knew I would get it to-day, because I had yesterday posted a letter for you. I think I could see you when you wrote it so much of you is in this letter. I am sorry I cannot write letters like that but my letters are so terribly cool and do not express what I want to say. And probably I say everything twice, because I speak so often with you, that I never know have I written you this in a letter or only told you while you were not here.

My questions what would you do if you could do what you want, was not purely curious, I very often think about the very similar question, what would you do if you came here and live with me, in the both possible cases, if I would have enough money for both of us or if I would not. But as yet I do not see a solution, not a clear one for the first case and none at all for the second.

Of course it is terrible to know about your future, but sometimes you have to decide for so long a time. So for instance I promised Pauli I would tell him half a year before I go away from Zurich, and as Simon needs the same time to find a position or more, I had to write now to Simon that I should like to go to Berlin next winter if possible.[43] Dear, I think I am much more lazy than you are, at least I am terribly lazy. Fancy what we would do together!

We seem to have the same idea about adventures and "morals" but I do not understand what you mean by "profanation"? Is not this highly conventional? Of course there is no need to discuss love-for-money. But there the point is not that it is dirty to buy something you want, but that you really will never get what you want. But if two people strongly want to be together a short time and have each other entirely, body and soul, and they are honest to each other, I mean tell each other the truth, where is the profanation? But stop discussing theories.

Do you really believe that the Maupassant story applies to us?[44] I think not, although there is no doubt, that the many feelings of

---
[42] Sabine Lee, Volume 1, page 132.
[43] See letter of September 13, 1930, page 45. -SL
[44] See Genia's letter of October 3, page 74.

strangeness between us are very important. But this is a strangeness of the whole person, of character, of opinion and so on, and the language is not important. In the contrary, I am so sorry you cannot speak Russian with me, because I enjoy so much hear you speak Russian. You have such a personal style of speaking that I could notice it although I only understand 10^{-3} of the words. Seriously, this is one of my greatest anxieties for the case you could ever come to Germany. To whom could you speak then? And I do not know whether it is possible at all to learn a foreign language, that one can speak it like you speak Russian. In March I hope I will know a bit more of the language, so that I will probably understand 0.3 of the words and perhaps already then your Maupassant story could happen, but will it? I think in one year I shall know enough Russian to understand even verses. But I have no great sense for verses. I like them very much, but it is not that I could not live without them, you understand?

Of course I remember that evening in Kislovodsk and I was very silly then, I must confess you now. It is nonsense of course when you believe I was not interested in the story. But I was terribly tired. You remember that we were in the railway the night before and that I went down to Vladikavkaz and back in the afternoon? But I did not want to be tired. I wanted so much to hear your story, and then I thought it would be so silly to sleep just when it was the first time we were together, first time that we could speak as we liked, and such a short time — and to spend this time with sleeping! But of course I could not help being tired, and I was silly enough not to tell you. And so some parts of your story are now for me like seen through fog, and you must tell me some more when we see us again (writing such things is so difficult).

With your engineer you are probably right, that it is better to love in vain than not to love at all. But still it is always sad; most sad, if he has some little hopes, which one takes so easily from a smile or a word, which did not mean anything. I seldom was so furious as on this evening in *Grusiya* and the morning in the train you speak of. This probably was the reason why I was not very polite to Sack then, and therefore again he probably was angry with me, and this was what *Ivanenko* understood as "being shocked."

By the way have you much time to read books and do you like to read? I should sometimes like to send you some books which you do not know and which would probably interest you. (At least I believe you told me you did not know them; for instance "Les climats" by Maurois (French) and "A Farewell to Arms" by Hemingway and "Liebe" by Wildgans (German) and many more.)

The included pictures are for *Nina Nicolaevna*. Tell her that Pauli has many faces but I send some of the nicest.

Now good night, Genia sleep well — oh I suddenly remember that it is now 2 o'clock in Leningrad so that you probably are long long asleep. Never mind — sleep well my dear!

Your Rudi

∽

Genia Kannegiser to Rudolf Peierls[45]

Leningrad
October 10, 1930

To day is my free day and in the morning I was in *Eremitage* with Nina and two my cousins *very nice, isn't it?* I was not there five months and it was so agreeable to look again on all "ensemble." You were charmed with it, but certainly one visit is nothing. What schools and painters do you prefer? I like very much van Dyck and *Rubens* — the portrait. (Do you remember these of *Eremitage* but to all I prefer the early Germans and Italians, von Eyck and his school and fra Bueto angelico. From Italian I love Sandro Botticelli awfully love his faces which are on the line between beauty and ugliness and Peruqino Madonnas — mothers and little girls together with childish eyes and wise lips and such touching trees — like big branches, and more genial painters of all peoples and all times — Leonardo da Vinci, but we have here only one his picture and I know him from books and reproduction only. But before Rafael I remain quite cool, more than cool I dislike him too quiet (ca/m, too "easy," too sweet) and I do not like Rembrandt, why I don't know, but he is nothing for me. We have now many new pictures — new French painters from *Moscow* one very nice Monet and Degas and Renoir. I like him very much I wrote you that in one of my letters. We will go together there in March. Yes?

After dinner we were sitting with Nina on our divan and we write verses about "Grusiya" and our trip, some of them quite funny, Oh,

---

[45]Sabine Lee, Volume 1, page 134.

Rudi if you will see here "Herr Dr. Sack" or write to him tell him that I am very angry, because he promised me to send photo to me and did not and more than that, I wrote him a letter with Bubi together and he did not answer. Perhaps he is angry that I did not see him before he went from *Kislovodsk* but I begged his pardon in the letter and really I was very sorry with that. Do you remember? On the railway station. So I am angry.

The day before yesterday I have a great walk quite alone in the Ostrova.[46] (You were there with Bronshtein, it is near the Buddist temple.) It was splendid gold parks and the lakes as looking glaces in the sun set and the sea white and rose two stone lions with wise eyes, and I was alone quite quite alone. Only the moon came after and laughed on the sky. *Nice*. I like sometimes to walk alone, this time I thought about you darling and wanted so much that you were near me.

And yesterday began the winter in the way to my laboratory. I fell into the snow storm and it was such wind that I cannot stay on my feet when I arrived at last in my room I was a bit of ice. To day is stormy and it rains and it is very high water in Neva — a typical autumn paysage, and it is cold...I speak so much about this "weather," there because my warm overcoat will be ready only in three weeks(!) and it is rather disagreeable to wait it such long time, but we must buy many things and the dressmaker is occupied. Oh, oh, oh!

But it will be splendid.

Yesterday I was half mad, because — every few minutes I read a splendid criminal novel from Edgar Wallace "The Face in the Night." Do you like Sherlock Holmes style? I awfully. In the evening I souped with one of my friends (this Kiebel of this incomprehensible wire)[47] the first time after my traveling in restaurant I remembered our last supper in *Kislovodsk* do you remember the gray sky and musik in the garden?

We finished it with a glass of chocolate, ou I adore it! Kiebel is very good boy, really very good, but a little "strange" too often too sad and don't know what to do next. When he is with me — all right in five minutes he begin to be unhappy again. It was a sad love story with him, but I think that it is simple "melancholic character." You know about this wire; all the town gossip about us. I make an ununderstanding

---

[46] Kamenny ostrov, an island in Leningrad. -SL

[47] Letter from Rudolf Peierls of September 29, 1930, see page 70. Ilya Efroimovich Kibel (1904–1970) was a notable Soviet mathematician and meteorologist. In the 1930s he was known to his friends as Kib and belonged to a circle of "astronomers" closely related with Genia, a group which also included Ambartsumyan, Gamow and Bronshtein. -SL

face and big eyes, but *Dimus* made all he can. For me it is nothing at all, but for you?

I got the abstracts from Gamow, from *Nature*. The last was with photo of Piz de Daint with you and Rosenfeld on the snow. Very well indeed, is it yours? You know that it was the Gamow's dream to inscribe under his work some mountain. Now I think he is quite happy.[48] I must write to him, but if you know how many letters I must write! Now it is 2h 30m and tomorrow I must go up at 7h 30m.

So good night my very dear. I wanted to write you one thing, but you would say it is for compliments. Good night! How many things passed and it is only less than 1.5 months. I am very glad of your letter. Very.

Yours, Genia

co

Genia Kannegiser to Rudolf Peierls[49]

Leningrad
October 14, 1930

Dearest,

I write you very high on the tower, where I work every day because it is of wood and glace and my Kohlhörster[50] is too sensitive for the strong walls, my writing table is below, but I must to measure every ten minutes and I wrote you with a pencil on a book in intervals. This tower stay right on the bank of the Neva and is the highest point of the district. So I can see very far around, and it is beautiful — three bridges, blue river, with *fabelhaft* buildings around, and on the horizon sea and forests. It is quite a pleasure to sit here.

Rudi, dearest some days before I wrote you a long letter, but when I came to post office I find that in my pocket I had enough money only for the simple letter and I send it so, I think that you will get it together with this.

---

[48] Peierls, Gamow and Rosenfeld went hiking in the mountains where Gamow conceived the idea for the paper. He wrote the paper in a mountain hut, took it and signed it on the top. See George Gamow, interview by Charles Weiner, April 25, 1968, Niels Bohr Library, AIP. -SL

[49] Sabine Lee, Volume 1, page 138. The letter was written in pencil and some passages are illegible as a result. -SL

[50] A set-up for measuring cosmic rays.

The day before yesterday and yesterday I got your both letters.[51] I was awfully glad. Awfully! You know, darling, I read in them many things I wanted to write you, it is very curious indeed. You know that every evening in my bed I wrote you long long letters, so long and complete that I also forgot always what I had write to you in my head and what on the paper.

Rudi, every day, every hour, it becomes [illegible] to wait. I thought that it will be better, but ... Really it is too hard to speak on such distance every evening! The Russian people says that love is how fire and distance how wind, it kills a little love and blows up a big one. It is true, I think. Yes? You wrote that your letters are cooly and I just thought why you have never tell me such "warm" things as you write! Perhaps, you simply haven't a time. Yes? Oh, *Rudi* how warm, hot. I think now, and now it's difficult to write! Yes, about "profanation" simply I forgot all less "pathetical" words and so I put profanation into " " to make it more "at home" but you understand.

Your Russian progress are "fabelhaft." Really, in two weeks I shall write you Russian letters for practice, but I am afraid that they will be too lyrical for a teacher!

Yes, I received yesterday a letter from Sack with four photos, two of them you in *Tiflis*. Sack wrote one half of his very short letter in Russian, and really very well, so well that I think that [...] wrote $\frac{3}{4}$ of this half. Thanks for your photo. I enjoy them very much. It is your Zurich house? Tomorrow is my free day (it is so nice in the *USSR* every fifth day you are free, poor boy with all your laziness you have the moral right to do nothing only each seventh day!) and I will photo myself and sent you in next letter quite new portrets.

I have been at the opening of the season of our *concert hall.* Our "Generalmusikdirektor" is nothing in comparison with Bruno Walter,[52] Kleiber,[53] or Klemperer,[54] but the orchestra is very nice and a splendid first violin. It was Skriabin[55] evening. Do you like it? I don't know him very well, but I love the things I know very much. The "poem of extase" is splendid, really splendid with its ill violins in the middle and joyful trumpets at the end. Skriabin is free and "windy." Yes? I

---

[51] See pages 78 and 80.
[52] Bruno Walter (1876–1962), at the time conductor of the Leipzig Gewandhausorchester. -SL
[53] Erich Kleiber (1890–1956), at the time Generalmusikdirektor of the Berlin State Opera. In 1934, he resigned in protest against the Nazi regime and emigrated to South America. -SL
[54] Otto Klemperer (1885–1973), at the time director of the Staatsoper am Platz der Republik, a branch of the Berliner State Opera. -SL
[55] Aleksandr Nikolaevich Skriabin (1872–1915), Russian composer and pianist. -SL

don't like Wagner very much, only few things the "fly of Walkiire," for instance, but only once in my life I fall asleep in the theatre and it was on Tangeuser.[56] It is true that I was little girl then, but it is true also that remembering it now I close my eyes. When he begin something he can not stop, fifty times the same thing! You know that my motive of my love for you is Carman of Schumann's "Carneval." I don't know why, but perhaps because when I in the first time thought of you very seriously, someone played "Carneval" in the ladies room in *Grusiya*.

I was so glad to read *"you"*[57] in your letter, it is stupid this English with his "you" in all conditions. Good bye, till to morrow! I must go downstairs and work without intervals till the evening and then go home — dinner, bath myself and sleep. Not a very interesting program! Yes good bye, darling.

Yours, Genia

∽

Rudolf Peierls to Genia Kannegiser[58]

Zürich
October 14, [19]30

My dearest,

Now the winter begins, and that means that the airplane Berlin-Leningrad no longer goes. So the distance between you and me increased by some days. That's terrible, isn't it? On the other hand, it means that 0.25 of the time till 1st March has passed. I just came from a concert of Stravinski, very interesting, but not easy to understand. Still I liked it very much, especially one piece where he played the piano himself. A very light "capriccio."

There is one man more now in our institute: Waller from Upsala[59] (Scattering of X-rays, theoretically). He is a sympathetic man, but terribly silent. I am usually not talking much either, but he is 1000 times worse. Still one can walk with him and so we did last Sunday. It

---

[56] Tannhäuser.

[57] ты. In Russian, there is a distinction between the formal (вы) and less formal address of you. Rudi, in his letter to Genia, used the informal term. -SL

[58] Sabine Lee, Volume 1, page 140.

[59] Ivar Waller (1898–1991), theoretical physicist and crystallographer; later professor of mathematical physics at Upsala University. -SL

is very nice, walking in the hills around Zurich, especially in this time of year. The fog lies on all valleys and covers them, so that one can only see the tops of all hills. In the background of course the higher mountains, which already are covered with snow.

I wrote you the other day of Heisenberg's letters in which he usually writes very wild theories to Pauli. Some time ago he had written a paper on the self-energy of the electron and connected it with the other difficulty, the zero-point energy of electromagnetic field.[60] We showed that if one solves the one difficulty, then the other will be solved, too. Pauli wrote to him, then, that this paper was like the old proverb: "If my aunt had four wheels, she would be an omnibus." Now some days ago, Heisenberg wrote: "I think the aunt is an omnibus," and gave a solution of both difficulties. But we did not like this solution, it has no physical idea in it, and we tried to find a mistake. And really today I found it. Fancy the face of the post-official, when he had to receive the telegram: "Tante ist doch kein Omnibus, Fehler gefunden, Hurra." Pauli was so glad about it that he promised me a bottle of champagne, which we will drink when he is over back from the Solvay Congress.[61] I am so tired, Darling, good night. Write soon!

Your Rudi

PS. How are you and how is your Kohlhörster?

∽

Genia Kannegiser to Rudolf Peierls[62]

Leningrad
October 15, 1930

It is $11\frac{1}{2}$ o'clock now and I go up to only 15 minutes before. Quite a "Pauli's" inclinations!

I am awfully jolly today, darling. It is so nice in the street — sun and blue sky, and warm wind. I still go at post office and after to islands what I shall do in the evening, I don't know yet. ("It is better not to

---

[60] W. Heisenberg, "Die Selbstenergie des Elektrons," Z.Phys., **64**, 4–13 (1930).

[61] Wolfgang Pauli was attending the 6th Solvay Congress in Brussels between 20th and 25th October 1930 where he spoke about the 'magnetic electron.' W. Pauli, "Les Theories Quantiques du Magnetisme: l'Electron Magnetique," *Le Magnetisme*, (Gauthiers-Villars, Paris, 1932), pp. 175–238. -SL

[62] Sabine Lee, Volume 1, page 141.

know the future" from the letter of ... (I don't know the date) October. Yesterday was really nice, also. I returned from my laboratoria on foot, and by the way had climbed on the top of St. Isaac's church' it is awfully high and difficult climbing indeed, but it is "fabelhaft." Will you, we shall go there together.

Rudi, if you knew how afraid I was this evening in Kislovodsk when I let you go alone in Mineralnaya! But I was so tired, so dreadfully tired that I thought that if I went with you you would be obliged to carry not only three trunks but me also.

My story, really, is not very interesting, but you must to know it — for to know me. It is difficult to write such things but I will try. I don't remember you or not, but I told you this evening that you are my second love, four or five men loved me, but I loved only once before this summer and my first love was Ivanenko. I was a little girl then, this story began four years ago and ended $2\frac{1}{2}$ years, and now seems to me that I read it in some book. But it was very "serious" and I awfully growed up during that time. It began with pity and something like mother's feelings. (He was very unhappy then) afterwards came the classical "first love" and it ended from both side in the same time. So I was Dimus first bride near one year. Bride not <u>wife</u>, you understand? Ouf! Rudi, perhaps it is very uncomprehensible but I shall relate to you the details when we shall see one another again.

Now we are in very cool relation, more than cool, because Dimus has an awful character and was so angry when this story ended, although it had a very "symmetrical" end, that we did not speak for half a year. Do you remember the way from Beslan till Mineralnaja, it was very strange my first and second love in one car! [...]

I will sent you the Russian book when you will understand them, certainly you have not read Bodin, Sredin and many many other new writers. Some one are very interesting, and for you will be easier to understand them after your traveling, and before another one.

Rudi, I forgot to tell you that I started speaking German. Oh! Es ist fabelhaft, ich spreche mit meinem Stiefvater und meinem Vetter. I speak awfully with a lot of English words but I can write that, Ich liebe dich sehr stark, sehr sehr stark!

Nina wrote these lines below before I begin this side.

So good bye, I want to talk with you more, but it will be without end and you will never receive this letter.

A Romance Taking Shape in Letters

Figure 4.1 Lev Landau, unknown (according to some sources, Genia Kannegiser), Dmitri Ivanenko, circa 1928.

I write you so long, long, long letters that you can write me not only answers. Yes? Ou, Nina telephoned that she is waiting for me so at last good bye.

Rudi, my dear. I miss you terribly. Good bye. Kiss you. Love you strongly. Genia

— Hello. Mr. Peierls! How do you do? Thank you very much for yours photos. Pauli sitting in the boat is really charming. I shall be very glad to see you here next March, then more. I hope till this time you shall speak Russian fluently. Once again thank you very much. Do you understand?

Nina Kannegiser

Rudolf Peierls to Genia Kannegiser[63]

Zürich
October 16, [19]30

My Dearest!

How is that possible? I thought you were so much used to be teased, and take such theories humorously, and now you seem to be offended by that harmless joke about "fishing for compliments"! Isn't that silly? No, seriously, you will write me what you meant in your letter without being afraid of jokes I would possibly make about it? Yes? You will get much more training from the side of Dimus, and so on, and it must not be agre[e]able for you. I, of course, do not mind it, I am far from Leningrad and then, much talk usually is less agre[e]able for a woman than for a man. But it must be terrible for you and I am so sorry to be the cause of such inconveniences for you. And I am afraid that your situation will be even worse in March and afterwards. Afterwards — a funny word; really I cannot think of the time after I will have left Leningrad again, it is quite impossible to imagine it now, that there is some time "afterwards." I agree with your taste as far as Raphael is concerned, I think him very tedious and pedantic. [...]

Today I had my Russian lesson again. *It is good to speak Russian.* My teacher is Russian, but she left Russia very early, she was only a child when her parents took her out from Russia. But she loves this country so much. Today I stayed there for two hours after the lesson and had to tell her about Russia and she became quite upset when I showed her pictures and told her many things I had seen there, things that made me also think of Russia with a kind of homesickness. She knows a lot of Russian people here who live here now for many years, are married here and have their profession and so on, but they all say that they can never feel at home here, that their home is still in Russia. But my room is not heated. It is very cold. (How cold will it be for you in Leningrad and without winter coat!) So I will go to bed and go on writing tomorrow. Good Night Genia. In Leningrad it is now 2 o'clock so you already will be fast asleep. Sleep well!

I am so glad your letter only went 5 days. So after all it is not so far to Leningrad!

I know that Gamow laid so much stress on undersigning his paper

---
[63]Sabine Lee, Volume 1, page 144.

in the mountains, and he wanted by all means a mountain with snow on it. Every time when we planned a trip he asked what was the name of the place and was there snow. I have the suspicion that he only came with us for that reason! But it is a pity that we could not go on "Piz Quatter Vals" — the name is so much more thrilling, but it was too difficult. Gamow had difficulties with his foot (he had had some accident in England before) and Rosenfeld's shoes went entirely to pieces and we got into fog and so on. But Piz Da Daint was also very beautiful.

Tomorrow probably my friend Thorner comes from Heidelberg for a weekend. In those two hours in Heidelberg we spoke also about "philosophy" you know: causality. Heisenberg's uncertainty principle, and so on. He is a doctor of medicine, but one of the few people who could possibly understand these things without being a physicist. But of course it will be hard work and one must always be very careful that these biologists do not take anything they want and claim it as "a consequence of the new physical theories." I will be very glad, for personal reasons, too, if he comes for there are not very many people in Zurich with whom one can speak reasonably. I am so sorry, I wanted to have a photograph of me for you as you wanted it and I made Waller take it. But it came out so dreadful, that I did not send it to you.

O dearie.

So as I said, Thorner was here and it were very nice days. Last night we went to the theater to see "How to become rich and happy," a very harmless but nice comedy, well played and well made up. Before and afterwards we had long discussions about the meaning and the use of the notion "time" in biology. He had some plans, some problems and ideas he was very fascinated by and he wanted to discuss them a long time, there remained no problem, no idea and he had to confess that he had had no clear thought. So in the end the effect of our discussion is only destructive. But besides this we had a splendid time. This morning we went with an early train to Einsiedeln, a place near the Vierwaldstatter See, climbed a 1900 meter mountain there, the most beautiful, although unknown point of this part of Switzerland, the great "Mythen." This is very sharp and steep rock, quite near the lake, and as it is the highest of the surrounding mountains, you have a beautiful view on the whole chain of the mountains behind. Beneath you you have many lakes in narrow valleys. Today like always in autumn, the valleys were filled with fog but on the higher mountains the sun was shining and one looked down on the fog like a sea with many islands, the other mountains above a dark blue sky. I had never seen this spectacle before so beautifully. I tried to make some pictures and when they are ready I hope I can give you a little impression of the beauty.

The end of the beautiful day was a bit exciting, because our train back to Zurich was late, and he had to get here the train to Heidelberg, as he must be in his hospital again to-morrow morning. Now his train to Heidelberg waited for the other one but the man to whom he had given his bag (which he could not carry to great Mythen and back) was not there. So he got his train all right, but his bag stayed here and I had to send it on to him and there were all kinds of complications. Now it is quite a general theorem that every letter has an end, and even this one must have it, although it sounds very improbable. I think often of you, dear, I dream often of you (far too often!) and I am longing for March.

Good bye, my Genia,
Your Rudi

One must always be careful with general theorems! I just got your letter of 14./15. and I was so happy.[64] Now this letter would certainly not have an end if I had the time. But sometimes one must work a little bit! I am longing for you, Du Liebes!

༄

Genia Kannegiser to Rudolf Peierls[65]

Leningrad
October 22, 1930

Rudi,

It is a too long monologue! Think only, that it is the third letter and I have not the answer to the first one. I know dear, that it is not your fault, but there is now two weeks between an letter and an answer. Horrible! Why aeroplanes cannot fly in the autumn? It is so stupid that the air post is only in the summer. Yes? But nothing to do. I think that I shall have your letter in four or five days, so the next one will be a little more "diagonally."

---

[64] See page 84.
[65] Sabine Lee, Volume 1, page 147.

Rudi I don't think that I killed somebody — this red spots are only Fe2O3,[66] I work with it now and I am tired to wash my hands every three minutes, it is awfully difficult with it so fine fine pulver, and I am now quite an Indian girl. Really! [...]

Rudi, tell me the things without which you "cannot live." For me verses, Musik, sunny and windy days and I awfully love the "white nights" perhaps you will see a bit of them in the April. At the concert I saw a lot of people between others Mrs. Frenkel who was in her blue dress of *Grusija*, Ivanenko with his wife. Dimus is here only few days and will return very soon in Kharkov. In Kharkov two weeks were Jordan, Fock and Ambartsumyan.[67] I like him (Ambarts) very much, such a nice boy without any tricks, and such a lot of people has them, Landau, Ivanenko, Bronshtein, this all are a little "mad" (too complicated). Only Gamow is always in good state of mind and it is very easy with him. Yes? Ambartsumyan relate me very many funny things, and I was glad to see him. Do you like him? He has a splendid head not only for the [illegible] but for all. You understand? He is a very clever boy and very "original." Rudi, a boat goes under my window and a long tail of white foam is behind it how beautiful it was three nights on *Grusija*. Do you remember?

It is so sad to write when you want to see and speak. Oh, Rudi, it is soon two months, one third only, a half of a year it is such long time. Think and you and I can be quite different people. We can "grow up." This letter is too dirty, but if I copy it it will be too clean and unnatural. So " beg your pardon"(!!!) and sent it so.

What do you now? How is Bloch? How are you with this girl? How is Pauli? It seems to me that I will never receive a letter from you! I am very interested now in cosmical rays, and of absorption of hard γ and so on. I think that you can have from absorption coef. an desirable wave length very easily. Yes? Here is only "..." Yesterday I was looking in "Berichte"[68] and meet your name there hundred times. I was glad to see your name, if I cannot see you. Good bye, darling! Good bye.

Genia

Dear Rudi,

I was so happy when I received your letter and the photos. Your

---

[66] Iron(III) oxide or ferric oxide is the inorganic compound with the formula Fe_2O_3.
[67] Victor Ambartsumyan (1908–1996) was a renowned Soviet/Armenian scientist, one of the founders of theoretical astrophysics.
[68] *Physikalische Berichte*, a German physical journal published in 1920–1978.

write Russian *fabelhaft*. Serious *fabelhaft*. I think that with my English you are not so satisfied. Yes? (It is not "fishing.") Darling, if I could be near you only half an hour! But half an hour is a lot of time. Do you remember the train between Vladikavkaz and Beslan, so many things passed in it, and it was only 40 minutes. In half an hour I could tell you at least five letters, because you know that I can speak 10^3 times easier than write. Many people says that I speak too easy. But I never thought that it was difficult for you to speak, at contrary I remember how you screamed writing letter. [...]

Darling, this story with your friend is really too "european." But the position of these people for my sight is not so bad. Why, they are young, healthy, loving, they had some position. You say that they can lost it, but in this world all is very uncertain. And, you know, for me the money is not a great thing. I am so habituated till the early childhood to the difficulties with money! Worn and homekeeping, but 0.9 of the women do so, and I think that they were very happy when all these stories turn their end and they were together. But poor girl, it is awfully hard to marry a man against his family, because now she certainly think that she must be for him all — and family, and position, and money (you understand?) and if she is not sure, that he is happier now with her and without all those things, than before, she certainly will be very unhappy. If he is a clever boy, he must be very careful and tender, because it is always easier to give something that to take and he gives her too many. But you know, although it seems to me very "unnatural" I can understand the parents (and you?), and perhaps they are the unhappiest in this story. Think only they loved their boy, they loves him also certainly they think that they are right, that he would be happier if he made all that they wanted. That now his life is spoiled, that he did not love them etc. etc. It is an awful thing to think such things (!!!). But the boy is right, because in 23 man must live with his own head, he knows in general what it is better for him and what is more honest for him. (Ouf, how many philosophy can one write, you are quite tired now my poor boy.)

Dear Rudi, dearest Darling, what I must do? I write this few lines in russian because they are not very interesting for you and you can pass them if it is difficult for you to read awful handwriting in strange language. Rudi, what a splendid little girl of five months I saw yesterday! She is a little daughter of one of my acquaintance, she has big round eyes, quite round and dark grey, she is rose, rose, and her nose is turned up and so small! She laughs all the time and begin to "speak" and she has curled hair. How I love the little children. I think that even more that boys. How are your nieces and nephew? Dear I think that the position of uncle and aunt is the best in the world. We have all the pleasures and none of the disagreements and responsibility. Yes? I have many little nephews (the children of my cousins) but they all live far away. And always the uncle for the child is the most splendid thing. Because he is always in good humour and brings nice and sweet things in his pocket, I remember how I awaited one of my uncles! Darling, I want to walk with you, to day in sun and blue river with boats in so near!

Thorner is a very nice boy, but very very boy. He is younger than you. Yes? How are your lectures? Last winter I read the lectures for the worker, who wanted to go into the high school. I learned with them mathematics and physics. If you knew how I was afraid the first time, the youngest of them was 27 minimum! And amongst them were quite old men (35!). But they were very nice and learned with great pleasure. Afterwards I had another group, with more younger people. They were very lazy and I was awfully severe — awfully! You don't know how severe I am!?

Darling, and you with Stradoba is very nice. I don't know why but old people and little children always like me, with the middle age it is not as easy, but with old one if I don't tease them it is always all right. I did not know that you have bad character. At contrary, I think that you have a very nice one in comparison with Landau for instance or even with Sack. Do you remember how easy he became unhappy, every insect desolated him and with you it was very easy!

Dear, I was very astonished in Odessa with you. You were so "uneuropean" and I like in you so much that you can make "nonsense," there are some people who cannot. Sack cannot certainly, and Mr. Simon, and Ramsauer(!)[69] and Bursian[70] (do you know him?) and even Dorfman,[71] but I think that Pauli can, perhaps Bothe, and any Russian physicist — Gamow, Frenkel, Landau, Bronshtein, Krutkow,[72] Fock.

---

[69] Carl Ramsauer (1879–1955) was a German physicist famous for the discovery of the Ramsauer-Townsend effect.

[70] Victor Bursian (1886–1945) was a Russian/Soviet physicist who worked on geophysics and physics of crystals. Bursian's seminar at the Ioffe Institute stimulated the development of quantum mechanics in the early days of Soviet Russia.

[71] Yakov Dorfman (1898–1974) was a Soviet theoretical physicist known for his works on condensed matter and history of physics. Yakov Dorfman and Yakov Frenkel were the first to develop the theory of domain structure in ferromagnets.

[72] Yuri Krutkov (1890–1952) was a Russian/Soviet theoretical physicist known for his contribution

You understand what I mean under "nonsense."

October 23

Darling, I think that I will never stop this letter. But I must end it because it will lost any interest in one week. So good bye my dear, dear, dear! I kiss you.

Genia

P.S. I beg your pardon, because this letter is so dirty (It is not fishing, really I am very ashamed with it), but I cannot help it because I write it in free places, and in my work room also with dirty hand.

Good bye.

Genia

∞

*****

I think that the above correspondence covering six weeks in September–October of 1930 gives a clear idea of the romantic feelings between Genia and Rudi, and their growing attraction. They openly shared their thoughts about love, physics, friends, music, everyday routine, and many other things. The exchange of letters continued unabated until early March when Peierls arrived in Leningrad by the invitation of Yakov Frenkel who had asked him to deliver a number of lectures on quantum solid state theory. On March 4, 1931, Yakov Dorfman, Frenkel's assistant, met Peierls at Leningrad railway terminal (Frenkel himself was visiting the University of Minnesota at that time). Dorfman told him right away that all theorists at Leningrad Fiztech were looking forward to his (Peierls') lecture course but would like it to be delivered in Russian. This was a surprise to Rudolf, since he had started learning Russian only five months before. Despite all odds his course in Russian was a great success. Peierls mentions in his recollections [5] that half a century later he met a number of former students who had attended his lectures. They praised him for his excellent introduction.

Below I will quote just a few excerpts from the letters of Genia and Rudi referring to November–December of 1930 and January–February 1931.

---

in the organization of physics research in the early days of Soviet Russia. Arrested in 1936, he spent subsequently, 11 years in Gulag.

October 25, 1930

— Dear Genia, send me your eyes that I may look into them half an hour; send me your hands that I may hold them in mine for half an hour, send me your lips that I may kiss them half an hour; I shall return everything half an hour later, well packed, recorded delivery and post-paid. No, you better send them not, for I would not return them, I would keep them and you would be uncomfortable in Leningrad without eyes, without hands and without lips.

October 28, 1930

— Dear Rudi! My laboratoria is awfully far from my home, 40 minutes on the tram, so I every morning have a long time to read but in tram I can only read the "stupid" novels, so I do. The last was "A Girl on the Boat" by P.G. Woodhouse it is funny, very funny. Rudi, I also cannot, cannot imagine the time after March. It is the "end of the world", but it is such a long time till this end! I think that I used a bottle of ink and now only two months are passed, and how many letters wrote with pencil! Four months — 120 days — 24 free days! Awful, awful, awful.

November 4, 1930

— Meine Liebe Genia! My room is about as large as yours and fortunately not the usual "chambre meublic". All the house is made up with a good and original taste. The best of my room is the view from the window — I include the picture — and also the view from the street if I walk to the institute. Then if the weather is clear, one sees the whole chain of the mountains. Now, they are already quite white. Sometimes are clouds above us, so that it is here quite dark, but the mountains are bright and sunny... I am at home usually only in the evening, and even then I am often out, but I generally have supper at home (or at least something which I call supper, for I make it myself) and I am bourgeois enough to spend half an hour or even an hour on this occupation, reading my newspaper and so on. From about 9 till 7 o'clock I am in the institute with half an hour or a bit more interruption for lunch. I do not do very much besides. On Wednesdays after our colloquium I have to drink with Pauli and Wentzel. I have the Russian lessons with Fania Moskovskaya. She is a nice young girl, an emigré from Russia studying law at Zürich University.

November 6, 1930

— Rudi! Yesterday was my free day, the worst are free days of all that time. I had a rendezvous with Nina and mama at my dressmaker, they were certainly late half an hour (the women always are late, I am the only exception) and the worst of all my dressmaker was not at home, and she lives awfully far and the weather was as bad as possible. And I was hungry (!!!) and wet and angry... I am angry now because I want a new dress, long dress, all my old ones are very short. I forgot to tell you — I've got a new overcoat. My old one I had for 8 years(!!) and I could not to look at it once more. At night the wind opened our window and while Nina and I were arguing whose turn it was to shut it I got so chilled that I think I have a cold anyway. Rudi, I don't think that you are a good cook. I always marvel at how thin you are in comparison to, say, Sack, not to mention Pauli. It's because you don't eat enough. You must eat lunch, and a lot of it, alright, darling? Yes, a pleasant news. I got the translation from English journals for one institute. I am very glad because it is very well paid, and I'll have some spare cash now. When you come I'll have enough to go to Novgorod and even to Moscow. I will be very occupied in the evenings now.

November 16, 1930

— Dear Genia! I very often think of the first days in Odessa: No you were not at all "exotic" to me, on the contrary, there was something like the feeling I had known you long before. I never dreamt of comparing you to any other girl I have seen but I do not think that I know any other like you. It is curious that you are at the same time typically Jewish (and therefore familiar to me) and typically Russian (and therefore strange). But all this is only the frame. When you speak or laugh you are sometimes like an exploding firework. Landau in the mountains on skis is a funny picture and I remember how once he was in some place and could neither forward nor backward, because he declared he would fall down. Then Thorner became very angry and called him a coward and then with little little steps he managed to get through. I must leave for a lesson with Fania Moskovskaya. Love.

November 17, 1930

— Dear Rudi, if you only knew how anxious I was these days, think two weeks without any word from you. I thought that you fell from some mountain and broke your leg like Bloch, or even worse that this was your head broken, or that you are ill and are lying alone in Zürich. Rudi, my darling, darling, I felt so sorry for you, when I thought of you lying there all alone, with nobody to care for you and look after you.

November 30, 1930

— My dear Genia,[73] Do you know that today is half time? Three more months, still a long time, although February only has 28 days. What will it be like in March, what will I be like, what will you be like? I am a great optimist, I think everything will be fine. I am glad that we have the letters. This way there was always a Genia close to me. But was it the same Genia who was in Odessa and whom I shall see again? The same Genia whom I am longing to see (and not only see, as you know, also kiss and caress, talk to and listen to, feel her hand in my hair, and more, more!) Dau wrote that he will be here soon. He is very lazy and he only answered after I had written him three letters, and only a postcard on which there was nothing but a big "?". I had to know whether he would be renting, as I was supposed to book a room. Now everything is o.k. Landau, Thorner and I will be in Arosa, in the same house where we stayed last year. Next week I will not have any Russian lessons. Fania Moskovskaya will be in London to meet a friend. To-morrow Landau will come and I am very glad, I have so many things to speak with him that probably we will have to talk 48 hours without interruption. You in the university without gown must be fabelhaft.

December 6, 1930

— Dear Rudi, I have in my mouth a very nice — fabelhaft — bonbon. So I think my letter will be very sweet. Rudi, for me it is very difficult to write when I want to speak, and now I want to speak with you. Every man, if he even a little interesting have many "faces". I have three groups of them. First — to everybody — you know it very well certainly, secondly, one for very few persons, I think that I show you all that I could in such a short time, of this second face, and the third nobody knows, even I do not know it till the end, if I shall write about it. I think you will understand nothing, it is very difficult even to speak about such things, and to write? In two words — I think I am awfully "girl" to everybody, and in the middle I am awfully "woman". You understand? The days run by awfully quick; all the day I run around the house and I think that I jumped so many stairs as the height of Yungfrau. I lost my purse will all my money (not very much, 1 rouble) but I must go home by tram and send this letter to you. Where can I get these 50 kopeks and key to lock laboratoria? Oh, I know, I will take them from the "portier". You can see what a wind is in my head. I

---

[73]This is Rudi's first letter written in Russian.

am hungry, but it is not so bad to be hungry if you think about dinner. I like sweet things (do you remember Odessa?). I ate chocolate only when I was a little girl — six years old, because after the war we have not then in Leningrad. I remember them but not very well. Bring, yes? In my last free day my cousin made of me and Nina seven pictures, five of which were spoiled. I will send you me in new coat, perhaps in new dress. I will have a new dress long, long till my end.

December 6, 1930

— My dear Genia, Today and tomorrow there is a physics meeting in Tübingen where Lise Meitner and Geiger and many others will be. I am supposed to go there and talk to them. Pauli wants to know something, and cannot go himself because there is a big ball of the Italian students and he will have to be there. Pauli begins his open letter addressing it to "Dear Radioactive Ladies and Gentlemen!" and then goes on presents his explanation of β decays by virtue of "little neutrons". My train left in the middle of the night. I had to get up today at 6 in the morning, and it was particularly unpleasant. Last night I was in the "Kanne" with Pauli and Dau and we drank very much. Kanne's landlord is drunk every night and then gives long speeches in mixed German, Schweizerdytsch, and English, very good philosophy mixed with indecent jokes. Therefore, he does not allow ladies to come to his bar. Pauli is a very good friend of his, and it is very funny to watch him in this milieu. Now there is some life in the institute, ardent discussions about everything, and if you would come to my room you would hear a dreadful noise, because at any time of the day there will be some people of different opinion trying to convince each other. All this of course is due to Landau. It is noon now, this is the time when the letters from Leningrad arrive. Perhaps your letter is now in the institute and I will only get to read it in the day after tomorrow.

December 11, 1930

— Darling, darling, darling. It is terribly hard to wait when you can't count days, as there are too many, but you have to count weeks or, rather, five-day weeks, because the days off are like milestones on the road from September to March, when you come. There remain 16 days-off, just as many letters and lots and lots of dreams and evenings when I lie down and speak to you. One of these days I had a dream as if you were sitting on our wardrobe and I was feeding you with buttered bread and honey. Honey flowed down on my fingers, they get sticky and sweet, sweet. It sometimes seems to me that I know you not only for three months, but for many, many years. Rudi, do you remember we were in Vladikavkaz and you asked me, "Do you want me to stay in Kislovodsk, too?" and I answered, "I do want and I don't want; it will be too sad to part. Now that I haven't got used to you it wouldn't be so difficult, but later?" It proved to be 1000 times more difficult. I can't help crying but I am very happy. If only they could invent the "Time Machine" (remember Wells' book!): I'd press the button and we would have March. Today I tried all the morning to put a "draht" in the galvanometer.[74] The diameter of "draht" is 0.003 mm. It is difficult as to a camel to go through a needle, but I did all till the end and in the very end I wanted to make a little better and teared this damn "draht" into five pieces. Yes, darling, "le mieux est l'ennemi du bien".[75] Now it is 10 o'clock in the evening and I am still in my laboratoria. I want to post my letter on the way home. Oh, yes, darling, please, tell to Das, if it is not difficult to him, he brings me a little thermos (one glass). It is for my stepfather. He has "stomach ulcers" and some days can only drink hot milk, and his thermos is broken.

December 14, 1930

— My dear girl! When you receive this letter it is almost New Year's Day. But I am longing, honestly. Dau philosophizes all day and that is perhaps too much. Now I spend almost all day arguing with him about theories of physics or theories of films or many more. This morning Dau and I looked at the paintings of an Italian-Swiss painter: Giacometti. Beautiful bright colors.

December 21, 1930

Genia, I desperately want to see you. I cannot imagine, what it will be like in March. Too good! I am often thinking about you, English, German, Russian and often without language. But I am still speaking Russian too slowly, and therefore it is still difficult to talk to you in Russian. It was useful to speak Russian with Dau, but we are now talking frequently and 'much about physics and we both speak German so quickly that it would take a week to say in Russian what we say in German in one day. We are both very much occupied with the "uncertainty relations." Every day we are improving the experiments, which we have done the previous day. This is very entertaining, but there are few results. Wednesday we all were invited by Wentzel and

---

[74] Draht (German) means wire.
[75] (French) Perfect is the enemy of good, or more literally, the best is the enemy of the good.

his wife and there again Dau made great impression. The main point why he makes such an impression on "bourgeois" people is not mainly that he has different and revolutionary views. This is not seldom and people are used to defend themselves. But he discusses not at all with reasons, he simply puts against them a system which is closed and as "obvious" as theirs. He does not try to be objective but is subjective so they are only with different prejudices and this amazes them awfully much. Pauli, Waller, Wentzel, Fania Moskovskaya, and others were amazed.

December 23, 1930

Dear Rudi, I will be in the hospital in the very beginning of January. It is the operation in the throat. You know I have a little toubercoulose and sometimes temperature jumps up till 38.2 °C.

December 25–January 1

Dear Genia! Today we went skiing for the first time. This was strange because I have not been skiing all year. Very strange when the skis all of a sudden move apart. Dau skis very badly and when we go together I always have to wait for him a long time. It is almost impossible to teach him. Dau will be in Leningrad in March, a little later than me. Kisses, many kisses.

January 1–2, 1931

The New Year — hurrah! Rudi, today is the first day of the New Year and the night was the New Year's night. I'd like to go on dancing, singing, and drinking (!) too. The party was very hilarious. Thank you for the books. I have not yet read them but I shall take them into the hospital. Hospital Brrr. I was once a fortnight with my appendix. Women in hospital rooms always speak about their children. All from the very beginning till the children of children. Ooooo! You know that I love the children awfully that I will be happy to have them, but these conversations... (January 2) I am ill today darling, high temperature etc. and I am in bed now because of that and my handwriting is even worse than normal. So, Dau, I am sure that I will never quarrel with him, because I think he is a little boy, quite, quite little boy and so I do not take seriously all that he says, all that he does. It is too childish you can only laugh, as you laugh when a boy of ten years speaks about "Weltproblem", "love" or something like that. He has the "heart" (you understand) of a man but everything else of a child, even an awfully theoretical child; he has the theories of all cases of life. But he is charming — and I love him awfully — like a "little brother" perhaps. I cannot bear when he is unhappy. Darling, I want you to be here now.

Rudolf Peierls to Genia Kannegiser[76]

Arosa
January 6, [19]31

Dear, my dear Genia!

I did not realize your operation was so complicated as that and that the tuberculosis affected you so much. My poor girl! Now while I am writing to you, you are probably already in the hospital and I imagine how painful all this will be. Poor dear child! Just while I am writing, I sit before the window and outside one sees Arosa[77] quite below, with all the strong and snowy mountains around, the morning sun is brightly shining on all and dark blue sky above. Very cold today, the frost crackling, everything is very beautiful.

The things I told you in my last letter became much simpler in the meantime. There is nothing at all "serious" in them. My relations to this Eva are only a play, for me as well as for her. And truly — a harmless play, impossible to be called an adventure. There is no doubt now that she is seriously in love with Bübchen and that she gave up the man in Berlin. Bübchen is more difficult to understand, he is much cooler, but it seems to be what Dau would call a symmetric situation and so everything seems simple and clear.

As for myself, I am not sorry for anything — I did, because I actually did nothing at all, but also for anything I thought. There was no adventure, but there was an opportunity to learn that I would probably take adventures if they would come. But I am not even sure about this. Adventures do not simply come so that you must take them, you must really want them. And in the present case I do not know whether I really wanted it, i.e. did all to get it. Well, anyhow, the situation is now clear. Some days ago we made a trip to the "Weisshorn" a peak about 2800 m high. We were very tired then and got to the "Weisshornhütte" a house on the edge of this peak, about 2500 m. Quite near of the house, a boy of 15 years fell and damaged his arm. Thorner immediately proved to be a good doctor. But he could not do anything without ether because

---

[76] Sabine Lee, Volume 1, page 203.
[77] Arosa is a tourist resort and a municipality in the Plessur Region in the canton of Graubünden in Switzerland.

the adjustment of the elbow was too painful. So the before-mentioned Bübchen and I went down to Arosa to the pharmacy to get ether and so on. As the patient had bad pains, we hurried and waisted only two hours to go down to a height of 1700 m and back.

In the meantime, Thorner had asked the people in the house to help him for the operation and for bringing the patient down. There were a lot of spokesmen, but nobody answered except three very nice English people, a man and two ladies. We all were terribly upset by all this. Later we met some friends of Thorner and they said: "O, you have been there when the accident happened? Yes we had also seen it, but we went away as soon as possible, we cannot stand such dreadful view!" So when Bübchen and I came up again — without breath of course, we had to assist for the operation and then we four (Thorner, the English man, Bübchen and I) took all our skies down. It was already dark and at every step we sank into the snow up to the knees or even to the hips. But at the end we got down safely, though late, and next morning, the patient could take his train home. Thorner made this operation very well (afterwards he confessed that it was the first time). He is really a good doctor. But personally the relations between Thorner and me are a bit cooler than before. This is mostly due to Dau. Not in such a way that he "separated" us, I would never give him this influence, but he is a very good "Reagenspapier." He makes people speak and tell their ideas, and although his own ideas are in many ways silly (for my taste) it is easy to use him and find out what other people are like.

And Thorner's ideas are really dreadful, especially in the question of relation between men and women. He has the ancient opinion of the two classes of women. One of them "decent" and the others have to be regarded as "Freiwild." It would be not too bad, if he had such views theoretically, but he uses them also practically. So for instance there is in Arosa a friend of that Eva Gabeler. She is a type of girl I do not like very much. She plays consciously with men. But she does it so openly that of course a man knows what he does if he joins the play. After all these plays are quite harmless. Now one day she complained that men behaved terribly against her. For instance one man had invited her for New Year's night and payed for her against her protest. Afterwards she evidently did not do everything he wanted and then he became angry and made her feel always again that he had payed for her and she was "ingrateful." Instead of being upset Thorner said that the man was not so perfectly wrong, that if a girl behaved like she did, she could assume that [s]he was not a "decent girl" and that he could behave to her as he liked. He even made herself feel that opinion. This of course made a great break in the atmosphere here.

I told you perhaps that Thorner in last time has close connections to France, he had very often been there and he likes French culture very much. Now this is a typical example of French culture and that is also why I do not like it.

As for the books you wrote me about, I wanted to read them long ago, but I never could make up my mind. But as soon as possible I will really do it.

Evening

To-morrow it is finished and we must go back, because to-morrow afternoon I must speak in the colloquium in Zurich.

My darling, I am so strongly looking forward to your letter which will show me how you accepted my last one. My dear, do you believe me that nothing at all is lost between us? Good night, I kiss you so strongly! You cannot even speak but probably you will be able to kiss and I will try to take off some of your pains, poor girl.

Good night.

Your Rudi

January 10, 1931

My dearest girl from Odessa, if I was with you I would be gentle and would be caring and I would do whatever I could so that you can forget your pain. But I am not there and instead I am writing to you such unpleasant things! You wrote me a letter lying in bed with a fever, and immediately I am answering in the same condition. But luckily I do not have a lot: a slight inflammation of the mucus "stomatitis aftosa". But it is not very bad, I just have to stay at home for a few days. As for Dau, dearest, I think one can take him a bit more seriously than you do. Of course you are right that he is very childish and what you say about his "heart" is even more true. But all this does not prevent that a lot of his ideas not only in physics but also in life are very interesting and important. I do not say, of course, that I usually agree with them, but it is difficult to say anything against them and that shows that they are new and necessary. Many people act very silly against him, because if he makes them angry (what he claims to be his main occupation or at least his main purpose) they bring the discussion from theory to practice and let him feel how little experience he has in life and especially in love affairs and they say (what is very easy) a lot of things which make him very unhappy.

January 15, 1931

My dear boy, nothing can be lost, nothing can be obtained. Life, love, they are "vectors." They have not only magnitude but and the direction also. And it is dynamic not static, every moment you lost something and obtain another and every moment the directions are changed. Yes? These "changes" are more or less big, certainly. So, my darling, after your last letters, because I got two letters together I will meet you in March not quite the same way, as if I did not receive them but it will not be less or more, it will be a little "another style". I think that you will even prefer it. And certainly it will be different from March, this sea of ink between us and these 6 months, it means something, I believe! Rudi, we are very good friends. I am sure of that now. Yes? Friendship is more difficult than love, I think that between us both it is very well. No, seriously, do you love me? All are charming with me, bring me bonbons, flowers, books. But I was lying and waiting your letter and did not know, who are wrong, you or the post. I have $37.8\,°C$ but really it is impossible to stay all the time at home! Good bye!

January 18, 1931

My dear Genia, how are you? Where are you now? In hospital already? Or has the operation already been completed and you are already at home? I am working with Dau all the time, but progress is very slow. Whenever we think we have understood everything, there is a mistake and we have to start all over again. At the moment there are difficulties, and we don't know what to do. Dau is rather melancholic at the moment — I don't know why and he does not seem to either. It is quite frequently like that with him. They are now showing the film "Im Westen nichts Neues" which is banned in Germany. It is supposed to be a very good film. Tonight I will watch it with Lilly Fenigstein. It was very difficult to get tickets, there was a long queue. The German ban was the best advertisement.

January 21, 1931

Rudi, my evening temperature is still 38 °C, and they make me swallow all kinds of stuff and mother makes me eat something every 20 minutes. I was palpated, checked over, X-rayed, in a word, I underwent all sorts of humiliation. If it goes like this, one of these days I'll go to the graveyard to choose a proper place. But enough is enough! Assez! My patience is pushed to the limit. In spite of doctors, medicines, etc., my spirits are up inappropriately, and I feel happy. If I keep on feeling like this, then, as it seems, at my funeral I'll rise up in the coffin and order to play a merry fox-trot instead of the Marche funébre. In the meantime I gave up working. Two days ago, though, I began to work a little, regardless of the doctor's anger. This job is at another Institute, much nearer to my home. You know, in the first time I thought that I am not for you a simple "adventure" when I met your eyes on that wet deck on that grey morning. And now you are alone, and I cannot help you.

January 28, 1931

My dear sick girl! What only is the matter with you. Seriously, Genia, you have to get better. Last week Dau and I finished the paper. I am very glad that it is completed now. Even Pauli, "der Theorienfresser" did not discuss a lot and Heisenberg even wrote that he agrees. At the moment I am reading Chekhov in Russian. That's great. And I am reading a lot faster and more easily than ever before and I can almost read without a dictionary. My dear girl. I kiss you strongly. 30 more days!

February 2, 1931

Dear Rudi, I write you in my laboratoria! It is great, darling! I didn't succeed to go (to sanatorium) in Detskoye. I was not allowed to go on a long leave of absence, etc.; so I made up my mind to begin working a little. I am still in the middle of my temperature 37.8–38.0 °C every evening, but it is in the evening and in the morning 37.2 °C only. If my temperature will spring on the higher degree I will receive ten days and will go to Detskoye in the middle of this month. My dearest, dearest, dearest, in that 4 days I wrote you a minimum of 10 letters (9 were in mind and 1 I began on the paper, but it was so incomprehensible, that I had no courage to sent it to you) and sent one telegram, have you got it, dear and did you recognize the author, I did not know how many cost the words in Swiss and had not money enough for signature. Only one month now, but I want that it will be this evening. Darling, certainly we shall organize our first "meeting". But how? Perhaps I will meet you on railway station? Dear, thanks for the letter and for the love: 3 and 4 pages are love in my sense of the word. You know the morning when I got your letter, I wanted to see you so awfully, I wanted you to be here so strong, that I think that I would burst into 1000 parts. Rudi, I dream every night that morning is March. Love you so much.

February 3, 1931

My dear girl, I was so happy when I received your telegram on Sunday evening. What a question whether I recognized the author of the telegram? Of course, I thought it came from Fock or Mrs. Frenkel! You are still not well? And you even cannot go to Detskoye... You know, Dau and I sent our paper to Bohr. He did not answer for a long time, but now we have heard that he disagrees with everything. We just do not know what he does not believe; it is very difficult for him to write. Therefore I'll probably leave on 22nd of February and will be in Copenhagen for a week. Do you remember I told you about my plans with Simon and Berlin? They are completely ruined now, because Simon is going to be professor in Breslau. I will probably not travel via Berlin, but by boat Kopenhagen-Helsingfors. The ship arrives in Helsingfors on March 2 in the morning, and I'll be in Leningrad on the 2nd or 3rd. And then I'll see you!

February 7, 1931

My dear boy! Behind me I can hear the coals cracking in the stove (it is so cold, that central heating is unable to make the room warm and we have to heat the stove). I like very much to watch the embers in the stove; I imagine towns, castles, ships appear and disappear... Nice thoughts come into my mind when I'm by the fireside, and daydreams too. I do so much want you to be here — even if we weren't talking.

February 9, 1931

Genia, I am terribly tired and I don't know why. Yesterday I went skiing with Dau, Waller and Motschan (a nice chemist). Impossible to tell you how beautiful all this, running down with great speed about one hour on splendid snow. I am feeling more and more sure on my skies and even Dau gradually learns to manage them. It would be awfully nice if you could meet me at the station in Leningrad, but I am afraid that Dorfman, Bronshtein etc. will also be there. They are all very nice people, but... And Dorfman has to know when I arrive, because he has to book a room for me. Now I'll go to my room and there will be a letter telling me where and when I can see you, yes?

February 13, 1931

Rudi, darling! This is the last letter that you get in Zurich, for the last time I'll write on the envelope "Gloriastr. 35". On the 11th I'm going to Detskoye for 10–12 days. I am still ill, and if I'm not in the laboratory or theatre, I should stay in bed it is mother who insists on staying in bed; I hate it). By the 3rd I'll come back to the city. By the way, you wrote some time ago, that when you receive my letters, our relations are reflected on your face. What Dau may know? He lacks the power of observation, even worse than Bronshtein. But still, what does he know? I have a purely "technical" interest. And what did he gossip?

February 15, 1931

Dear Genia, soon we'll have the opportunity to talk. And we shall talk, talk very much! Another 15 days! The time that passed was so short, but these 15 days will be so terribly long. I won't talk to you in my sleep, I can really be together with you. It is terrible though that we cannot be together the entire time.

February 18, 1931

Rudi, darling! This is my last letter before your departure, and it will be very short because I left the paper at home and now have to write on some scraps. I am writing from Detskoye. It is snowing, and right before my eyes high fir trees are swaying, very high and all covered with snow. In the distance some dogs are barking, and I can hear the yard-keeper cutting firewood. I'm lying on the verandah thrust in a fur sack and covered with a blanket. Only my nose and eyes are outside and at this moment also the right hand with which I'm scribbling this. I arrived here yesterday in the afternoon. It is so wonderful here! A white house located in the woods; the interior being decorated in the style of the English Modern. You know these wooden panels in the lower part of the wall with damask in the upper part. There is wooden ceiling in the dining hall and the smoking room. Amazing armchairs and sofas, a big piano, carpets — all this very cozy. The bedrooms accommodate several people each, so I had to share my bedroom with other women; only one of them snores in her sleep, not too loudly, so it is quite tolerable. I haven't made my acquaintance with everybody yet. The local doctor discovered I had endocarditis (a heart disease). If they go on and on, they may well find puerperal fever and melancholic insanity!!! March 1 is quite close, oh Rudi, 12 days! It is good we aren't in our usual surroundings: you are in Copenhagen and I'm here in Detskoye — it is easier this way.

References

[1] Edoardo Amaldi, *The Adventurous Life of Friedrich Georg Houtermans, Physicist*, (Springer, 2012).
[2] *Physics in a Mad World*, Ed. M. Shifman, (World Scientific, 2016).
[3] M. Shifman, *Standing Together in Troubled Times*, (World Scientific, 2017).

Chapter 5

Flashback

In this Chapter, we will travel back in time and learn about Genia's and Nina's family. The Kannegisers belonged to a branch of a huge Russian-Jewish family tree of Mandelshtams, who played an important role in Russian culture and science for over two hundred years. Unfortunately, it is almost extinct now in Russia, after the cataclysms of the 20th century. This will be discussed at the end of this chapter. A significant number of the documents presented here are new, they appeared in the public domain recently and are hardly accessible to the Western reader.

***

Not much is known about Maria Abramovna Levina-Kannegiser, Genia's and Nina's mother, except for the fact that she had a brother, David Levin, whose granddaughter Maria (Masha) Verblovskaya was mentioned more than once in Rudolf Peierls' letters. Even the year of Maria Levina's birth is known only approximately (circa 1888). She died in 1953 in Kazakhstan, in exile. Genia's and Nina's father, Nikolai Samuilovich Kannegiser, on the other hand, is more familiar to historians because he belonged to the family tree of the Mandelshtams, which was thoroughly studied in connection with various famous people who belong to this tree as well. At least three of them — Osip Mandelshtam, Leonid Kannegiser and Vladimir Arnold — left indelible marks in Russian culture and history.

Nikolai Kannegiser, Nina's and Genia's biological father, was born in 1863 into the family of Samuil Kannegiser who was a medical doctor. He became a medical doctor too, an acclaimed gynecologist in St. Petersburg, a graduate of St. Vladimir University in Kiev. He worked at the Clinical Institute of Grand Duchess Elena Pavlovna in St. Petersburg. In 1904 he started teaching at the Women's Medical Institute and authored, among others, the treatises: "Surgical treatment of uterine fibroids" and "Lectures on operative obstetrics." In 1909 Nikolai Kannegiser died of septicemia.

In 1912 Maria Levina-Kannegiser remarried. Her second husband, Isai Benediktovich Mandelshtam (1885, Kiev-1954, Alma-Ata, Kazakhstan), treated Genia and Nina as his daughters. Nikolai Kannegiser was Isai's cousin (see below).

***

Isai B. Mandelshtam: My Stepfather[1]

Nina Kannegiser

Isai Benediktovich Mandelshtam, in his time a well-known translator of western-European literature into Russian, was born in Kiev in 1885. His father, Benedikt Yemelyanovich Mandelshtam, an ophthalmologist in Kiev, died when Isai was 9 years old.

Benedikt Mandelshtam's elder brother, Max, became the guardian of his three children (Isai was the youngest). Max was an outstanding ophthalmologist and a well-known Jewish public figure who also lived in Kiev.[2] Isai's mother Anna, née Jeannette Gurevich, moved with the children into an annex of Max Mandelshtam's house. Throughout his childhood and youth, Isai was in constant close contact with his uncle and uncle's family.

At the age of ten, Isai enrolled in the first form (in the nomenclature of the time) of a secondary school. With equal ease he studied both art and science subjects.

---

[1] Published in *Minuvshee. Istoricheskiy Almanakh*, 1991, №11, (Atheneum, Paris), pp. 382–413. Publication by L.I. Volodarskaya, Courtesy of M.B. Verblovskaya. Translated to English by Wladimir von Schlippe. In R. Peierls archive (Bodleian Library, Oxford University) one can find Volodarskaya's letter to Peierls and his reply. Volodarskaya asks Peierls about Isai Mandelshtam. Peierls recommends her to address the last surviving relative, M.B. Verblovskaya, see page 218.

[2] Max Mandelshtam was an activist of the Zionist movement in Russia. He came from a family of prominent figures of Jewish enlightenment, Haskala. Max Mandelshtam studied medicine at German university in Dorpat (now Tartu), and then completed his studies at Kharkov University (1860). In 1864–68 he specialized in ophthalmology in Berlin, then defended his doctoral dissertation at the Medical and Surgical Academy in St. Petersburg. In 1869, he became a Professor of Ophthalmology at St. Vladimir Imperial University in Kiev. He also founded his own clinic, the best in Ukraine. In 1880 Max Mandelshtam left a successful university career because of the anti-Semitic decision of the Kiev University Council to cancel his election to full professorship, which was an almost unanimous decision by his department. Shocked by the pogroms in southern Russia in the early 1880s, Max Mandelshtam founded the Committee for Assistance to Victims of Pogroms (1881). Later he realized the futility of the struggle for the emancipation of Jews in Russia. For the rest of his life he dedicated himself to advancing the idea of mass Jewish emigration from Russia.

In 1903, Isai graduated with a Gold Medal (distinction). His mother and uncle offered him the choice of one of the three traditional intellectual professions: medicine, law or engineering. In 1902 a shipbuilding department was opened at the St. Petersburg Polytechnic Institute. Isai greatly wished to be in St. Petersburg, and so he decided to become a ship designer.

The ship design courses at the Institute of Technology did not greatly appeal to Isai, where early on he had to expend much time and effort on technical drawing. He found solace instead, in the theaters of St. Petersburg. When, in connection with the revolutionary students' unrests of 1905, the Polytechnic Institute, together with other institutes of higher education, was closed for an indefinite period, Isai went abroad and enrolled at the Liége University Faculty of Technology. He chose to study electrical engineering, as this was closer to physics and there was no technical drawing, which made it much more interesting than shipbuilding. His lively and cheerful disposition, his talent for comic acting, and his good-natured interest in people made Isai popular among the students of Liége university, both Belgians and Russian émigrés. And as for Isai himself — the time spent in Liége, and his travels in Belgium and France left a mark not only in his memory but also in his world view and even in his character.

In Belgium he mastered the French language to perfection, which previously, he had known only enough to understand the lectures and to get along with his Belgian peers. (Isai knew German from childhood: one spoke German in the home of his uncle Max Mandelshtam, who was born and bred in the Baltic provinces and had studied medicine in Germany. Isai used to say that in Germany he was usually taken for a German from the Baltics.)

From his time in Liége, Isai acquired a love for modern French literature, and in particular for the work of Anatole France, to which he was attached for many years.

Figure 5.1   Isai Benediktovich Mandelshtam. Courtesy of Natalia Alexander.

After graduating in 1908 from the Faculty of Technology, Isai returned to Russia. For over a year he worked as an engineer in Nikolaev at a ship-building wharf, then moved to St. Petersburg and joined the "Allgemeine Elektrizitäts Gesellschaft" — the Russian branch of the German firm AEG. The Liége university diploma did not give Jews the right to live in St. Petersburg, and so he enrolled in the Faculty of Law at St. Petersburg University.

In 1912 he married Maria Abramovna Kannegiser (née Levin), the young widow of his cousin Nikolai Samuilovich Kannegiser, who had been an obstetrician and professor at the Medical Institute for Women. Maria had two little daughters from her first marriage — I and my elder sister Genia. Isai brought us up without adopting us; he did not have children of his own.

Engineering, legal studies, and family commitments — all this left Isai with little time for leisure, the source of fulfillment. Even so, the love of the theatre he kept for life. He went to the theatre as often as money and free time would allow, and frequently went to concerts — he was very musical, and knew and loved serious music. He read much, having wide interests in philosophy, sociology, physics, and, naturally, literature.

He found an outlet for intellectual activity in literary work as a translator. In school, he had already started translating German and French poetry into Russian. At that time he was also writing his own lyrical poems, as well as humorous verse and epigrams. Isai never published his own poems, and later considered them as initial practice in versification. His first translations were published in one of the journals in Kiev.

In 1910, three of Isai's translations of poems by Detlev von Liliencron[3] were published in the journal "European Messenger" (*Vestnik Evropy*). These translations are included in the second edition of poems by Osip Mandelshtam, published in 1964 in Washington, and erroneously attributed to him, with a comment that, unlike later translations of, for example, Barbier[4] and Petrarch,[5] these translations are distinguished by their accuracy and closeness to the original. The same edition includes Heine's "Disputation", also in Isai's translation which was first published in 1917 in the "Annals" (*Letopis*) and later repeatedly reprinted in various Russian editions of the works of Heine.

Isai himself considered his translation of E. Hardt's play *Tantris der Narr* (Tristram the Jester) to be the beginning of his literary work; it was published in 1910 in the journal "Theatre and Art" (*Teatr i iskusstvo*). After this publication, Isai joined the Society of Writers of Drama and Music.

In the 1910s (including the first years of the revolution) Isai continued translating German poetry of the end of the 19th and beginning of the 20th centuries: Detlev von Liliencron, Gustav Falke, Richard Dehmel, Otto Bierbaum, and Christian Morgenstern; he also translated lyrics by Heine (he continued translating Heine until his death), and tested himself translating Baudelaire and Verlaine. He made little effort to have his translations published: the very process of translating

---

[3] Detlev von Liliencron (1844–1909) was a German lyric poet and novelist.
[4] Henri Barbier (1805–1882) was a French dramatist and poet.
[5] Francesco Petrarca (1304–1374), anglicized as Petrarch, was a great poet of Renaissance Italy, one of the earliest European humanists.

poetry gave him great pleasure and satisfaction throughout his life. In 1917, Isai made his debut as a translator of prose, translating Anatole France's "Les Opinions de M. Jerome Coignard," which was published by the "Universal Library" in 1918.

On August 30th, 1918, my cousin Leonid Kannegiser, a member of the Socialist Revolutionary Party, assassinated M.S. Uritsky, the head of the Petrograd *Cheka*.[6] Leonid Kannegiser was a son of the engineer Ioakim Samuilovich Kannegiser, a cousin of Isai's, with whom he was on friendly terms and whom he frequently visited. In the night of the first of September Isai was arrested together with other regular guests of that house (described by Marina Tsvetaeva in her essay *An Otherworldly Evening*). The investigation established the absence of any involvement of Isai and all the other arrested people in the assassination, and they were all released after a few months. Isai spent four months in prison (the Deriabinsky barracks). This false arrest, that had apparently no immediate adverse consequences, played a fateful role in Isai's life at a later time, when the corresponding authorities reconsidered their erstwhile "mistakes" as a pretext for new repressions.

After his release from prison, Isai continued working as an engineer for the former AEG, and later at Svirstroi. Around 1920 his work as a translator acquired a more professional character: he started working on contracts with the "World Literature" publishers. For these publishers he translated "Die Leiden des jungen Werthers" (The Sorrows of Young Werther), published in 1922, *Wilhelm Meisters Wanderjahre* (Wilhelm Meister's Journeyman Years) — unpublished — for the planned collected works of Goethe, and *Illusions Perdues* (Lost Illusions) by Balzac, published by the Academia publishers in 1930.

In the early 1920s, Isai became a professional man of literature without leaving his work as an engineer. Translations were his main interest, and royalties a significant part of his earnings. He was a member of the "House of Literary Men" from its foundation to its abolition, frequently participating in disputations, concerts and various soirées held there, and he was socializing almost exclusively with men of literature —

---

[6]The Soviet acronym for the name of the secret police at that time. The Soviet secret police has changed acronyms like a chameleon. It started out as the Cheka, and then became the GPU, the OGPU, the NKVD, the NKGB, the MGB, and finally the KGB. Even today, however, people often simply refer to the secret police as "the Cheka" and the secret agents as "the Chekists." Currently, it is known as FSB, the Federal Security Service.

A.G. Gornfeld,[7] M.A. Kuzmin[8] Yu.I. Yurkun,[9] P.K. Guber,[10] B.M. Eichenbaum,[11] B.K. Lifshitz,[12] and M.E. Loevberg.[13] After another 2 or 3 years, Isai left his engineering job and for some time worked as a freelancer, as in those days people of various arts who were not in State employment were called.

The time of NEP (New Economic Policy) was accompanied by a rapid flourishing of publishing activity. Private publishers — Antenei, Petrograd, Time, Thought, Book, Sower, Book Corner — appeared one after the other, and they all published translations of literature in huge quantities, sometimes almost to the exclusion of new work. Getting hold of the latest foreign literature was rather difficult (but all the same was possible by the most enterprising publishers and translators), and it frequently was at odds with the requirements of the censors. Therefore, there was a demand for works of older writers, that were forgotten by Russian readers or not known to them at all.

Isai translated with equal ease and precision both from German and French; he was in perfect command of literary Russian. He worked enthusiastically, even with fervor, and was very hard working — he would translate half a printer's sheet per day. He had neat handwriting, and his manuscripts would be type-set directly, especially when there was pressure of time. He met deadlines precisely, bringing his working habits as a project engineer to the freelance profession. Isai was very resourceful in searching for little-known or forgotten books. Thus he translated Eugene Fromentin's *Dominique*, Henry Murger's *Scénes de la vie de Bohéme* which had not been translated since 1892, Abbé

---

[7] Arkady Georgievich Gornfeld (1867–1941) was a Russian/Soviet literary critic, translator, publicist, and journalist.

[8] Mikhail Alexeevich Kuzmin (1872–1936) was a Russian writer, poet, playwright, translator, and composer of the Silver Age.

[9] Yuri Ivanovich Yurkun (Juozas Jurkūnas, 1895–1938) was a Russian writer and graphic artist of Lithuanian origin. He and Mikhail Kuzmin were in a homosexual relationship. Shot by NKVD firing squad in 1938.

[10] Pyotr Konstantinovich Guber (1886–1941) was a Russian/Soviet writer and literary critic. Arrested by the NKVD in 1938 and died in Gulag.

[11] Boris Mikhailovich Eichenbaum (1886–1959) was an outstanding Russian/Soviet literary critic, one of the key figures of the so-called "formal school" (Society for the Study of the Theory of Poetic Language which existed in 1916–1925 as a scientific association created by a group of theorists and historians of literature and linguists). In 1949–1953 he fell victim to Stalin's anti-Semitic campaign.

[12] Benedikt Konstantinovich (Nakhmanovich) Lifshitz (1886–1938) was a Russian poet, translator and researcher of "futurism." Arrested on October 5, 1937 "for participation in the anti-Soviet Trotskyist organization" and sentenced to death on September 20, 1938. Shot by firing squad the next day.

[13] Maria Evgenievna Loevberg (1894–1934) was a Russian/Soviet writer, poet and translator.

Prévost's *Manon Lescaut*, short stories by Arthur Schnitzler, novels by Leo Perutz, little known novels by Balzac, Claude Tillier's *Mon oncle Benjamin*, Gustave Flaubert, Sir Arthur Conan Doyle, Paul Verlaine, Charles Baudelaire, and so on. More than 50 books of his translations (not counting further editions) appeared from 1920 to 1930. In this time, he was one of the recognized masters of translation. Early in 1930, a teaching seminar was organized in the translators' section of the Leningrad branch of the Writers' Union. The head of the seminar was A.A. Smirnov, a scholar of English and of Romance languages and literature who later became a professor at Leningrad University. He himself was running the classes of translating from French and Spanish, and invited Isai to take the classes for translators of German literature.

The gradual demise of private initiative in everything concerning books, and also the financial instability of independent editing houses who were appallingly slow in paying for work done, were producing a permanent state of destitution. This led Isai to extend his translating activity from pure literature into books on technical subjects, and soon he became a staff editor of the KUBUCH publishing house[14] which specialized in technical literature. This gave a certain minimum stability to his material situation.

In 1931 Genia married Rudolf Peierls, a young theoretical physicist of German descent who was working in Zurich under the celebrated Pauli. Six month after their marriage was registered, she was allowed to renounce her Soviet citizenship and she left for Switzerland to join her husband. This event, which seemed to be purely a family matter and an entirely joyful one, later turned into a tragedy for Isai.

In 1934 the Peierls family (who by that time were firmly settled in England) came to Leningrad to show their year-old daughter to her grandparents.[15]

At the end of 1934, Isai was commissioned by the Leningrad Section of GIKhL (State Fiction Publishers) to translate "Wilhelm Tell" and "Don Carlos" for a one-volume collection of Schiller's plays. He started on this task with great joy, as he had until then worked for a considerable time almost exclusively on technical translations, moreover he

---

[14]KUBUCH (1924–1936) was a publishing house founded by the Commission for the Improvement of Students' Life of the Leningrad Soviet of Workers' and Peasants' Deputies. It published educational literature on various fields of knowledge, guides, and reference books for the "proletarian students."

[15]Report entitled "Names of arrivals from USSR including Rudolph (sic!) Peierls, physicist" (File KV 2/1658, serial 4a in the British National Archives) mentions Peierls' return from the USSR on September 26, 1934. Courtesy of Frank Close.

had not done any poetry translations for a very long time, apart from Schnitzler's play "Casanova in Spa."

In March of 1935 he was arrested again and spent about ten days in prison. He was not accused of anything, during his interrogations all questions concerned his family relationship with the socialist-revolutionary Leonid Kannegiser and Genia's marriage. This was during "Kirov Days"[16] when new practices were established, and thus Isai was exiled from Leningrad by administrative order, without a trial. The destination of his deportation was Ufa, and the length of exile was five years. My mother, Maria Abramovna Mandelshtam, and I were also exiled with him for the same duration. We left Leningrad in a special rail transport — at that time several such transports were sent eastwards every day.

Ufa was already overcrowded with people sent into exile from Leningrad. The Mandelshtams succeeded in finding a small room (14 square meters) in a village cottage at the edge of the town. In spite of the crowded conditions and of the extreme practical discomfort, Isai immediately started work on translating Schiller. Work was for him the only distraction from life's woes. His deportation had not been aggravated by a deprivation of civil rights, therefore the contracts agreed with publishing houses remained in force, but it was difficult to rely on new commissions: publishers were either lazy or wary of initiating out of their own free will any collaboration with such "faraway places," and it was not really possible to demonstrate inventiveness in Ufa.

As early as 1934, Isai had begun to experience strong sporadic pain in his feet and legs, this sometimes made walking difficult. This was the beginning of endarteritis, which was not diagnosed immediately. Attacks of these pains became ever more frequent, and once the family had arrived in Ufa, they became very prolonged, sometimes Isai could not go out of the house for several days on end. This was already a sufficient reason for him to give up any thought of a regular job. Moreover, although Ufa lacked specialists at the time — especially at the beginning — people sent into exile were offered employment almost exclusively for manual work.

---

[16]Sergei Kirov, the head of the Leningrad Communist Party, was assassinated on December 1, 1934. This assassination triggered the onset of mass terror in Leningrad. Two days later a law was enacted according to which all investigations of terrorist acts had to be completed within ten days, indictments delivered to the accused just one day before "court trials" without participation of the parties and with no possibility of appeal. Death sentences were ordered to be executed immediately. On February 28, 1935, a mass deportation from Leningrad of the "bourgeois and other undesirable elements" began. No trials were held. Within a month, 39,000 people were exiled to Siberia and other remote areas.

Isai had left his library (and all furniture) in Leningrad, distributing them among those people of his acquaintance who were the least vulnerable in respect of deportation. Only one trunk was brought to Ufa, containing books chosen in a great hurry: the family had been given only five days to dispose of everything that it had collected over the 25 years in Leningrad.

Among the books brought to Ufa was a volume of Shakespeare in English. Isai did not know English as well as French or German, as he had studied it independently, only from books, and one of his methods was to read Shakespeare in English, in parallel with Tiek's German translation,[17] which was famous for its exactitude. Tiek's translation had also been brought to Ufa.

Isai had wanted to apply his translating abilities to a Shakespeare text. Once he had completed his Schiller translation in Ufa, and in the absence of any other commission, he had enough free time for such an attempt. He started to translate "Richard II" — a chronicle which was not as well known or as frequently translated as "Richard III". It was somewhat shorter than the others and it had a very intense plot.

He worked with immense enthusiasm. The intensive verse of Shakespeare, in addition to the great number of monosyllabic words in the English language, demanded that he create a previously unknown technique if he was to achieve at least a tentative equivalent of lines. The search and development of this technique gave him a pure creative pleasure. He urgently needed readers of a level at which they could serve as critics — and, amazingly, he found them.

Among the many people exiled from Leningrad he found the young Levinson siblings (a brother and a sister) — I had known them in Leningrad. They were exceptionally original — not to say peculiar — but no less gifted. Evgeniy Maksimilianovich Levinson was a talented and widely known mathematician, Natalya Maksimilianovna was (I think) an archaeologist. They were both passionate lovers of comparative linguistics and philology, they were also brilliant connoisseurs of English language and literature. They had brought to Ufa their special English dictionaries of Shakespeare's language.

Within a short time they became the closest friends of Isai — they brought their dictionaries, and visited him very often indeed. Isai would read out to them his translation of Shakespeare as it was completed,

---

[17] Ludwig Tieck (1773–1853) translated Shakespeare to German in the first half of the 19th century. He was one of the founding fathers of the Romantic movement in the late 18th and early 19th centuries.

scene by scene. They would listen, following the English original text, paying exceptionally close attention to and discussing the Russian version in great detail. Isai used to say later that it is thanks to their passionate approval of his version of "Richard II", based on their careful analysis of all detail, that he felt for the first time truly capable of translating Shakespeare at a high level, although he was far from perfect in his knowledge of the English language. (Natalya and Evgeniy Levinson lost their lives during the war, presumably during the German occupation of Pushkin where they had been living after their return from Ufa.)

Six months after arriving in Ufa, Isai and his family moved to a separate flat, consisting of two rooms and a kitchen, which had been made out of one large room in a private house situated on the high shore of the river Belaya (Salavata street). In spite of the great distance between Ufa and Moscow (40 hours by train), several friends and relatives of the Mandelshtams visited them during their very first summer there. Such visits happened also in later years, and many of their visitors used the tourist shipping route along the rivers Volga, Kama, and Belaya. His mother,[18] who was already in her eighties, visited him twice, accompanied by his sister. In spite of a good climate, of relatively bearable practical arrangements and of the psychological support created by his work on Shakespeare, Isai's physical condition in Ufa was poor: the pain in his feet and legs prevented him from taking exercise, and in any case, there was nowhere to go, apart from the police station for "registering his presence" and the library. He put on weight and became short of breath.

The slogan "Children are not answerable for their parents" which Stalin suddenly produced at the start of 1936 immediately granted freedom to all young people who had been exiled from Leningrad as "members of the family", and I was one of these. At the end of April I returned to Leningrad.

In the summer of 1936, Anastasia Mikhailovna Kharitonova (Nastya) arrived in Ufa to live with the Mandelshtams. She had been the Mandelshtams' domestic help from 1919, and had long since become a very close and very faithful member of the family.

Autumn 1937 saw the start of arrests among people exiled from Leningrad. Isai was arrested in March 1938. He spent 8 months in an Ufa jail, in an overcrowded prison cell, and in the psychological

---

[18] J. Gurevich, see Fig. 5.5 on page 139.

atmosphere of those years which we cannot imagine nowadays, among prisoners who had already been tortured under interrogation and those who were still only awaiting this fate. Isai lay on the floor under the two-storied pallets which were occupied by those who had been arrested earlier, in conditions so cramped that if one person turned over, all the others had to turn with him — and he would translate Pushkin's verse that he knew by heart into German. This was the time when he translated "Hymn to the Plague", the monologue of the Miserly Knight, Salieri's monologue, "The Prophet", "On Georgia's Hills" and some other poems. He also translated some lyrical works of Heine which he knew by heart.

He was interrogated twice: a repeat interrogation about the murder of Uritsky which had happened 20 years before, and on the "spying activities" of Rudolph Peierls, who by that time was already a physicist of world renown. No "physical methods" — that is, no torture — were used on Isai, only threats, screaming and extreme foul language.

Without any trial, by decision of a "special council", Isai was sent to a labor camp — Solikamskbumstroi — for a three-year sentence. He spent this time working as a qualified engineer. He wrote down the translations that he had made in prison. This camp was not a "special regime" one, the inmates were allowed to receive a parcel or two every month, and they were occasionally granted a visitor. I visited Isai in summer 1939, and in spring 1940, when the sentence of exile to Ufa came to an end, my mother went to see him. Both of us visited him in autumn of the same year.

Under (internal) passport rules, Mother was not allowed to return to Leningrad after the end of her exile sentence to Ufa. She chose Malaya Vishera as her designated address. Isai left the labour camp in March 1941. He arrived "ready for departure with all chattels" in the office of the head of his camp, where (internal) passports were completed and handed out, and was asked to make a choice of his new place of permanent residence. Excluded from this choice were: all capital cities, industrial centers, railway junctions, holiday resorts, seafront and frontier zones, etc. For the purpose of facilitating the choice, a map had been hung on the office wall. Isai looked at it, inspected the stretch between Leningrad and Moscow, then the Bologoye area, and he noticed Ostashkov. He remembered that this small town was a favorite holiday destination for the Leningrad arts elite. Isai decided that the elite would not "choose a bad place" — and picked up Ostashkov.

> *From a letter of Mandelshtam to B.M. Eichenbaum, 15th July 1943...*
>
> *Maria Abramovna works as a nurse in our local hospital. The workload is huge, with nightly and sometimes round-the-clock duty, and with pitiful wages, equal to about 10 litres of milk, i.e. enough for 10 days. Nina earns the remaining 20 litres. For all other needs, we have managed so far by selling our belongings on the market? For four months this winter I taught German in the senior forms of our Middle School. But I found these pupils to be so ignorant that I despaired of being able to raise them to the standard of the course and was afraid that I'd have to fail 2/3 of them, and hence dropped teaching? We have sunk into such a "misery" as we have been never before. From our Genia we had only a few telegrams during these two years but not a single letter...*

The NKVD (Peoples Commissariat of the Interior) allowed Isai to spend a week in Leningrad to sort out his literary and family affairs. He made a preliminary agreement with the "Life of Remarkable People" publishers for a book on Beaumarchais, whose extraordinary character and biography had long since attracted him. Isai and mother moved to Ostashkov in March 1941. Isai never returned to Leningrad. The war started on June 22, 1941.

Isai's translations could not be published under his own name while he was in a prison camp. A.A. Smirnov did him a great service when he decided not to give up the publishing of his Shakespeare translations,[19] but rather published "The Merchant of Venice" and "Julius Caesar" purportedly translated by B.D. Levin, while "King Henry VIII" had no translator's name attached at all, but rather had been edited by A.A. Smirnov. (Boris Danilovich Levin, an orientalist and law specialist, who occasionally did some translations, was a nephew of my mother. He died in the Leningrad blockade.)

The German advance towards Toropets-Ostashkov was swift. At the end of July, a happy accident allowed Isai, my mother and A.M. Kharitonova, who would not abandon them, to leave Ostashkov. They ended up in Melekess,[20] a tiny town in the Kuybyshev region. The Leningrad Public Library funds were being evacuated there. At that time Melekess was a godforsaken place, a tiny regional center with a population of less than 15,000. However, the flood of refugees soon brought this total up to 50,000.

---

[19]W. Shakespeare, Selected Works in 4 Volumes, (Detgizizdat, Moscow-Leningrad, 1940), Vol. 3.
[20]Now Dmitrovgrad, Ulyanov Region, 800 km east of Moscow.

In 1940, while in prison camp, Isai had survived serious pneumonia which had developed into pulmonary tuberculosis. His endarteritis also deteriorated. He was 56 now, but he looked an old man, he was practically semi-disabled. In spite of it all, he did try to work: he taught German in school for a while, then he taught German to graduate students who were taking correspondence courses while also teaching at the local teachers' college. My mother returned (after an immensely long break) to the profession of her teens and worked as a nurse in the local hospital. I too arrived in Melekess in late November, as I had again been sent into exile from Leningrad. I started working immediately, so that our family's standard of living remained more or less stable compared to wartime.

> *Maria Verblovskaya, unpublished*
>
> *I remember how he and Maria Abramovna appeared in Leningrad in March 1941. When the issue of my summer was raised, they offered to take me to spend the summer with them, in Ostashkov. My nanny Frosya brought me there on May 1st. It was still snowing in Leningrad. It was the day when I last saw daddy and remembered his sad face, swam away behind the moving car window. The war began on June 22nd. The Mandelshtams, who were tied up with the NKVD, did not know what to do. Germans were already approaching Kalinin (Tver). The echelon with the funds of the Leningrad Public Library was being evacuated to the city Melekess, in Kuybyshev region. My mother, Library's employee, was with this echelon.*
>
> *In Bologoye, my mother begged the director to give her the evacuation precept with the final destination address (which was being kept secret from the employees). She managed to reach Ostashkov. Germans were already close, and if it were not for my mother who took us with her to Melekess, all of us would end in the same way, in accordance with the German plan of the total extermination of Jews.*

Isai used the time during his prolonged enforced periods without paid work by translating poetry. In these years he translated Shakespeare's "Macbeth", "Othello", the second part of "Henry IV", "Richard II", also Gasso, Goethe's "Reineke Fuchs", and "Hermann and Dorothea". He completed the Pushkin translations into German that he had started while in Ufa prison: "The Miserly Knight", "A Feast During the Plague", "Mozart and Salieri", also he translated his "Scenes from the Age of Chivalry", "The Stone Guest", and some short poems of Pushkin. At that time, he also translated Victor Hugo's cycle of poems "L'année terrible" — he said that in his thoughts he dedicated this to the besieged Leningrad. He also continued to translate lyrical poems by Heine and Goethe's poems. Almost none of these translations have ever been published.

In the early spring of 1945 Isai suffered a major flare-up of pulmonary tuberculosis, which nearly ended his life. He did however recover and a year after the War ended, in 1946, he decided to move closer to Moscow. All administrative restrictions remained in force; he could not live within 100 kilometers from Moscow, and on his friends' advice he chose Maloyaroslavets. His wish for closer links with Moscow was due to his perennial hope of collaboration with publishing houses, his desire for visits — however fleeting — to a large town ("I am an urbanist", he used to say) and moreover, his closest friend (and cousin), Professor A.G. Gurvich,[21] who had lived in Leningrad before the war, and now lived in Moscow.

> *From a letter of Mandelshtam to B.M. Eichenbaum, October 12, 1943*
>
> *...On what I am working at present? — alas, I cannot work at all. I cannot overcome the single and insurmountable obstacle to writing — the lack of paper. "The fingers are asking for the pen", but there is no paper. Therefore, I cannot even think about working on a translation...*
> *However, just recently I spent a whole month translating the poems from Victor Hugo's "Année Terrible". I could do this on scraps of paper and on an old ledger which I was given, writing between the lines of numbers...*
> *From our Zhenechku we had the other day a telegram ending with the words: "hope worst is behind and soon all meet again?"*

Isai's hope for paid-for literary work turned out to be fulfilled to some extent by his move closer to Moscow. He was given the task of translating Shakespeare's "Pericles" (Lengosizdat, Vol. 7, 1949). His translations of "The Merchant of Venice" and "Julius Caesar" were accepted for a new edition, and his translation into German of *"Pir vo vremia chumy"* (The Feast During the Plague) was published under his name [1].

---

[21] Alexander Gavrilovich Gurvich (also spelled as Gurwitsch, 1874–1954) was a Russian/Soviet biologist who discovered superweak radiation from living systems (mitogenetic rays, also known as biophotons. Non-thermal in origin, the biophotons are technically considered as a type of bioluminescence).

Isai did not visit Moscow frequently. These visits were actually illegal, since by the rules of those times the passport restrictions did not allow spending the night in Moscow, and the travel from Maloyaroslavets to Moscow took four hours, so that it was not possible to go for a visit to Moscow and return on the same day. But all the same he did visit Moscow, occasionally going to a theatre or concert, or meeting with a few of his friends. Visiting publishers in search of work was very distressing for him — during the 10–15 years of his exile from Leningrad, new people had appeared everywhere, people to whom Isai's name meant nothing, but to whom his Maloyaroslavets address did mean a lot.

In those days half of the population of Maloyaroslavets were people with passport restrictions, i.e. people who had served prison sentences. In 1949 arrests began among people of this category in Maloyaroslavets. The number kept increasing, and in March of 1951 Isai was arrested, again.

He was already 66 years old. As Maloyaroslavets had no prison, he was sent to Kaluga.[22] Once again, he spent eight months in prison, was interrogated once and sentenced to exile for 10 years. The Dzhambul region of Kazakhstan was assigned to him as his place of exile.

The transport to Dzhambul[23] was done in stages. This took over three weeks. The regional NKVD administration did not consider it possible to leave him in the administrative center of the region and sent him to the village of Mikhailovka,[24] the center of the Sverdlovsk district, 16 kilometers from Dzhambul.

---

[22]Kaluga is a city located on the Oka River 150 kilometers (93 miles) southwest of Moscow.
[23]In 1997, Dzhambul was renamed and is currently known as Taraz (also spelled as Talas). Taraz is a city and the administrative center of Jambyl Region in Kazakhstan, located on the Talas (Taraz) River in the south of the country near the border with Kyrgyzstan. Starting in the 1930s, Dzhambul, along with other places in Kazakhstan, became the destination for large numbers of the deported peoples who were subject to internal exile. There were millions of Volga Germans, Chechens, Ukrainians, Koreans and other ethnic minorities, along with other persecuted persons (former *kulaks*, bourgeoisie and aristocracy, families of convicted "enemies of the people," etc.).
[24]Currently, the village of Sarykemer, 530 km to the west of Alma-Ata. Today, a comfortable highway connects them. In 1951 there were just a few dusty dirt roads in this area.

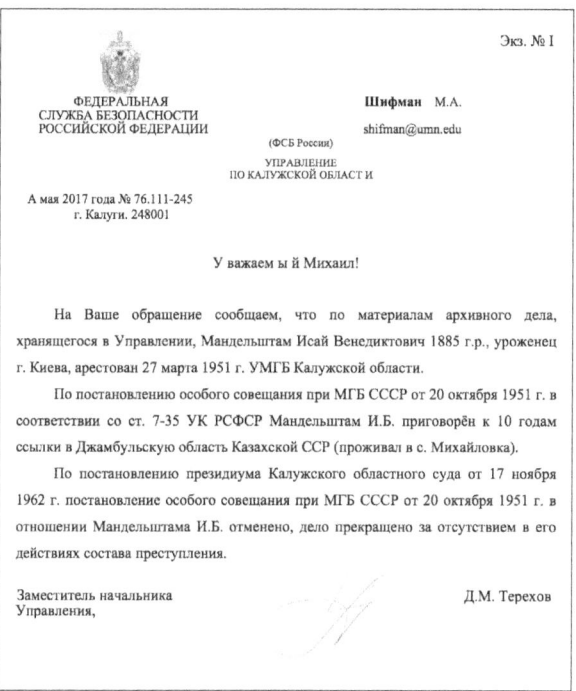

Figure 5.2 Reply to my inquiry to the Federal Security Service (FSB), Kaluga Branch, about the circumstances of Isai Mandelshtam's arrest in 1951. The letter of May 5, 2017, reads: "At your request, we inform you that, based on the archival file kept in our Office, Isai Benediktovich Mandelshtam, born in 1885, a native of Kiev, was arrested on March 27, 1951 by the Kaluga Region Branch of Ministry of State Security (MGB).

According to the decision of the Special Committee of the USSR MGB of October 20, 1951, in accordance with Article 7-35 of the Penal Code of Russian Federation I.B. Mandelshtam was sentenced to 10 years of exile in the Dzhambul region of the Kazakh Soviet Socialist Republic (to reside in the village of Mikhailovka).

According to the decision of the Presidium of the Kaluga Regional Court of November 17, 1962, the verdict of the Special Committee of the USSR MGB dated October 20, 1951, concerning I.B. Mandelshtam was canceled (posthumously), the case being dismissed for lack of corpus delicti in his actions."

Article 7-35 of the Penal Code prescribes the following sentence: citizens found to be "socially dangerous elements" are to be limited in residence to specially designated areas for a period of 3 to 10 years. Special Committee (1922–1953) was an extrajudicial body deciding the cases in absentia. It had the right to sentence the accused either to exile or to Gulag.

> *From a letter of Mandelshtam to B.M. Eichenbaum, May 4, 1944*
>
> *...I cannot write — so difficult and hopeless is our life...*
>
> *December 3, 1946*
>
> *...three years ago Maria Abramovna was dying of typhus, and the doctor told me one evening that there is no more hope...*

My mother went also immediately from Maloyaroslavets to join Isai at Mikhailovka. They saw in the New Year of 1952 together.

Mikhailovka is a village in a forestless arid steppe. The streets are lined with poplars and elms. The houses are one-storied, built of adobe brick. Each house has its vineyard, vegetable garden and small fruit trees. Irrigation canals run along the streets. The river Talas runs round the village, and beyond the river lies the steppe, and in the distance are mountains. The winter is short and with snow but no deep frost, the summers are hot. The population, by the Kazakhstan standards of those times, was large — 3,000 to 4,000. The native population were the descendants of the Cossacks of the Seven-river region, who stayed there after the conquest of Turkmenistan, and Kazakhs. But in the early 1930s, many Ukrainians moved to these places, escaping from collectivization, and in the late 1940s there was another influx of peoples moved there for their ethnicity: Chechens, Ingushs, Karachaevs, Greeks from the Black Sea regions, also Germans from the Volga. Isai's health had suffered badly from the prison and transport, but during his first time in Asia, which he had never seen before, he was greatly interested in observing nature, peoples' way of life, the "mix of garments and faces". However, this time he was completely cut off from cultural life. There was even no constant electric lighting in Mikhailovka — the hydroelectric power station on the Talas river worked only from time to time, and the village had no cinema: occasionally a mobile cinema would visit. Among the inhabitants of Mikhailovka there was not a single intellectual; Isai had nobody with whom to converse.

He got hold of a radio (on batteries). Through the pitiable district library, he could order books from Moscow, from the Lenin library. The orders took a long time, but eventually the books arrived, and the librarian, violating the rules, allowed him to take the books home. Sometimes he received books sent by friends from Moscow, and from Leningrad he once received nearly an entire bookcase on the history of French culture in the 16th to 19th centuries.

During his stay in Maloyaroslavets, Isai had met Argo,[25] a translator of poetry and a satirical poet. They did not see each other frequently but kept up a correspondence. At the same time Isai met in Moscow his old friend E.Yu. Gerken[26] — the "King of Operetta".

After Isai's move to Kazakhstan, they both showed to him their warmest concern, not only by keeping up their correspondence — which in itself was an act of great civil courage — but also in their constant efforts to find literary work for him. They both worked with him: with Argo he translated Prosper Merimée's *Le Théâtre de Clara Gazud* (which was later published by Goslitizdat), and with Gerken he translated Anatole France's *L'affaire Crainquebille* and other plays. These translations earned him little money, but it was work, which gave the illusion of a continuation of a professional life and took his mind off sad thoughts. In 1953, Isai's term of exile was reduced to five years — he was due to be released in March 1956. At the end of November 1953 my mother died unexpectedly. Isai continued living in Mikhailovka by himself. His household was managed by a neighbor, the wife of an exiled Greek, who had helped them while mama was still living.

Isai forced himself to read and to continue the translation of the play "The Noisy Secret" by Calderon.[27] In the beginning of May he received — at last — permission to move to Alma-Ata, where I had moved to in 1951, i.e. from the beginning of Isai's exile to Kazakhstan. Isai moved to Alma-Ata as soon as he was permitted.

On June 29th, 1954, Isai died in Alma-Ata.

In 1962 he was rehabilitated posthumously.

---

[25] Abram Markovich Argo (real name Goldenberg, 1897–1968) was a Russian/Soviet poet, playwright and translator. In the 1920s he participated in the creation of the first review shows of the Moscow Theater of Satire.

[26] Evgeny Gerken (1886–1962) was a Russian/Soviet poet, playwright and translator. He was known for his operetta "Serf Girl" and was a close friend of Mikhail Kuzmin. In August 1933, Gerken was arrested in connection with the "case of Leningrad homosexuals" accused of "counter-revolutionary activities" and espionage. In 1934 he was sentenced to 10 years in Gulag.

[27] In the English translation this play by Pedro Calderón de la Barca (1600–1681) is entitled "The Secret Spoken Aloud."

The Mandelshtams[28]

Maria (Masha) Verblovskaya

In the late 1920s, their home was a gathering place for talented young people, friends of Genia, who was a student of the Department of Physics and Mathematics, and friends of Nina, a student of the Biology Department. The house at 26 Mokhovaya Street was a welcoming place, and the young people who got together there came up with a nickname for themselves, appropriate for the time and their youthful exuberance: the "Jazz Band." The young physicists included George Gamow, Andrei Anselm, Lev Landau, Nikolai Kozyrev, Dmitri Ivanenko, Valentina Ioffe,[29] Irina Sokolskaya, the sisters Natalia and Anna Gurvich,[30] and the brothers Elephter Andronikashvili and Iraklii Andronikov.[31] Their house was perhaps the first place where Iraklii performed as a parody storyteller. Later, in 1927, they were joined by Matvei Bronshtein. This group shared not only an interest in physics, but also a passion for history and literature, especially poetry. Maria Kannegirser [Genia's and Nina's mother] ran the house. A book from her collection, *Noon in Croton*[32] by B. Lifshitz, contains the inscription:

"To the enchanting Maria Abramovna, from Benedikt Lifshitz on bended knee."

Genia worked at a factory from the beginning of the WWII. She sent her children, a daughter and son, to the north of England, far from the bombings, and then to Canada. Genia claimed that she always carried poison with her, in case the Nazis ever landed in England.

In March 1941, when Isai Mandelshtam [Genia's and Nina's stepfather] was released from the camp, he was offered several places to settle, and he chose the town of Ostashkov. I recall that around that time, he and [his wife] Maria[33] came to see us in Leningrad. When the subject of my summer plans came up, they offered to take me to spend the

---

[28] Unpublished notes by Maria Verblovsky, also known as Masha Verblovskaya. From Natalia Alexander Family Archive. Courtesy of Natalia Alexander. Translated from Russian by Alice West. Translation of poetry by Larry Bogoslaw.

[29] The daughter of the famous Russian/Soviet Academician Abram Ioffe, known as "the Father of Soviet physics."

[30] Daughters of Alexander Gavrilovich Gurvich, see footnote on page 125.

[31] Elephter Andronikashvili (1910–1989) was a renowned physicist in the areas of quantum hydrodynamics and low-temperature physics (see page 363). Since 1933 he lived and worked in Soviet Georgia. His brother, Iraklii Andronikov (1908–1990), received national acclaim for his oral stories and TV programs.

[32] Кротонский полдень, Москва, Узел, 1928.

[33] Maria Verblovsky's great-aunt.

season with them, and my nanny Frosya took me there on the first day of May. It was snowing in Leningrad. This day was the last time that I ever saw my father, and I remember watching through the window of the moving train car as his sad face receded into the distance. On June 22, the war began.

I remember how Isai was overjoyed by the news of the Soviet Union's alliance with England, because it meant they would finally hear from the Peierls family, who had lived in England since 1933. He already had two grandchildren there, a boy and a girl.

In the first days of the war,[34] the Public Library was ordered to evacuate its most valuable collections. In early July, a special train carrying the precious cargo moved out, accompanied by 15 employees, including my mother, Olga Vraskaya.

The Mandelshtams, who were under the watchful eye of the NKVD, didn't know what to do. The Germans were approaching Kalinin (Tver). At this time the train carrying the Public Library's collections was *en route* to the city of Melekes in Ulyanovsk Region.

In Bologoye, mother begged the director to give her the evacuation order showing the final destination of their trip, which had been kept secret from the employees until then. She made it to Ostashkov. The Germans were close, and if mother had not taken us away, it was quite likely that only one outcome awaited us all, according to the Nazi plan to exterminate the Jews.

In 1946, the Mandelshtams received permission to settle in Maloyaroslavets.[35] Isai Mandelshtam wrote:

> My Maloyaroslávets has won my honest
> Love, although it's been mine just half a year.
> In golden autumn, it's a young Adonis;
> In winter, it's a grandpa with a beard.
> O let me live till spring fulfills its promise
> And see this town's bright greenery appear;
> Then let the Maloyaroslavets days
> Of summer take this poet's breath away.

In March 1951, Isai Mandelshtam was arrested again. He served about eight months in a prison in Kaluga without a trial, as with his previous arrests, and then was exiled for ten years to the village of

---

[34] The war in Russia started with the German invasion on June 22, 1941.
[35] A small town 130 km to the south of Moscow.

Mikhailovka, in the Dzhambul Region of Kazakhstan.[36] But even there, cut off from civilization, he never gave up: he translated plays, for which he was paid a pittance, and even managed to check out books from the Lenin Library in Moscow. In late November 1953, a terrible tragedy befell him: his wife Maria died.

> How could one so young
> Suddenly cease to be?
> How did I get stung
> By this evil bee?
> Her heartbeat beat so strong,
> Then it ceased to beat?

Nina, following her parents to Kazakhstan, made her way to Alma-Ata in 1951. She worked on expeditions studying hemorrhagic fever and earned her candidate of science degree in Alma-Ata. After her mother's death, she petitioned for a reduction of her father's term of exile in Mikhailovka. When the request was granted in May 1954, she took Isai home to live with her. But on June 29, 1954, Isai Mandelshtam died in Alma-Ata.

In the mid-1950s, Nina returned to Leningrad. By then it was possible for her to be in contact with Genia and her family. At an academic conference in Moscow, Rudolf Peierls asked Nikita Khrushchev: "Why won't you let my sister-in-law come to visit us?" Khrushchev replied: "Why do you say we don't let her?! We let everyone go anywhere!" Later, Nina visited England on numerous occasions, and Genia started coming to Leningrad in the early 1980s, until 1985, when my mother died. After Nina died in 1982, Genia and mother looked through the Mandelshtam's papers, and Genia wept while reading Maria Abramovna's letters.

Nina translated specialized literature, mainly books on biology and microbiology. She visited us often. She lived in a neighborhood that was quite remote back in the 1960s. Members of the *intelligentsia* who had survived the repression were settled in tiny, ten square meter rooms in concrete panel buildings occupied primarily by workers from nearby factories. Her only possessions were a cot, a kitchen table, a bookshelf, and a pair of stools. She had no interest in things. In a suitcase she kept a few of Isai Mandelshtam's manuscripts, letters, and books.

---

[36] See footnote on page 126.

Of her childhood friends, Nina remained close with Valentina Abramovna Ioffe and Stepan Borisovich Vraskiy, my uncle. But in her heart, of course, she longed to be in England, with Genia and Rudolf Peierls, who by then had three daughters and a son. In the summer of 1982, already critically ill, she went to England and passed away there. Genia passed away in 1986.

*****

Letters to Genia from Russia, circa 1936[37]
Letter From Nina and Mother

May 1936

My little Genia,

I'm very sorry you won't have any letters from us for several days but, by God, I have been busy as hell. I work for 10 hours a day plus tram trips, meals, sleep, talks with the parents — that covers 24 hours! Thank you for fine pictures of Gaby. I implore you to send me for personal use the one where she is sitting on a chamber-pot with a trumpet and the one where she is getting from bushes with the hand outstretched (you and Rudi are also very well seen there). We also received the Berlin pictures. Gaby there is a bit worse seen, but your *Schwieger* is amazingly clear-cut; he has the face of an actor and resembles Isai. The family picture would be *sehr nett* if you and your nephew were excluded — this is an apotheosis! apotheosis! The faces look foolish like hell! Gaby is politely trying to kick Use's head. Rudi looks best of all from the viewpoint of expressiveness. In general he is photogenic and when it will be everything composed and clear what's what in physics, he may become a film-star (even in the sound movies with [...] *Sabinchen*). Now it is very late and I'll have to get up at 8.30 tomorrow. It is terrible! Mother is a bit unwell — the flu. She is in bed, sneezing, and the temperature is 37.4° [C] today. Isai is struggling with "Don Carlos." Ernest works like mad: he leaves at 8 in the morning and comes back at 12 at night. He and I have concluded peace and he feeds me with candies. My foot hurts a bit — I have to put on the shoes without heels. [...] the birds are singing and there is water up to the knees. I imagine how soon you will come and how you, Gaby and I are traveling by train and fellow travelers in our compartment are annoyed

---
[37] Courtesy of Gaby Gross.

and we are annoyed why they are. Yesterday I saw off Bubi: he had come here and we had heard his girl sing; there was a competition of young singers and musicians. The girl is impossible, to my mind, but probably I'm not versed in it. For the whole month I haven't been to a theater or concert. I have become savage, I don't even visit bathhouse.

Real honest, I cannot write any further, I'm falling asleep. I kiss you all affectionately.

N[ina]

[Added by Maria Kannegiser]

My dear,

Though I lie in bed and I'm not feeling well yet, I'd like to write to you; when I learned that Nina wrote to you, I stole her envelope and read this letter. It is a pity she didn't write anything about her marriage. [...] I'm getting weaker. I have been ill all through the winter and haven't gone out. This story will be the end of me. Please, don't betray me, otherwise Nina will again be very annoyed. Now that will do about it; it is the most awful event in the whole life that I have lived... thank you, my dear, for the photos, it was the greatest pleasure for me. I spend whole days looking at you all and I'm happy for you, my darlings. The Granddad is very much like Isai. Now that I'm in bed we speak a lot with Isai, and the conclusion we came to is very sad: we should have had more children. I was a fool being afraid to have them, only because I thought my daughters will feel the difference in Isai's attitude toward them. And now we feel very lonely: you are so far off, Nina is just like a lodger, nothing more. Well, you will pardon me for these grievous words, but I have long thought about it and I'd wish to tell someone. I wish you good health, my darlings, and kiss all four of you.

Mother

Postcard from Mother and Stepfather

[Oufa bureau de poste No. 139]
To: England, Chilcompton, Long Road, Mrs. E. Peierls, Cambridge
From: M. A. Mandelshtam, Ufa, 27 Salavata str.

May 9, 1936

My darlings,

I haven't written to you for quite a long time, I was in low spirits after Nina had left. I don't know where she is staying now, I know only that she is in Leningrad. She may even got married Lev Semyonovich,[38] but she wouldn't write to me about that. As you see, my dear Gerda, I cannot live a calm life, and when I look at little Gaby I shed endless tears. Her picture is on our table; Eric made a large portrait. My dear, if only this isn't a great trouble for you, please send Isai a light summer jacket[39] and a cap. You will excuse me, will you, if I put you to expenses. Now Isai is about to write to you, I kiss you, my darlings.

Mother

My darlings,

I kiss you affectionately and thank for congratulations and for the books (Shylock and Romains). On the cover of the book by Romains I noticed an ad of a new Bernard Grasset's edition of *Mémoires Du Chevalier d'Eon: Le Mystère De Sa Vie* (chez Flammarion, 12 fr.) and

---

[38] On April 4, 2017, Gaby Gross wrote me: "[Nina] sometimes mentioned a blind man (a mathematician) that I thought she may have had a relationship with." The letter of May 9 and the last paragraph of the previous letter (written by Maria Kannegiser, Nina's mother) solves the mystery of this romantic story. The reference to blind mathematician Lev Semyonovich allows one to identify L.S. Pontryagin (1908–1988), who was one of the greatest mathematicians of the 20th century representing the Moscow school. I quote here a passage from Rosa Berri's memoir (https://taki-terrier.livejournal.com/13683.html): "Realizing that it will be difficult for a blind young man to find a girlfriend at the age of 18–19, Pontryagin's mother took into her family an orphan girl, Tasia (of his age) when her son was still a boy, and, when time came, literally put her in bed with him. Tasia was an ordinary good-looking girl, and not stupid, completely undeservedly doomed to a bitter fate. She loved Lev Semyonovich from childhood. Having started studying at the Faculty of Mathematics at Moscow University, Lev Semyonovich got into the intellectual elite and soon announced to Tasia that he would not live with her, because he would seek 'true love,' and she was too simple for him. Tasia studied biology at Moscow University; after graduating, she left for [Soviet] Georgia, where she did not marry, because she did not cease to love Lev Semyonovich. And Lev Semyonovich began to fall in love, I will not list the names of his 'objects' — the name to them is a legion — I will add only one funny detail: they were all Jewish." Nina never married.
[39] For God's sake, nothing that is more expensive than lustrine. -MK

I'd like to read it. There is no hurry, though. I've just finished the second course of medical treatment with *padutin* and I'm feeling much better. Thank you for it, too! What news there is in physics? Anything sensational?

Yours, Isai

☙

Postcard from Genia to her Parents Exiled in Ufa, 1936[40]

Figure 5.3 To M.A. Mandelshtam, ul. Salavata 27, appt. 3, Ufa, USSR. — November 25, 1936 — Airmail — My dear, I have no stamps to send you a letter but I want to mail you something today. Sending you this postcard with Gaby's photo. Others will follow. Look what a wonderful granddaughter you have. This snapshot is ideal. I kiss you madly. Genia — PS. Gaby looks exactly as her grandma. Rudi

Apparently, in 1937 their correspondence was interrupted by the advent of the Great Terror.

*****

---

[40]Courtesy of Gaby Gross.

Family Tree

As was mentioned, the extended Mandelshtam family gave Russia many glorious sons and daughters. Their contributions to Russian art, literature, science and engineering are impossible to overestimate. The Mandelshtam family tree is traced to the beginning of the 18th century [2]. In the 1730s a Mendel Mandelshtam moved to a *shtetl* Zhagory in the Duchy of Courland and Semigallia.[41] A family legend says that Mendel Mandelshtam's ancestors were descendants of the Sephardim, who fled to Holland and northern Germany from Spain in the late 15th century from the persecution of Isabella of Castile. The family name "Mandelshtam" (sometimes also spelled as Mandelstam) was given to him instead of his Hebrew name in the Duchy of Courland.

This family was huge, probably hundreds of descendants live today. In a sketch below I will present only two branches, due to their importance to this narrative: they lead us to the main characters. The very fact that the Mandelshtam family tree is so thoroughly reconstructed is due to Osip Mandelshtam (bottom left corner of the tree), one of the greatest Russian poets of the 20th century, who perished on the wake of the Great Terror in 1938. Kannegisers and Gurviches (see pages 125, 157), and Gureviches (see page 112) also belong to side branches of this tree not shown below.

It is a sad story of a large, beautiful, and once happy family practically extinct in today's Russia.

Also of interest are *three* lines descending from Emelian Emmanuil Mandelshtam (the upper right corner in Fig. 5.4). One will lead us to Isai Mandelshtam, Genia's and Nina's stepfather. Another to Vladimir Arnold, one of the greatest mathematicians of the 20th century. The dashed line indicates that Isai Mandelshtam was raised by his uncle Max (Fig. 5.5).

---

[41]Currently Žagarė in Lithuania. The 1914 census of this part of the Russian Empire shows that Zhagory's population was 14,000, of which 8,000 were Jewish. See also page 32.

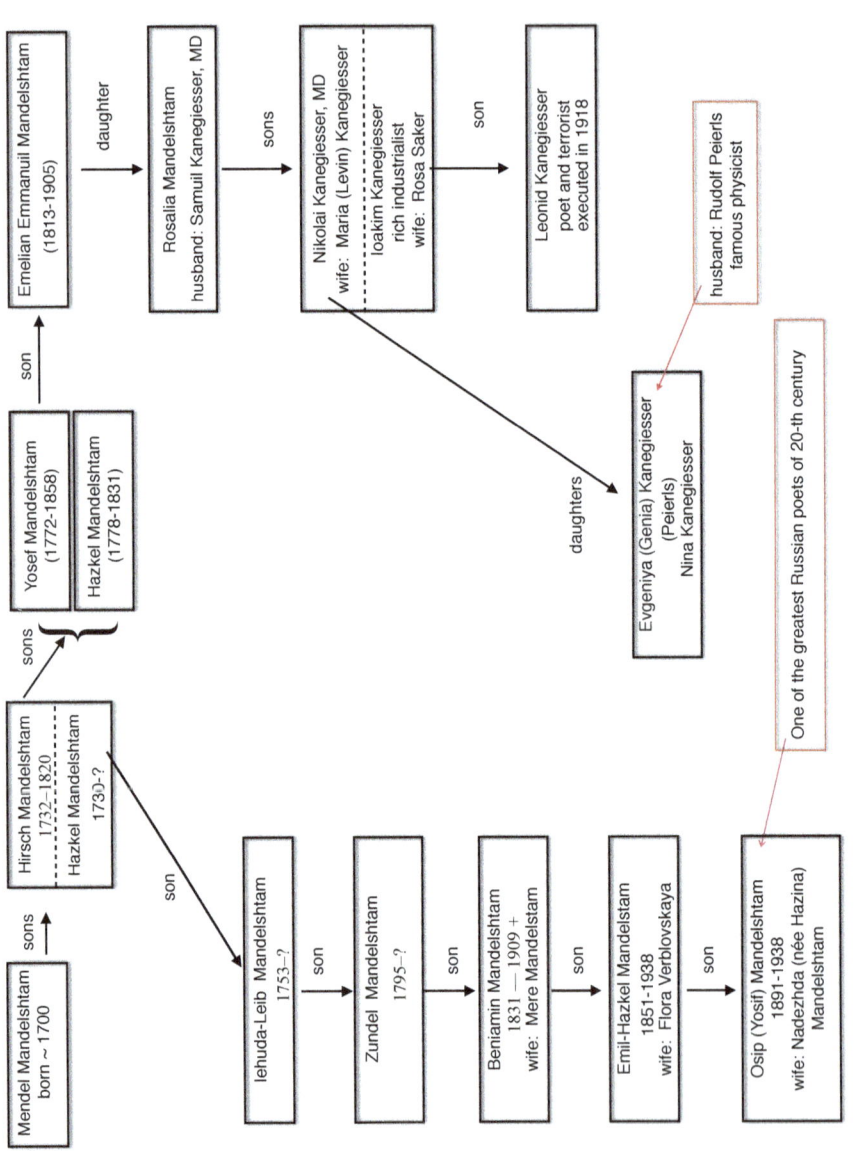

Figure 5.4 Mandelshtam family tree. I

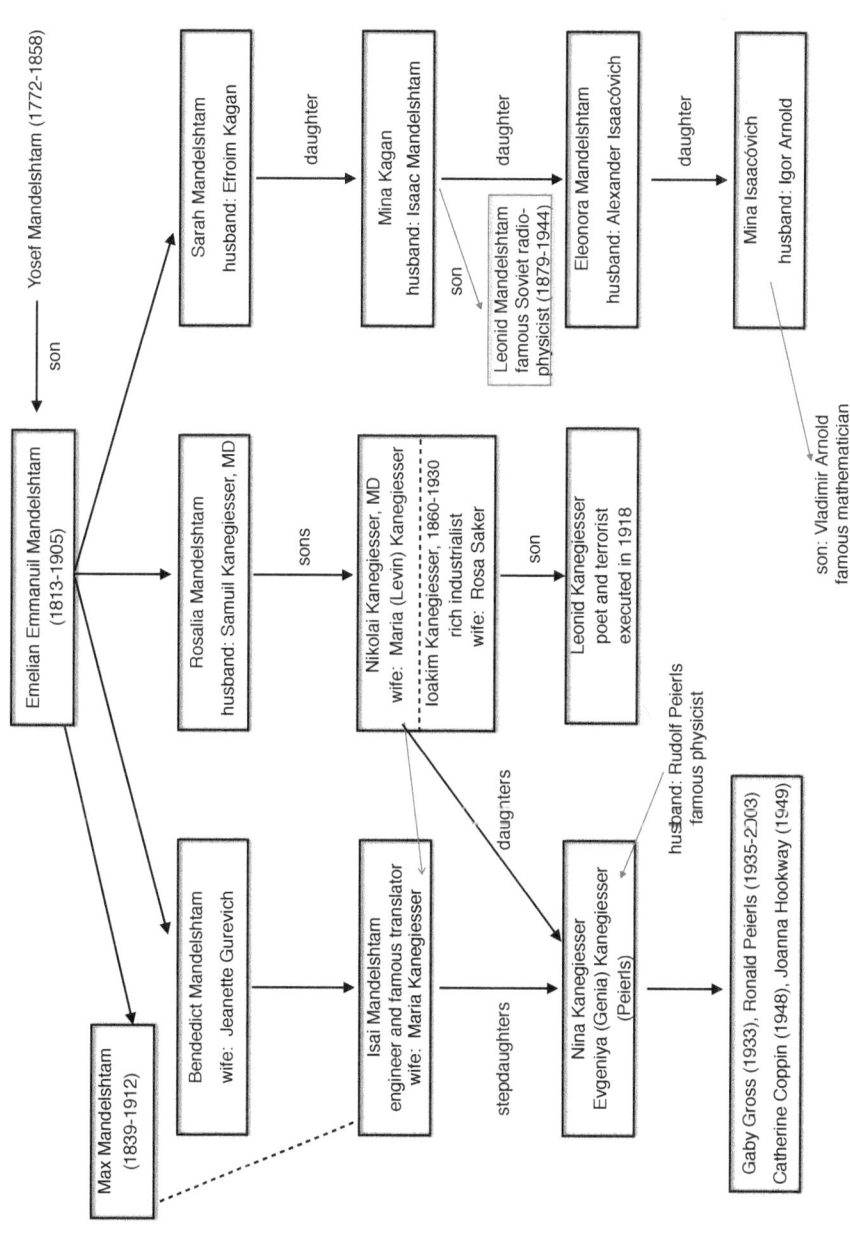

Figure 5.5 Mandelshtam family tree. II

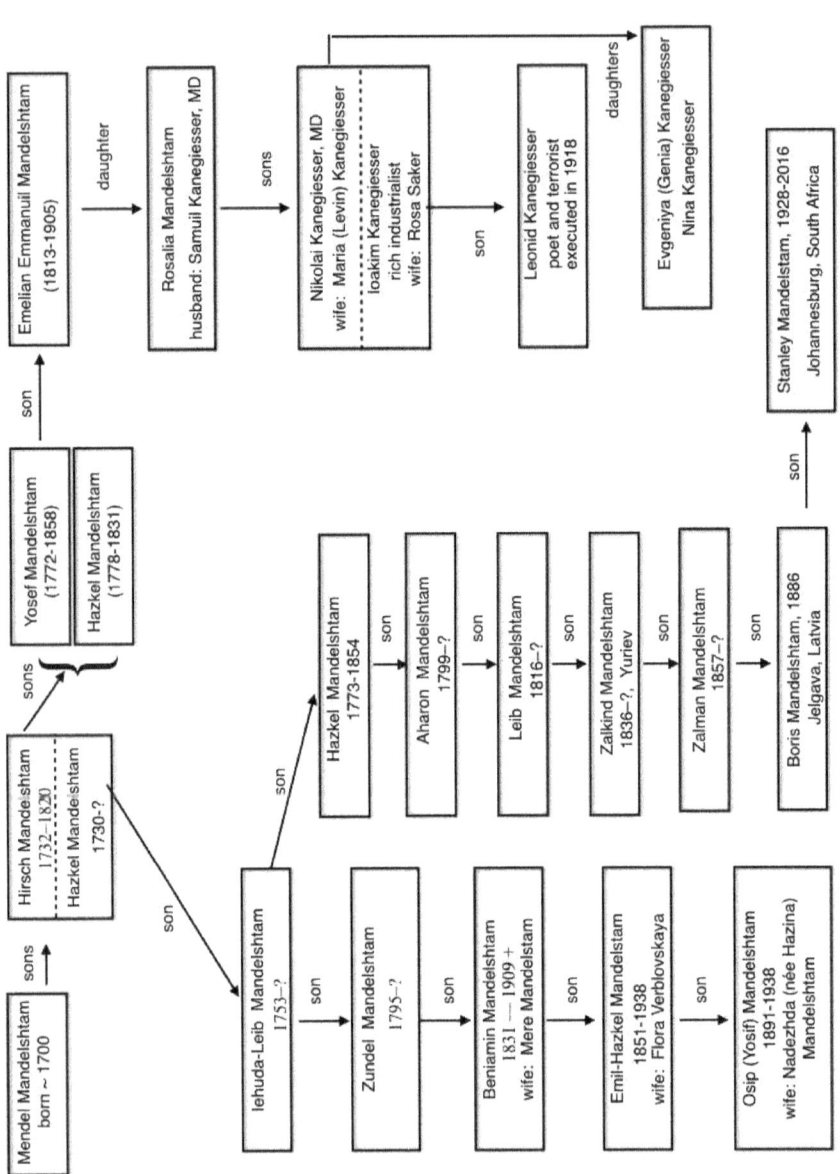

Figure 5.6 Mandelshtam family tree, III. Stanley Mandelshtam's branch.

*****

This digression is intended for physicists. Is there a modern theorist who has not heard of Stanley Mandelstam? I doubt so. He invented Mandelstam variables, suggested the bootstrap program in the Regge theory, was one of the earliest pioneers of string theory, put forward the idea of the dual Meissner effect as the mechanism of quark confinement, and so on. At Birmingham University he was a student of Rudolf Peierls (PhD in 1956) and then his colleague (1960–1963).

On August 30, 2017, Gaby Gross (née Peierls) wrote to me:

> My mother suspected she might be related to Stanley Mandelstam who was at Birmingham for a few years and called him her "grandson" but she never expected that the relationship could be demonstrated.

Indeed, this hypothesis could not be verified at that time due to insufficient data on the Mandelshtam family tree. Now we do have such data.

I think that Genia and Rudolf Peierls would be amused to learn for certain that Stanley Mandelstam belonged to a branch of the same Mandelshtam family tree as Genia, albeit a different branch. This curious fact was observed by Dolores Itkina [3]. I present her findings in a graphic form, see Fig. 5.6. Stanley Mandelstam is in the right lower corner. He represents the tenth generation of the tree, counting from Mendel Mandelshtam, the Patriarch. Figure 5.6 is an extension of Fig. 5.4 on page 138.

References

[1] A.S. Puschkin, "Ausgewählte Werke", Bänden 1-4, (Verlag für fremdsprachige Literatur, Moskau, 1949), Translated from Russian by I.B. Mandelshtam.

[2] Maxim Arnold, unpublished; Dmitri Vainchtein, unpublished; Pavel Nerler, "Osip Emilievich Mandelshtam, Rozhdenie i Semya," Znamya, №12, 2016,
http://magazines.russ.ru/znamia/2016/12/osip-mandelshtam-rozhdenie-i-semya.html

[3] Dolores Itkina, *Known and Unknown Branches of the Mandelshtam Family Tree*, in Russian, 543.ru/people/ways/morti/arnold/Mand_Neizv_i_izvestn.pdf

Chapter 6

Marriage and Trepidation. March 1931

We are fortunate to have the opportunity to look at events preceding Genia's and Rudolf's wedding and the wedding itself through his eyes, as Peierls briefly described it in his recollections in *Bird of Passage*:

> From Copenhagen I went on to Leningrad via Stockholm and the Baltic. The boat from Stockholm to Finland ran into thick ice, and although it was built as an icebreaker, it had to turn back and find another route where the ice was not so thick. We eventually reached Finland and I waited for a day in Helsinki for the night train to Leningrad.
>
> In the Physico-Technical Institute I was taken in hand by J. Dorfman,[1] a senior experimentalist, since Frenkel was in America. Dorfman suggested I give a course of lectures on solid-state physics, and he further suggested I should give it in Russian. I agreed, perhaps unwisely. I greatly enjoyed the process, but for the audience it must have been an ordeal. However, I still occasionally meet Russian physicists who say that they learned the beginnings of solid-state physics from my course, so they must have understood something.
>
> But much of my free time was spent with Genia, and after ten days or so we decided we would get married. At that time there was no obstacle to Russians marrying foreigners, but whether Genia would be able to leave the country was another question. [...]

---
[1] Yakov Grigorievich Dorfman (1898–1974) was an outstanding Soviet physicist in the area of magnetism.

Figure 6.1 On the day of Genia's and Rudolf's wedding. Seated: Genia's stepfather Isai Mandelshtam, her mother Maria Kannegiser and Rudolf Peierls. Standing: Nastya (household help), Nina and Genia. From *Bird of Passage*.

I want to emphasize Peierls' last remark in the paragraph above. This was a visionary observation, indeed. Then he continues:

> Our plan was to marry on Friday, the 13th of March, to mock superstition, but this plan was not realized. At that time, a compulsory smallpox vaccination was in effect, as there had been some cases of the disease, and I had a reaction to the vaccine. On the 13th I was in bed with a high fever, and the marriage was postponed to Sunday the 15th. Sundays had no significance in Russia at the time. The week had been officially replaced by a five-day period, with every fifth day a day of rest.
>
> So on Sunday morning we set out for the register office. But we found it had been moved and had to get directions from a policeman. The procedure was very simple (it has become much more ceremonious by now) — we just signed a form that had been filled in by a clerk. The clerk did spoil two forms, because first she could not spell my name and then she could not spell Genia's. When

she finally got it right, and had correctly entered our professions as "scientific workers," she put her pen down and said, "You are both scientific workers? And both so cheerful?"

At that time I felt strongly that getting married was the private affair of two people, and not an occasion for crowds to celebrate, but I consented to our being toasted by Genia's family. I had of course met her family by this time; in fact, I had called on them during my first visit to Leningrad. Her mother, very gentle and feminine, had unbelievable energy when it was required — and bringing up a family during revolution, civil war, and hunger had certainly required it. Genia's father, a famous gynecologist, had died when she was a baby; her stepfather was a wonderful person, trained as an engineer, but also a talented writer and translator, a great raconteur, and a man of outstanding integrity. Genia's sister, Nina, a biologist, completed the family. Another important member of the household was Nastya,[2] who for many years had been the help and friend of the family. A cousin, who was family photographer, also stayed in the flat. Without an extra person, the flat would have been too large and would have had to be shared with strangers.

*****

In 1969, in an interview conducted by Charles Weiner for the American Institute of Physics Oral History Project, Rudolf Peierls described his pre-WWII visits to Russia in 1931, 1932, 1934, and 1937 [1]:

I was in Leningrad in the spring of 1931 for two months by invitation. I gave some lectures and, incidentally, I got married. At that point my wife did not immediately get permission to leave the country, and so when I left at the end of that two months, it was not clear whether she'd ever be able to come out, or, for that matter, whether I'd ever be able to get back, because you needed a visa for every trip, which wasn't a matter of routine.

Obviously, on summer vacation, as soon as I could get away, which was, I think, at the end of July, I went back to Leningrad and stayed until I had to get back to my job

---
[2] Anastasia Mikhailovna Kharitonova.

in Zurich, which was presumably the end of September. And obviously I must have been to the Kharkov Conference that summer. I remember being in Kharkov, though I wouldn't be sure of the date.

We — Genia and I — met actually the previous year at the conference in Odessa and saw a good deal of each other then; and, in fact, there was after the conference a small party that went on. Well, the whole conference, all those who wanted, were taken on the boat across the Black Sea, and then a smaller party traveled through the Caucasus, and then in fact my future wife and I stayed a little longer there. And then we started to correspond. And so when we got married we'd seen each other for a total of two or three weeks, but we'd corresponded in the middle for six months.

At the last minute the papers came through two days before I had to leave or something like that.

In 1932 I spent the summer [in the USSR], essentially a vacation, though I visited some institutes as well. And, yes, then I went on a walking tour through the Caucasus with Landau and a friend of his,[3] and I think on the return visited Kharkov where Landau then had a job — spent a few days there. That was 1932. I came back from Leningrad earlier than my wife, who was staying longer.

The next journey [to the USSR] must have been in 1934.[4] And the last time I went before the war was in 1937 when there was a conference in Moscow and when the chance of foreigners to go there was already deteriorating — the mass arrests had started. This was heading for Stalinism.

The last time in 1937, everybody's fear of arrests and so on was very noticeable. I remember two remarks which perhaps characterize the climate there. It was said as a joke that those people with private telephones were trying to get it disconnected because they thought that people to be arrested were maybe picked from the telephone directory.

---

[3]Styrikovich, see page 376.
[4]The Peierls took their one-year old daughter Gaby (born on August 20, 1933) with them, to show her to Genia's parents.

Of course, it was quite an irrational remark, because as the telephone directory was published only every three or four years, getting your telephone disconnected wouldn't get you out of the list very quickly. [...]

The 1937 conference [in Moscow] was slightly halfhearted because people were too preoccupied with what was going on. I don't remember much in detail about the conference. It was a time when work on cyclotrons in Russia had started. People were reporting on the progress. I don't think they had a working cyclotron yet. [...] They had, of course, quite competent work going on in radioactivity, and they had excellent theorists. I remember talking with Landau, not in the conference but on that occasion. I have a stronger memory of his worries about how the situation was deteriorating and how unhappy everybody was rather than about physics.

Things were changing critically about that time. [In 1934] for Soviet physicists to get out was rather difficult and rather exceptional. On the other hand, it was rather easy for people from the west to go to the Soviet Union. And in fact, any reputable physicist who was willing to pay his expenses as far as the first stopping place in the Soviet Union, or, if he was really keen to save money until the border, they could arrange for some rubles to be sent him somehow to the border, could go more or less for the asking. It was quite easy to get an invitation and to spend a few weeks there and have all your expenses paid, and be welcome in some laboratory. Not everybody did that, of course, but quite a few people, particularly theoreticians, made use of that. People like Weisskopf and Placzek and so on did a lot of traveling in that sort of way. This was completely cut off in 1937. The nuclear physics conference in Moscow in 1937 was the last occasion where people were invited officially. I think the plans for that had been made some time ahead; otherwise it wouldn't have come off. After that it was practically impossible to go.

*****

After Genia and Rudolf were married in Leningrad on March 15, 1931, Genia applied for the exit visa. In fact, she wanted to denounce her Soviet citizenship altogether. This was the most problematic point of their plan. They were not sure that she would be allowed to leave the country. Rudolf's concerns about this are seen also from the remark in Heinrich Peierls' letter to his son of March 19, 1931 (page 149), which mentions Rudi's consideration that at a "later time it would be difficult to get Genia out of Russia." They clearly realized the rapid deterioration of the political situation in Russia.

In fact, already in 1931 exit visas were rarely granted to Soviet citizens, to put it mildly. It was good that the young couple apparently did not realize the scale of the problem, which kept their spirits high. My former PhD adviser, Boris Ioffe, who in 1933 visited Poland with his mother, testifies [2]:

> At that time to get permission to travel abroad was very difficult, almost impossible, but we were lucky: an uncle of mine had an acquaintance — Artur Artuzov,[5] who was at that time the Head of the Foreign Department of the OGPU (NKVD).

From the same letter it is clear that Rudolf informed his father of his imminent marriage in the very last moment. Heinrich was outraged. He accused Rudolf of "irresponsibility," tried to talk Rudolf out of "such stupidity" and informed him that he — his father — would cut off his financial support and would not resume it under any circumstances. Later this promise was forgotten: Heinrich Peierls resumed his financial assistance to the (limited) extent that Nazi regulations allowed money transfers to other countries.

The last important point in Heinrich's letter cautions his son that in the future, authorities would hold a foreign wife against Rudolf. Remarkably, 20 years later this remark proved to be visionary.

*****

Rudolf Peierls arrived in Leningrad on March 4, 1931 and spent almost two months with Genia. He left for Zurich in the last days of April.

---

[5] In 1937 Artur Artuzov was arrested by the NKVD and executed by firing squad.

Heinrich Peierls to Rudolf Peierls[6]

Berlin
March 19, 1931

Dear Rudi,

Although I could not completely overcome the suspicion that I expressed in my registered letter sent to Zürich, I must admit that the content of your letter of the 13th has by far exceeded the worst apprehensions and has been in the strongest contrast to what you have said when we last met. I do not want to repeat all my arguments against such an early relationship, since you are sufficiently familiar with them. But I had not thought you capable of such an irresponsibility as wanting to engage in marriage now, before you something [words missing -WvS] and before you are able to support a wife even in the most modest way. How can you, as a reasonable person, make your decision dependent on the consideration that at a later time it would be difficult to get your future wife out of Russia? Don't you see that both of you would have to completely "proletarize"[7] and that it would soon be the end of your career which is just beginning? Are you so ignorant of the opinion of the authorities on whom university appointments depend, to believe that they would not hold a foreign wife against you? Where are your life, your plans with America and so on? Do you want to leave your wife behind or do you intend to carry the cost of all her journeys?

You can, of course, make such a stupidity which will ruin your life, and I cannot forbid it. But I feel obliged to do everything in my power to prevent an economic suicide of this scale.

I must therefore say with utmost certainty that I would feel obliged not only to refrain from any additional support, but also to withdraw the current one. However difficult it is to me, I must also tell you that you could not expect under any condition to receive even a single pfennig in the economic complications which will probably (that is certainly) arise. If you want to run into your disaster, and if you are not receptive to reasonable advice, then smash all china to pieces, leave all culture behind and drink your cup to the dregs. Then nobody can help you. Should there be no suitable life companion for you in the whole of Germany, then at least test yourself for a while before you take unalterable decisions, and then marry once your economy allows it. I can

---

[6]Sabine Lee, Volume 1, page 243. Translated from German by Wladimir von Schlippe.
[7]"Proletarisieren" in the German original.

see how scatter-brained you are, considering that you want to see us in Berlin at the end of April, although I have told you repeatedly that we would be away at that time. We will go on the 10th of April first to Lugano, Park-Hotel, following destinations uncertain. At any rate, at the end of April or beginning of May we will not yet be at home. My telegram has probably prepared you for our position. And now do after mature and sober consideration what your reason, that surely has not yet completely left you, tells you to do.

With best wishes,

Your father

Dear Rudel!

Your sudden decision gives us great sorrow. You seem to be in such a state of being in love that you throw away all your principles, your career and your entire future. When one is as gifted as you are, one has duties towards oneself, and when one has a father with a rich experience of life, a father who is not narrow-minded nor old-fashioned but kind, generous and of clear vision, then I think one should follow one's father's advice! Take your time and regain your senses. We all have experienced things which we had to overcome. Such experience does enrich one if one overcomes it and if one does not expect of oneself something that one cannot live up to.

In well-meaning love,

Else[8]

---

[8] Rudolf Peierls' stepmother.

On August 15, 1931, Rudolf Peierls arrived in Leningrad to collect Genia. However, her emigration papers were not yet ready, and they decided to go on a "suspended honeymoon" for 2–3 weeks in Teberda, a town in the Caucasus Mountains 65 miles south of Cherkessk. Teberda was, and still is, the gateway to the Teberda Nature Reserve, an area known for its natural beauty. Rudi and Genia planned to descend from Teberda (its elevation over the sea level is about four thousand feet) down to Sukhumi on the Black Sea shore. In the 1930s there were no roads connecting these two towns, only narrow trails. However, this part of the plan had to be abandoned because Rudi's arrival to Leningrad was delayed by a week or so. His Soviet visa came later than expected.

On the way back they stayed in Moscow to make inquiries about the status of Genia's application. At the railway station they were met by Genia's relative, also a physicist, Leonid Isaakovich Mandelshtam. He drove them to his home. From there Genia made a call to the office which was in charge of her exit documents. The exchange went something like this:

– Once again, please, your surname, name and patronymic.
– Kannegiser, Evgenia Nikolaevna.
– With two ens or one?
– With two.
– Well ... wait a minute. Your request is approved. Documents sent to Leningrad.

In the evening of that same day they left for Leningrad to prepare for the departure. A week passed, then another, Rudolf became nervous since he was supposed to return to Pauli in Zurich by the end of September. "It is impossible to wait any longer, said Rudy, tomorrow I am going to Moscow and will try to settle everything."

What happened next is described in Peierls' book [3]

> In the Moscow office I was not kindly received: "Your papers have gone off; we have nothing more to do with the matter!" I persisted. To what address had they been sent? To the right one, I was assured. Could I speak to the person who dispatched them? No, but they would check with her, if I insisted. I did insist, and it turned out the papers had been sent to another office in

Leningrad. I wanted to communicate this to the family in Leningrad as quickly as possible, and used a new picture telegram service for the purpose. I wrote my message on a form, and the same afternoon the facsimile was delivered in Leningrad. Genia had a cold, so her sister went to the other office, and after some search a clerk found the papers in a drawer, and commented "Oh, that Maria Ivanovna; she always misplaces things!" But at last Genia and I could depart together.

Rudolf Peierls and Genia — now Frau Doktor Peierls — arrived in Zurich in late September. On September 29, 1931, Peierls received a letter from Pauli instructing him to clarify and perfect the arguments in the Landau-Peierls paper and to study a recent work by Bethe [4]. "In addition, he writes, please, look into the question of radiation in a time-dependent magnetic field, in particular, the line widths and the degree of polarization of the resonance radiation. You must and can find general answers. But this question is not very interesting."

Pauli was in the United States at this time. His colloquium at Princeton was scheduled on Thursday, October 1, later than was expected. In his letter, Pauli informs Rudolf that he would leave the US on October 2, and arrive in Naples on October 11, just in time not to miss the beginning of his (Pauli's) lecture course in Zurich on October 20th.

References

[1] https://www.aip.org/history-programs/niels-bohr-library/oral-histories/4816-1.
[2] Boris Ioffe, *Atom Projects, Events and People*, (World Scientific, Singapore, 2017), page 7.
[3] Rudolf Peierls, *Bird of Passage*, (Princeton University Press, 1985).
[4] H. Bethe, "On The Theory of Metals, Eigenvalues and Eigenfunctions of the Linear Atomic Chain," *Z. Phys.*, **71**, 205 (1931).

Figure 6.2 Skiing vacation in 1931. Ilse Thorner's brother, Genia, Rudolf, Max Delbrück, and Ilse Thorner. From *Bird of Passage*.

Chapter 7

Flashback 2

Nina, the Sister

Nina Kannegiser, a year younger than Genia, was born in 1909. They say, destination defines destiny. This was indeed true for Nina. While Genia left the USSR, the paradise country of workers and peasants, Nina stayed, and her fate prepared for her quite a few grim pages in her life. Despite that, everyone who knew her described her as a joyful and optimistic person.

A rare glimpse into Genia's and Nina's lives in the early 1920s is provided by Nina's posthumously published article [1]:

> My stepfather — Isai Benediktovich Mandelshtam — got acquainted with Yuri Yurkun[1] while he [Isai] was in prison, in 1918. Somewhat later he met Mikhail Kuzmin[2] in the apartment of my uncle, Ioakim Kannegiser. At that time we lived in someone else's apartment — the owners fled to Ukraine to save themselves from starvation — in a room with a plush ottoman, with beaded tassels on all lampshades and photographs in frames on the walls, but there was a large genuine stove, rather than homemade "burzhuyki."[3] Steam heat had long ceased to function in almost all but government buildings. My stepfather had two engineer jobs in two distinct places which doubled his ration cards. In the evenings he translated Balzac and France, and my mother was bartering

---

[1] Yuri Ivanovich Yurkun (1895–1938) was a Russian writer of the so-called Silver Age and a graphic artist. He was arrested by the NKVD and shot in 1938.
[2] Mikhail Alekseevich Kuzmin (1872–1936) was a Russian poet and prose writer of the Silver Age, translator, and composer, see J. Malmstad and N. Bogomolov, *Mikhail Kuzmin: A Life in Art* [2].
[3] A primitive heating device.

with "inspiration" her wedding ring for a carp fish, her jacket for a couple of logs of firewood, and her wedding set for 24 people for 15 kilos of rye flour. And as a result, the four of us were somehow fed, warmed and could enjoy the opportunity to warm up our wandering guests and treat them with millet porridge and boiling water with breadcrumbs.

Figure 7.1 Nina and Genia Kannegiser, circa 1929. Courtesy of Natalia Alexander.

[In the winter of 1919–1920][4] Kuzmin and Yurkun would come simply to warm up, most often in the early evening, sometimes before my stepfather's return from work.

---
[4] At that time Nina was 10 years old.

Once Yurkun said to my mother: "I have brought Mikhail Alekseevich a bit earlier today. I'm afraid that he might freeze to death." One had to understand that literally.... The sub-zero temperatures that year were harsh, and because of hunger, the fierce cold in their rooms, and constantly being in the same insufficiently warm and dilapidated clothing (wool coats and felt hats), they would arrive numb from the cold.... Mikhail Alekseevich, taking tiny little steps (he had a very characteristic gait), would go into the dining room, lean against the tiled stove, and gradually thaw out: he would turn slightly pink, cough, have a cigarette. Mama, carrying our only candle, would disappear into the kitchen (soup made out of cereal and an additional course — the hateful rutabagas), having instructed my sister (a year older) and me: "Sit with Mikhail Alekseevich, only don't pester him." He was very courteous to us, used the polite form of "you," and we would staidly chat about ways to cure frostbitten fingers and about the exceptional talent of our cat.... He was extremely thin — his clothes hung on him. A puffy face (from edema?)... His face and hands were a yellowish blue gray from the cold and dirt — a smoky fire supplied the light and heat in their apartment at the time, and water had to be carried out of the cellars.

The friendship between Nina and Mikhail Kuzmin lasted for life. Sometime after Kuzmin's death in 1936, Nina wrote an essay devoted to him which was hidden and published only in 1990 [1]. Nina received a university education in Leningrad, majoring in biology. One of her teachers and advisers was Alexander G. Gurvich (1874–1954), a relative of Isai Mandelshtam and a well-known biologist.[5] Since 1931 and until her first exile in 1935 Nina conducted research at the Leningrad branch of the Institute of Experimental Medicine.

---

[5]In the late 1920s, A.G. Gurvich was nominated to the Corresponding Members of the USSR Academy of Sciences, but this nomination was vetoed by the university Communist Party cell. In his "Memoirs" A.A. Lubishchev writes [3]:

" ... the regime of any dictatorship was absolutely unacceptable to him and to the question of the Rector of Moscow University Vyshinsky (Andrei Vyshinsky, the Chief Prosecutor of the USSR during the Great Terror -MS)

– Professor, can't we build a bridge between us?

Gurvich replied,

– Between us, is the abyss."

Since 1930, he worked in Leningrad at the Institute of Experimental Medicine.

As was mentioned, in 1935 her stepfather was arrested and the Mandelshtams were sent in exile to Ufa. On March 22, 1938, in Ufa, Isai Mandelshtam was sentenced to Gulag. Trials and tribulations of the family until 1954 are vividly described in detail in Nina Kannegiser's essay *Isai Benediktovich Mandelshtam. My Stepfather*, see page 112.

In 1951, Maria and Isai Mandelshtam were sent to exile in Kazakhstan. It seems likely that Nina had arrived in Kazakhstan around this time, on an epidemiological expedition. The exact date of her relocation to Alma-Ata (then the capital of Kazakhstan) remains unknown. Eventually, she defended her PhD thesis in Alma-Ata.

Since 1937, Alma-Ata became home for Moscow and Leningrad *intelligentsia* exiled for political reasons. In 1941, after the German invasion, it received in addition a large number of evacuees from the western parts of the USSR. Almost all film studios and major theaters were relocated there, with all their personnel. It suffices to mention Sergei Eisenstein (Battleship Potemkin!), Vsevolod Pudovkin, Grigori Kozintsev, Leonid Trauberg,[6] etc. All of a sudden, the city became, in a sense, an intellectual and cultural center.

However, close friendships even among these people were rare. Being afraid of denunciations they had to exercise great caution in conversations with newcomers. Even innocuous phrases could be deliberately misinterpreted and reported to the NKVD. Any frank word was too risky. From some hints, glances and gestures, people recognized soul mates in others. Of the really close friends of Nina Kannegiser one can name two: Boris Sergeevich Kuzin and Pavel Zaltsman, a Leningrad artist and writer, sent to exile in Alma-Ata in 1942 just because his father was ethnic German.[7]

Food was in short supply in Alma-Ata, and so was housing. The ration cards provided only basics, such as bread, and even that in limited amounts. On lamp posts, handwritten notes such as "Trading a loaf of bread for an egg" were commonly posted. A few cantines for starving evacuees from Leningrad were open. They served soup of bran (in exchange for a ration card).

In Kazakhstan, Nina worked as a epidemiologist. Boris Kuzin, an entomologist, poet, and a great friend of Osip Mandelshtam, was in exile in Alma-Ata since 1944, and this was not the first time for him, as it was not the first for the Mandelshtams. Kuzin and Nina Kannegiser

---

[6] Famous Russian filmmakers.
[7] His mother Maria Ornshtein was Jewish. In German his name would be written as Salztmann; Zaltsman is a Russified version.

were connected through their common interests in biology. Pavel Zaltsman was a frequent guest with the Kuzins. They spent hours talking about the causes of the catastrophe that struck Russia in 1917, the extermination of the *intelligentsia*, the Gulag .... Their friendship was deeply rooted in general poetic preferences: Anna Akhmatova and, of course, Osip Mandelshtam were idols for both.

It was Boris Kuzin who introduced Zaltsman to Nina Nikolaevna. They became close friends. Nina spent hours reading by heart, poems of Akhmatova and Gumilev to his daughter, Lotte Zaltsman. The memory of this friendship lives in the family. Here is what E.P. Zaltsman wrote in 2010 [4]:

> [...] In Alma-Ata Zaltsman met Nina Nikolaevna Kannegiser, also a biologist, who came to Kazakhstan in voluntary exile following her stepfather, Isai Mandelshtam, a well known translator, and her mother. Prior to that, she was wandering between small Kazakh towns in the middle of nowhere, bear's dens so to say. Compared to them Alma-Ata seemed quite an acceptable city, especially since here she was able to find employment in her trade. She worked at a semi-classified Anti-Plague Institute and traveled around Kazakhstan, taking part in localization of epidemics of plague and hemorrhagic fever. These diseases, officially exterminated in the Soviet Union, have repeatedly exploded in the WW2 and post-war period in Kazakhstan, and tuberculosis and syphilis were almost normal in some areas. She belonged to the elite of the Russian *intelligentsia*. She recited by heart Boris Pasternak, Anna Akhmatova, Nikolai Gumilev, Irina Odoevtseva.[8] She lived in Alma-Ata until 1958, but even after 1958 her contacts with Zaltsman did not stop.

I will also quote a touching paragraph from the memoirs of Maria Verblovskaya,[9] Nina's relative:

---

[8] Boris Pasternak, Anna Akhmatova, and Nikolai Gumilev were great Russian poets of the 20th century. Irina Odoevtseva (real name Iraïda Heinike (1895–1990)) was a Russian poet and prose writer, Nikolai Gumilev's favorite student. In 1921, Gumilev was arrested on suspicion of participating in an anti-Soviet conspiracy and soon after, executed. Irina Odoevtseva managed to emigrate to France in 1922. Both became "non-persons." Even spelling out their names was dangerous in the 1930s–1950s.

[9] M. Verblovskaya, *Flow of Time*, 2014, unpublished.

Figure 7.2   Pavel Zaltsman. Self-portrait, 1940s.

After her return from Alma-Ata in the mid-1950s, Nina was involved in translating special literature, mainly books on biology and microbiology. She often visited us. She lived in a neighborhood of sloppily built five-storied "Khrushchev" houses, which at that time, in the 1960s, was a remote suburb. These block houses were inhabited mainly by workers of surrounding factories. Some of the *intelligentsia* who survived the repressions and were returning to Leningrad with nowhere to live were settled in these houses in small rooms — "cells" — of ten square meters.[10] They were supposed to share the "main" apartment with other families. All Nina's belongings consisted of were a cot, a small kitchen table, a bookshelf, and a couple of primitive wooden stools. She did not want even to hear about buying other things. She kept in her suitcase manuscripts, letters and some books by Isai Benediktovich Mandelshtam.

---

[10] About 100 square feet.

In the late 1960s Nina Kannegiser co-authored a book of recollections devoted to her beloved teacher Alexander Gurvich [5].

In the Annual Letters (to friends), Rudolf Peierls mentions Nina several times. For instance, in the letter dated December 1981 we read (page 363): "In the winter [of 1980–81] when we were at the *Institut des Hautes Études Scientifiques* in Bures-sur-Yvette, Genia's sister Nina, who had visited us in Oxford for the summer, spent the first month with us in Bures, and spent much time with or without Genia, exploring Paris." Nina died in 1982 while visiting Genia. In the Annual Letter of November 1982 the Peierls write (page 364): "The year was overshadowed by the death in October of Genia's sister, who had been with us since July. She was already unwell when she arrived, with what was diagnosed as spastic colitis. She was also very depressed, and suffered from loss of appetite, so that she had become very emaciated. With care and medication these troubles improved, but she was still very weak and anemic. A stay in hospital for observation did not show the cause, but in October she had to go to hospital for further tests, which showed cancer. She died from an internal hemorrhage, probably caused by the tests, but this was a blessing, as it saved her a period of suffering and misery. We also were relieved it happened here, where she could get better care and comfort than in Leningrad."

Later, in his memoir *Bird of Passage* [6] written in 1983–84 and released in 1985 Rudolph Peierls recollected:

> Genia had expected to accompany me, but Nina had, once again, been able to accept our invitation to come to Oxford. She had been ailing for some time. When she arrived she had lost much weight and was very depressed. Genia gave her all possible care, but the doctors could not do much, and the recovery progress was very slow. Eventually her depression lifted, and she seemed better, but her physical condition deteriorated, and she had to be taken to hospital. I left for Munich with a heavy heart. Five days later Nina died. She was a remarkable person, with a phenomenal knowledge and memory, and always willing to go to trouble for others. She had a hard life, and it was with some slight consolation that her last days were spent in the comfort of an English hospital, and without much suffering.

***

Figure 7.3 Nina Kannegiser in the 1960s. Leningrad. Courtesy of Natalia Alexander.

Katya Arnold's Recollections

We lived in Moscow and the Kannegisers lived in Leningrad. Our apartment was on Arbat, Staropeskovsky Lane. Now, the neighboring house is the American Ambassador's residence. My paternal grandmother, Vera Stepanovna Arnold, had been Lenin's assistant, that's why our family did not have to share the apartment with other families, as was the norm at that time ("communal apartments"). I was born in 1947. Nina Nikolaevna was our relative from the Jewish side, as were the Mandelshtams and the Isacoviches. The Arnolds are the Russian line. Those who lived in the communal flats were afraid of hosting foreigners and dissidents who went through Gulag. It was too risky. Therefore it was in our apartment that Genia, who visited the USSR around 1956 for the first time after her departure in 1931, could embrace her sister Nina.[11] Rudolf Peierls and Genia met each other at a conference in Odessa in 1930. They fell in love with each other, and Rudi proposed. Genia answered: "If you learn Russian and in a year return here to pick me up, I will marry you!"

And indeed, he learned Russian and came to marry her in Leningrad in 1931. At that time it was still possible. Just 4–5 years later, marrying a foreigner would be completely inconceivable. The Peierlses left, and soon all connections were cut off. People were afraid even to write to their relatives abroad. And here it was, the first encounter after 35 years in our apartment.

When Nina's mother and stepfather were sent to their second exile in Kazakhstan, she left everything that belonged to her in Leningrad, her apartment and all, and went with them voluntarily.

In Kazakhstan she worked as an epidemiologist. On donkeys and horses she was roaming between the distant Kazakh villages in the mountains and treated the Kazakh tribes for diseases which are now eradicated from Earth's face, but back then they still existed. Needless to say, for me it was incredible exotics. I was not even 10, but when Nina visited Moscow once in a while (perhaps, once a year), I could not miss her stories. She always stayed with Klara Efimovna Papaleksi on Kirov Street. Klara Efimovna was an academic lady, a widow of the well-known physicist Papaleksi. Each time Nina Nikolaevna would bring her one and the same present: a huge red apple, over a pound in

---

[11] In fact, Genia visited her parents and Nina in Leningrad in 1934. Gaby Gross, née Peierls, Genia's and Rudolf's daughter, believes that she (Gaby) was the first from the Peierls' family to visit the USSR after Stalin's death, in 1955 or 1956. Private communication, 2018.

weight, called Alma-Ata Apport. I was allowed to go by bus alone from Arbat to Kirov Street. We always had tea at first, and then strolled through Moscow, visiting various art museums and little shops nearby, buying there post cards with classical paintings on them. She seemingly knew all of them and right at the entrance she would loudly direct me: "Buy the postcards!", and I would buy one, two, five, ... ten... It was Nina who "infected" me with arts for the rest of my life. I was a child but she treated me as an adult which made our friendship closer. She cherished my uncle, Mikhail Aleksandrovich Isaacóvich. They grew up together in Odessa. They — my uncle and Nina — were great pranksters and joyful people. They loved to laugh. Even the gloomy Soviet regime failed to outroot this love, to suppress it. And, gosh, grim it was: my grandfather was shot in 1937 by the NKVD, Nina's parents died in exile.

Sometimes we spent time together, three of us, sometimes it was just I and Nina Nikolaevna. When I turned 12, my parents allowed me to go to Leningrad on winter vacation. Nina Nikolaevna invited me. She said: "If you swear that you won't get sick, come, visit me!" So I did, my parents put on me *valenki* and *kaloshi*,[12] and saw me off to the train station. Nina Nikolaevna met me in Leningrad at a railway terminal. She had a room in a communal flat in a suburb called Martyshkino. This happened about two years after her return from Kazakhstan. She took me to the Opera. Then there was New Year Eve celebration. We met on New Year Eve with her old friends, the previous generation of Petersburgers, who all but a few had died by this time. Her friends had a room in a communal flat with enormously high ceilings.

They would never say "Leningrad" only Petersburg or Peter or Petrograd arguing between themselves which of the three names were better. They drank vodka, and I was given a glass of Champagne, after which I did not feel well because I got drunk. The next morning she said: "Katya, if you cannot drink, do not drink!" She made no difference between a child and an adult.

She showed me the Pergam Altar which was stolen from Berlin after the war. It was kept in the basement of Hermitage. Although it was not accessible to general public, she had acquaintances there who let us in. I was so impressed that I remember this day till now — the Pergam Altar in Hermitage's basement. Nina Nikolaevna was passionate about my education. She had no children of her own, nor was she married. Most probably, by that time she was already retired.

---

[12] Felt boots and galoshes.

When it became possible, she started visiting Genia in England, each time staying with the Peierlses for a month or so. Once they traveled in America — Boston, New York, and the West Coast, from Los Angeles to Seattle. She brought back to Russia remarkable slides and shared her unbelievable impressions with us. Her stories were always very detailed.

Figure 7.4 Nina in 1977 in Brookhaven, NY. On the right is Ronnie (Ronald) Peierls, Nina's nephew. Courtesy of Natalia Alexander.

I remember her telling us that in America one can find excellent collections of modern art, while classical art is better represented in Europe. She told us all she knew about her nephews and nieces, Genia's children.

Nina Nikolaevna died in 1982 while visiting Genia. She had liver cancer[13] at an advanced stage and went to England hoping for a treatment. But it was too late. At that time we were in Israel. I called and Genia answered: "She no longer leaves her room, but I will ask." When I heard the voice of Nina Nikolaevna I told her that I would come immediately. "No, Katya, don't do that, I can't, it is too hard..." She died two weeks later.

---

[13] See R. Peierls' recollections on page 161.

In 1979 in New York I met Genia for the last time. She tried to advise us on how to start life in a new country, knowing how hard it is from her own experience, to encourage us.

***

Leonid Kannegiser, a Cousin[14]

A Russian writer, an anti-Bolshevik activist and later an emigré, Roman Gul wrote in his diary:

> At around 10 am on August 30, 1918, a handsome young man, a Jew of bourgeois origin, left his apartment on Sapyorny Lane. This was a young poet Leonid Ioakimovich Kannegiser. He mounted his bike and rode to Winter Palace. He stopped near the Ministry of Foreign Affairs where Uritsky, the Head of the Petrograd Cheka, had an office.
>
> "Is Uritsky seeing visitors?", he asked a doorman, a remnant from the czar time.
>
> – He has not yet arrived.
>
> – I will wait.
>
> He sat on a window-sill and looked out onto the Palace Square with a few pedestrians here and there. Finally, after 20 minutes he saw that a car — another remnant from the czar time — carrying Uritsky, had arrived. Arriving from his private apartment on Vasilievsky Island, a freak on short, curved legs, ducking like a duck, Uritsky ran into the entrance of the Palace. They say Uritsky liked to brag about the number of death sentences he signed. How many did he have to sign today? The young man in the leather jacket stood up. And while the chief of Cheka was mincing on his short legs to the elevator, Leonid Kannegiser who was six steps away fired at Uritsky and killed him on the spot.
>
> Kannegiser rushed out to the streets. Besides the doorman there were no other witnesses. In a state of shock, he forgot his cap and did not even bother to hide the revolver. Instead of mingling with the crowd, Kannegiser

---

[14] Following [7].

jumped on a bicycle and quickly rode away. Since a lonely bicycle rider on a huge square could be easily seen, quite soon he was captured and arrested.

The writer Mark Aldanov, who knew Kannegiser quite well, characterized him[15] as an exceptionally talented poet, exalted, living in an imaginary world of poetry rather than in the real world. Until the early 1918 he was indifferent to politics. However by April or May of 1918 his mindset turned upside down — he became an ardent hater of Bolsheviks. "The death of a friend, — concludes M. Aldanov, — turned Kannegiser the poet into a terrorist."

Many sources mention that Leonid had a personal motive for Uritsky's assassination. For among those executed by Cheka on August 21, in the very beginning of the Red Terror, was his close friend Vladimir Pereltsveig.

Uritsky, in his official capacity in the Cheka, signed hundreds of execution orders of the so-called counter-revolutionaries and innocent hostages. Thus, September's newspapers announced mass executions, approximately 500 people per day, two days in a row.

Cheka officers established family relations of Leonid Kannegiser and his friends and acquaintance — from his telephone notebook. There were 467 names in the list. His relatives, friends and other people from this list were arrested. Ioakim (Akim) Kannegiser, his father, was arrested that same day, August 30. Isai Benediktovich Mandelshtam was arrested the next day. Leonid Kannegiser was executed by firing squad in October of 1918. The exact date is unknown. His parents, after a number of interrogations which lasted from August till December 1918 were released. Ioakim Kannegiser was arrested again in 1921. Around 1924 he, his wife and his daughter were allowed to leave the country. The family settled in Paris, where Ioakim Kannegiser and his wife Rosa published poetry and diaries of their executed son. Their daughter Elizabeth (1897–1942) — the only child who survived the Bolshevik revolution (Genia's cousin) — was arrested by the French police in Nice in 1942 and through the transfer camp in Drancy was deported to Auschwitz where she was killed. That was the end of this line of the Kannegiser family tree.

---

[15] Mark Aldanov, *Assassination of Uritsky*, in "Sovremennye Zapiski," Paris, July 1923, № 16, page 350; Recent Edition: M. Aldanov, *Collected Works in Six Volumes*, (Pravda, Moscow, 1991), Vol. 6, pages 486–516.

Leonid Kannegiser was a promising poet, a close friend of Sergei Esenin, an acclaimed Russian poet. He was a member of the People's Socialist Party which vehemently opposed Bolsheviks and pursued moderate socialist ideology. This party was banned in 1918.

***

Leonid Kannegiser was born in 1896 into the family of Ioakim (Akim) Kannegiser (1860–1930), an outstanding Russian engineer and industrialist. In 1881, Ioakim Kannegiser graduated from the mathematical department of St. Vladimir's Imperial University in Kiev, and in 1884, from the St. Petersburg Institute of Railway Engineers. He was the Director of the Board of the Society of Nikolayev[16] Plants in St. Petersburg since 1907, and in 1910 he headed the Society of Nikolayev factories and shipyards. At that time the family lived in St. Petersburg, on Sapyorny Lane, No. 10, apt. 4-5, which served as a meeting place for the technical and financial elite of St. Petersburg. Moreover, this apartment was a major artistic and literary center of the northern capital. Numerous writers of the Russian emigration were to remember it in their memoirs. Marina Tsvetaeva saw a great deal of the Kannegiser family, including Leonid. In her essay *An Otherworldly Evening* [8] she vaguely hints that Sergey Esenin and Leonid Kannegiser were lovers at the time of her visit.[17]

Leonid's elder brother, Sergei Kannegiser (1894–1917), committed suicide in spring 1917. Leonid's mother, Rosa Saker (1863–1946), was a medical doctor. Ioakim Samuilovich Kannegiser was the brother of Nikolai Samuilovich Kannegiser, the biological father of Nina and Genia. A more detailed sketch of the family tree is presented on page 139.

Even after relocation from Nikolaev to St. Petersburg, the whole family spent every summer at their *dacha* in Odessa. This was an old family tradition. Genia and Nina spent their summers there too. During the Provisional Government in 1917 Ioakim Kannegiser was a member of the Presidium of the Military-Industrial Committee.

---

[16] Nikolayev, currently Mykolaiv, is a city in southern Ukraine. Mykolaiv is arguably the main shipbuilding center of the Black Sea.

[17] The only other evidence of Leonid's homosexuality I could find in the published literature comes from an N. Blumenfeld's (a family acquaintance from Odessa) who noted in passing that Leonid Kannegiser loved to show off and did not hide his homosexuality. At the same time it is also documented that in 1915, Leonid had a romantic relationship with the poetess Pallada Bogdanova-Belsky and in 1915–1917, an affair with the actress Olga Hildebrandt (according to her recollections, they were close to engagement, but in March 1917 their relationship faded away).

*****

Fragments of Leonid Kannegiser's Poetry[18]

From a review of A. Akhmatova's "Rosary" poems

For you, for the last time, it may be,
I've found fresh motion in my pen –
I see no more to agitate me
in your sweet form, my Harlequin!

I gifted you with hours, gave years,
my powers at full flower, my peak,
but melancholy is my nature –
I only plagued you with ennui.

I prayed, and now my wits are settled,
because the Maker heard my plea:
breaking off passions, ending battle,
I'll live in wisdom, finally.

ON REVIEW

In sunlight, with bayonets gleaming –
foot-soldiers. Beyond, in the deep –
Don Cossacks. In front of the legions –
Kerensky upon a white steed.

His weary eyelids are lifted.
He's making a speech. No one stirs.
O voice! To remember for ages:
Russia. Liberty. War.

Then hearts become fire and iron,
the spirit – an oak green with life,
and the Marseillaise eagle comes flying,
ascending from silvery pipes.

To battle! – we'll beat back the devils,
and through the dark pall of the sky,
Archangels will gaze down, jealous
to see us rejoice as we die.

---

[18]Translated from Russian by James Manteith.

And if, staggering, aching,
I fall upon you, mother earth,
to lie in a field, forsaken,
with a bullet hole near my heart,

on the verge of the blessed gateway,
in my jubilant dying dream,
I'll recall it – Russia, Liberty,
Kerensky upon a white steed.

June 27, 1917, Pavlovsk

SNOWY CHURCH
The arctic time and architect find amity
so full, the snow and walls fit undistinguished
where, modestly investitured in blizzardry,
the Lord's church stands – a wife impoverished.
And she sleeps amid white plots of burial,
her glass gleaming with mica's humbleness,
and with even what gold she wears as natural
as beads strung on a youthful country wench.
The copper sang, and reticence and ravishment
abruptly shuddered off their pious rest,
and it seems the very voice of snowdrifts
has melted in the belfry's resonance.

Nizhniy Novgorod, March 1918

To Yesenin

"With a fair friend, with a dear brother,
Crossing Volga in a boat…"

***

References

[1] Nina Kannegiser, *On M.A. Kuzmin*, published by N.G. Knyazeva and G.A. Morev, Iskusstvo Leningrada, 1990, No. 9, pp. 65–67.

[2] John E. Malmstad and Nikolai Bogomolov, *Mikhail Kuzmin: A Life in Art*, (Harvard University press, 1999).

[3] A.A. Lyubishchev, *Memories of Alexander Gavrilovich Gurvich*, in *A.A. Lyubishchev & A.G. Gurvich: A Dialog on Biofield*, Eds. V.A. Gurkin, A.N. Marasov, and R.V. Naumov, (Ulyanov Pedagogical University Press, 1998), in Russian,
http://www.mat.univie.ac.at/ neretin/misc/biology/b1/
%5BLyubishev_A.A.,_Gurvich_A.G.%5D_Dialog_o_biopole(BookFi).pdf

[4] http://www.pavelzaltsman.org/
index.php?option=com_content&view=article&id=17&Itemid=46

[5] L. V. Belousov, A.A. Gurvich, S. Ya. Zalkind, and N.N. Kannegiser, *Alexander Gavrilovich Gurvich*, (Moscow, Nauka, 1970).

[6] Rudolf Peierls, *Bird of Passage: Recollections of a Physicist*, (Princeton University Press, 2014), pages 337–338.

[7] http://tverdyi-znak.livejournal.com/1338622.html (in Russian).

[8] Marina Tsvetayeva, *Nezdeshnii Vecher*, in *Selected Works in Two Volumes*, (Kristall, St. Petersburg, 1999), Vol. 2.

PART 2

WAR AND PEACE

Chapter 8

Glimpses

Figure 8.1 The Peierls family, 1961. Left to right: Rudolf Peierls, Kitty, Jo, Genia (half-hidden behind Jo), Nina Kannegiser, Gaby, and Charlie Gross, Gaby's future husband. Courtesy of Gaby Gross.

1940: Selected Correspondence[1]

Genia Peierls to Hans Bethe

38, Calthorpe Road,
Birmingham, 15
June 17, 1940

Dear Hans,

Things are moving so quickly now that I don't know where we shall he and what we shall be when you get this letter. At the moment our children are with their school in the country, and we are both rather in the front line of ARP[2] and would be in exposed positions during a raid. It is quite possible that we may both get killed and that the children would then be left in a rather awkward situation.

If things should get very bad here I suppose the United States will organize some large scale scheme to get children removed and if they are really left without us perhaps you could use your influence to get them taken along. If this country [the UK] wins the war they will, of course, be alright. We have good friend who will look after them whatever happens to us. But if we should lose the war their existence here would become impossible. In that case I imagine the Germans would shoot us even if nothing happens to us during the air raids. But perhaps the children will be spared.

They are very nice children, very jolly, easy-going and bright. They will got all the possible scholarships and so on and be a lot of fun to the people who take them. If they do manage to get taken into some kind of refugee camp in USA perhaps you could occasionally keep an eye on them and bring them up in the good traditions of theoretical physics.

You know, Rudi has a lot of relations in USA, some of them even are not so bad. But I would much rather that the children would go to people who take them because they want to and not because they are the children of their brother-in-law or so. On no account would I let them feel like "poor relations" and that always happens in these conditions.

---

[1] Bodleian Archive, Oxford University, and Sabine Lee, Volume 1.
[2] Air Raid Precautions (ARP) was an organization in the United Kingdom dedicated to the protection of civilians from the danger of air raids.

I am writing a similar letter to Fankuchen. We are very well, terribly busy and in good spirits. *Qui vivra, verra*, and if even *"ne vivra pas,"* we had lots of fun and on the whole a long and interesting life. Nothing can be worse than death because you can always have that choice.

I hope to see you once more somewhere and some time.

Love to you all,

Genia

Genia Peierls to the Fankuchens[3]

38, Calthorpe Road,
Birmingham, 15
June 23, 1940

Dear Fankuchens,

Life is racing on with such a speed that I have not the slightest idea where and what we shall be when you get this letter. It is partly a business letter, too.

So far we are very well. We are in Birmingham (we are now British subjects) Rudi is a part-time fireman, and I am a full-time nurse. I learned a tremendous lot and am a very good nurse indeed. It is very hard work, but I like it and can stand it alright. Rudi does his University work as usual, and with the most amazing hours of duty which we have, we sometimes meet in the bathroom and have a nice chat between 7.30 and 8 a.m, and that is all for a day or two!

The children are evacuated with their school. We were very lucky, because they stay in the most lovely country seat of a Lord, with a deer park, a river, a water fall, pheasants on the premises. The house is full of Reynolds, but they do not appreciate this part very much yet.

They have grown terribly. Gaby is quite a lady. She learns history and geography, and is extremely bright and very popular. She is together with 8 and even 9 year old children.

Ronnie is a man, wears real suits and ties, just learned to read, and writes very funny letters. He has had his tonsils cut and now is not so

---

[3] Bodleian, A120.

dignified (looks less like a bank director). He also is everybody's pet and people spoil him, but the roots of very strict education are strong and they are still very nice children. They never had any homesickness and are the easiest to manage in all the school.

Now don't wonder why all this self-advertisement, because I am coming to business. In case (and quite a probable case) that we get both killed in an air raid (we shall both have to work during a raid) or shot by the Germans if they manage to invade this country, I would like to ask you a favor.

If the war is won, and if it does not last too long, I would like the children to stay here [in England]. We have many friends here and they will look after the children. But in the care of an invasion or hunger if we are no longer alive I would like them to go to USA if possible. I think in both cases people would start a kind of "Save the children" scheme. In such a case could you see what you can do for them to get their names considered for such a scheme, and also have a look at them occasionally when they come over. Perhaps you will even find somebody who will take them.

They are very nice healthy, cheerful and extremely bright children. They will probably get all scholarships (at any rate Gaby will). I shall enclose some of their pictures.

You know Rudi has a lot of relatives over in America. But I don't want the children to go to somebody just because they are uncles, brothers-in-law or cousins of their father and feel obliged to take them. I much prefer them to go to people who take them because they want to. I am writing a similar letter to Bethe, we shall leave his address and yours with friends here and they will let you know if we get killed.

Very possibly your country's turn will be next, if we lose the war. Let us hope there will be a decent, place left in the world. I hope you won't mind this letter. Even more, I know you won't mind it. Thank you very much in advance. Would you like a bit of local gossip? With utmost determine everybody is saving children. Mrs. Dirac — a daughter, Bretschers — a son, Woosters — not yet born, etc. Cambridge is simply a maternity home. Such is life. I hope you are having a good time in science and in general.

Yours sincerely,

Genia

Genia Peierls to the Bethes[4]

38, Calthorpe Road,
Birmingham, 15
July 16, 1940

My dear Bethes,

I am not quite sure which you will get first: this letter or the one which Gaby will write to you from Toronto.

After about a fortnight of red tape and bother I am now writing to you in a whirlwind of socks, jerseys, suite and dresses, coats and shoes. Everything is very small, everything is marked, some bags are "wanted" others "not wanted," but I think that one really wants everything for a journey on the high seas starting with a cool English July and finishing in a Canadian August.

We are going round the children looking for some more suitable spots on which they can he labeled. They are wearing identity disks, university badges, school blazers, and I think I shall just stick some labels over them, and varnish them over to make them waterproof.

In the meantime we are stuffing Gaby with wisdom and knowledge. They just came home from school, the time is very short, and Gaby gets slightly confuted. For instance, after I warned her about the dangers of constipation and ordered her to go to one of the mothers traveling with the party and ask for a remedy, she was also warned about the dangers of falling into the sea and the necessity of listening to what the sailors told you. Next morning she told me that "If I or Ronnie haven't been to the lavatory I must go and tell a sailor". Such a lot of pitfalls occur in every problem.

Rudi is writing you all the business side, I will tell you about the children.

Gaby is very jolly, very quick and very unsentimental. I think she has what is called a masculine mind. She is absolutely logical, very clear and unemotional in all her thinking. She loves arithmetic and one can explain to her everything. She will understand a very abstract problem, and she thinks a lot in this direction. For instance, she told Rudi already some time ago that every book should have an even number of pages because every leaf has two sides. Nobody had asked her about it. She is already, and will be even more, a book-worm. She reads

---
[4]Bodleian, A120.

everything and is quite capable of understanding easy geography and history books. She is mentally much older than her age. She actually enjoys learning and thinking and solving problems. As other people play tennis she plays arithmetic — for the sake of mental gymnastics. But I do not think she will be able to "invent" anything. She has not much imagination and is terribly *terre à terre*. She is never afraid or abashed in the mental or sentimental sphere. She will go and talk to everybody and to any amount of people; she is very self-confident (even cheeky) and free. But she is a coward in a lot of physical things, she does not believe in herself in that respect at all. She cannot ride a bicycle or climb, but loves enter and probably will soon learn to swim. She is bad at sewing, knitting and all handwork, and her handwriting is positively inherited! She is very much my daughter in very many respects (bicycle) but much more a woman than I ever was. She loves frocks and ribbons, can spend hours in front of a mirror, and adores babies and small children. She is actually very good with them and very reliable.

She tremendously enjoys a life and is very easily contented. She is a chatterbox, but has a lot of tact and *savoir vivre*.

Now about Ronnie or, as he is called at school Podgkin (he is still very fat). He is very much Hans' spiritual son. First of all, food is very important to him, and he can always eat as long as there is something to eat. But when it is good he enjoys it tremendously. Ronnie is a real gourmand.

Then he has the same sort of reckless and senseless bravery as you have. He will climb the most impossible tree, he will try and get on the bicycle until he is dark green and ultra-violet. He screams in between, but wipes his tears and goes on. He was swimming in Yorkshire last year in a real surf so that he was turning somersaults in the waves for minutes on end, but he never minded it. But if he is afraid of something (dogs!) nothing can be done.

And then he needs a lot of petting, an adores it. He is very affectionate and every woman loves him. But he must be dealt with very firmly and he must be encouraged to consider himself a grown-up.

He was extremely good when his tonsils were taken out. He stayed for three days in the nursing home and never even cried or rang the bell. Just because he wanted to show off. He can very easily be sorry for himself. He is very jolly and has a big cense of humor. He has a lot of imagination, but is not as quick and as logical as Gaby.

If he will ever stay with you, your mother will adore him and spoil him. And he will love her and tell her that "she is the best woman in the world, that he loves her more than everybody else, and that he will marry her when he grows up." He says that to a lot of people. He can read and write a little, and can do a bit of adding, etc. but he does not like arithmetic. He is a little lazy and easygoing.

The two are very fond of each other. When I told Gaby that in Toronto they might be separated, she went pale (you need a lot to make her feel anything) and said "but he is still so young and I must look after him!" I hope they will put them together. They both are quite independent, can dress, wash, Gaby can make her bed and wash up a bit, or dust a room. They will eat everything, sleep everywhere, and play or work in the middle of the most terrible din (Gaby particularly); they learned that at the boarding school.

They had whooping-cough and were inoculated against diphtheria. Ronnie had his tonsils out. He must be stopped when he eats too much; I he does not stand too much of eggs and of fat things.

Now you know about them nearly as much as I do. Gaby is a very simple problem. She is clever enough to understand another point of view and if she will not be frightened by something or somebody, she will always be happy and adapt herself to every situation. But she must be kept busy and set hard problems to think about. She did very badly at school when she was put in too low a form and work was too easy for her.

Ronnie is more complicated. He must be loved and petted or he will be unhappy, but not too much, or he will never grow up. He must be treated and talked to as a "big boy" and he will learn to play up and live up to it.

Thank you very much once more,

Yours sincerely,

Genia

Genia Peierls to Mrs. Blend[5]

38, Calthorpe Road,
Birmingham, 15
July 31, 1940

Dear Mrs. Blend,

I hope you are already settled and rested and enjoy Toronto and privacy. We got your letter from the boat on Wednesday almost on my birthday, and I never enjoyed any present as much as that letter. It immediately settled my worst worries. I never hoped you would take your responsibility for the children on your passport so whole-heartedly, and I pictured Ronnie having three helpings of everything four times a day for ten days on end and shivered. (Gaby always looks after him well on Christmas parties etc., and allows him "one of everything and two pieces of bread and butter" (her own description) but I was afraid that she will be too distracted on the boat to keep on this marvelous system.

I am so glad that you like the children. I always tried to bring them up not for the nursery, not for the home or school but for the world, so that they would fit in with unfamiliar surroundings and people. I also tried to make a "treat" out of everything new, even if it was an operation, castor oil, or the dentist. It is very useful if they have learned this trick.

I think they enjoyed their journey tremendously. Thank you so much for that, and for the letter for us, and also the letter to Mr. Blend, which he let us see; we had also a telephone message from Mrs. Russell. So very few people would have found time for all this, and it was such a relief to me. I knew that on the boat the children had somebody who really did care.

Very many thanks again

Yours sincerely,

Genia

---

[5] Bodleian, A120.

Hans Bethe to Rudolph Peierls[6]

38, Calthorpe Road,
Birmingham, 15
August 15, 1940

My dear Peierls,

Our telegram to you was so much obvious thing to do [...]

We are practically living with you all this time. Whenever there are news in the radio we listen anxiously. There is seldom reason for jubilation. Still, we are happy enough to know that so far the German planes have been repulsed every time with the greatest losses and that the RAF obviously has learnt a great deal from the mistakes made in the Belgian and French war. We are happy to know that the British ships still are everywhere on the sea and that England's sea power has not been seriously challenged, that the French fleet did not fall in Hitler's hands and many other things. And we hope for the best.

Yes, this life was beautiful and full, and we are enjoying it at the moment in spite of everything. The last year alone was worth living for (this concerns my personal affairs only), and if there should be an end soon, we still have had our life. If England should be defeated, our turn will be next — everybody knows that here [in the US]. And I am very much surprised how indifferent I am to the possibility of death. I dislike the idea of dying because there would be one less person to fight Hitler. If somebody had told me that three years ago, I should have considered him crazy.

Good luck — and I hope that we can soon write you something more positive about the children.

Yours,
Hans

---

[6]Bodleian, A120.

UNIVERSITY OF TORONTO
WOMEN'S WAR SERVICE COMMITTEE
SUB-COMMITTEE FOR BRITISH OVERSEAS CHILDREN

98 St. George Sreet
Toronto, Canada
August 23, 1940

Professor R.E. Peierls,
38 Cawthra Road,
Birmingham 15,
Warwickshire, ENGLAND

Dear Professor Peierls:

This will notify you officially of something you may already know. Your son, Ronald, has been placed by our Committee in the home of Dr. C.M. Jephcott, 323 Rosemary Road, Toronto; your daughter, Gaby, has been placed in the home of Dr. Jephcott's sister, Mrs. A. C. Sanderson, Old Yonge Street, York Mills (a suburb of Toronto), Ontario, for the duration of the War, or until we receive other instructions from you.

We regret the considerable delay in making these placements, which was caused in the first place by confusion on our part as to whether the children should go to Professor Bethe of Cornell, or whether they should be placed in Toronto; then when we had definite information from Professor Bethe, the children, having come in contact with measles, had been placed in quarantine for two weeks. However, this quarantine ends this week, and as neither child has contracted measles they will go to their new homes on Saturday.

We shall be very glad to receive any communications regarding your children that you may care to send, and can assure you that we will be only too pleased to answer any questions or to fulfill any commissions that are within our power.

Yours sincerely,

Karl S. Bernhardt

Chairman,
Placement Committee

Los Alamos

After about six months in New York and a brief vacation at Cape Cod (near Boston) the Peierlses headed to New Mexico. It was decided that Rudolf was needed in Los Alamos. General Leslie Groves, the Head of the Manhattan Project, did not allow the key participants to travel by air because of the risk of accidents, hence they traveled by train. The train *California Limited* which served Santa Fe, the nearest town to Los Alamos, circulated between Chicago and Los Angeles. Even before the war it was infamous for its delays; after the outbreak of the war quite often trains were held up for troop trains to pass by, which made *California Limited* even slower.

When the Peierlses got to the station in Chicago they saw hundreds of passengers on the platforms, tired from missing connections or searching for lost luggage. This reminded Genia and Rudi of their travel in the Caucasus in 1930, the Russian railway stations with their omnipresent turmoil and confusion. Friends of theirs helped them escort the children through the crowd.

New Mexico is probably the most otherworldly destination in the the US. Descending from the north one sees the vast farmlands intertwined with rivers and creeks gradually giving way to arid half-deserts with sparse vegetation of sagebrush and piñon, juniper-speckled hills, and mountain villages with Indian adobe homes. The vistas of the Sangre de Cristo mountains whose snowcapped peaks glow dusky red at sunset are breathtaking. Canyons and mesas, red with shades of crimson, are spread beneath deep blue skies. Evergreen forests on hills alternate with jagged sandstone cliffs on which sunlight and shadow perform a never ending dance. The beauty of this land casts a powerful spell. It is no accident that this land was chosen by Groves and Oppenheimer for a "secret town on the Hill."

All the Peierlses knew about their final destination was contained in the "Arrival Procedure Memorandum" instructing them to get off the train in Lamy, a railway station 18 miles to the south from Santa Fe, find their way to Santa Fe, and report to the US Army Corps of Engineers office at 109 East Palace Avenue. Los Alamos was not mentioned, as usual. Nor was Rudolf authorized to mention his profession. Just a Mr. Peierls with his family.

The Peierlses arrived at the small adobe train station in Lamy. They were mesmerized by the deserted landscape dotted by tumbleweeds that blew back and forth in the high wind. A young woman in the WAC[7] uniform greeted them: "Welcome, Mr. Peierls! Good afternoon, Mrs. Peierls. I am your driver. I will take you to Santa Fe." They were important visitors, indeed.

Figure 8.2   On the way to Lamy, NM.

The office at 109 East Palace Avenue was a front for the classified Manhattan project site on the "Hill." After a brief stop for formalities and further instructions, the Peierlses shook hands with Dorothy McKibbin who ran the office: "35 miles to your new home, Mr. Peierls, and I wish you a smooth journey." It took them almost two hours of being driven by the usual army car on a narrow road full of dips and sharp turns. In every direction they looked, there were dirt roads leading off into nowhere. The final ascent to the site was steep and boulder strewn and the sheer drop right beside them was unnerving.

---

[7]Women's Army Corps.

Figure 8.3 Aerial view of Los Alamos road on mesa. Courtesy of Bradbury Science Museum (Alan Carr).

The last 10-mile segment of the road was unpaved and riddled with deep ruts left by dry creek beds called "arroyos." The creeks flooded each spring turning the road into an almost impassable bog.

The site for Los Alamos was perched on the cone of an extinct volcano, on a broad tabletop of the mesa, two miles long and 7,300 feet above sea level. Construction on the site started just before 1943, the first scientists arrived in March of 1943, 14 months before the Peierlses. In these 14 months a lot had been achieved but most of the streets in Los Alamos remained unpaved. Rudolf Peierls recollects [1]

> The view of distant mountains gave the town a beauty marred only somewhat by the army huts that made up most of the town. It had been a fashionable boys' school, and the log houses of the school were left, though they were dwarfed by the drab prefabs. The old log houses had the distinction of containing bathtubs, as opposed to mere showers, and the street on which they stood was

known as "Bathtub Row." We were told, apologetically, that no room was left for us in Bathtub Row or in any of the houses nearby in the more desirable part of town. There was only a flat on the "other" side of town. We said, "Never mind; when we live there it will become respectable." Sure enough, when the Fermis arrived a little after us, they moved into the flat above us.

On our first evening there, the Oppenheimers invited Genia and me for dinner. This started with dry martinis, and we each had two. We did not then realize that the altitude substantially reduces one's tolerance for alcohol. After dinner we found we had great trouble getting up from the table and walking home. This was only slightly aggravated by the fierceness of Robert's martinis.

Figure 8.4 Oppenheimer's parties. Courtesy of Bradbury Science Museum (Alan Carr).

Then Peierls continues,

> It was a strange sensation to meet so many old friends from various phases of our lives in such an outlandish place as Los Alamos. Hans Bethe was the oldest friend there. [...] Other old friends included Enrico and Laura Fermi and Emilio Segrè whom we knew from Rome.

For security reasons they all were given "new" names: Niels Bohr was known as Nicholas Baker. All physicists on the Hill called him Uncle Nick. Emilio Segrè was Eugene Samson, and Enrico Fermi was Henry Farmer.

Fermi found his all-American code name quite funny, especially since as soon as he pronounced it in his heavily accented English, it immediately aroused the suspicions of the Los Alamos security guards. He was always stopped and questioned, and once the guards demanded he show letters addressed to "Mr. Farmer." A funny (and probably mythical) story about Niels Bohr vs. Nicholas Baker is narrated in many memoirs. Els Placzek before her divorce in 1943 was known as Els von Halban (see e.g. [2]). One day she met Bohr in a hotel lift. She greeted him, "Good morning, Professor Bohr. I do not know whether you remember me?" Bohr replied, "I am not Bohr, I am Nicholas Baker. But I do remember you; you are Mrs. von Halban." — "No, I am Mrs. Placzek."

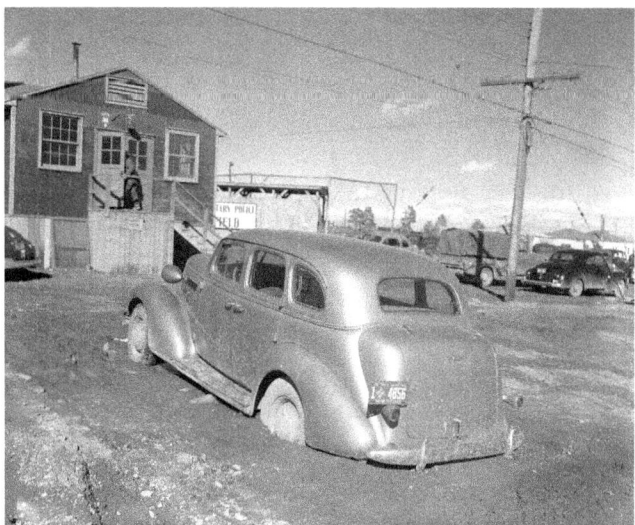

Figure 8.5 On one of the streets of Los Alamos. Courtesy of Alan Carr.

The Peierlses in Los Alamos[8]

A number of British Mission wives — Hanni Bretscher, Peggy Titterton, Winifred Moon, Els Placzek, Genia Peierls, and Elsie Tuck — accompanied their husbands to Los Alamos. Most arrived with their husbands, but James Tuck had to agitate the authorities for some time before his wife, Elsie, was allowed to join him. "I shall blow up this place proper, you know, if my little wife doesn't get oveh heah," he once said. "She's a tiny little one, no bigger than a minute, won't take up much war space." Eventually, a frightened Elsie Tuck boarded a ship bound for New York City, where she was met by a complete stranger. The man placed her on a train to Chicago (she did not know her destination) and three days later she was finally reunited with her husband.

Laboratory officials tried to utilize the wives' skills whenever possible. Winifred Moon worked as a secretary, and Peggy Titterton, trained as a laboratory technician, worked several months in the Tech area. Other women, such as Jane Wilson, Alice Smith and Ruth Marshak, taught in the Los Alamos school system. On the other hand, physicist Genia Peierls and mathematician Hanni Bretscher did not work in the schools or laboratory; instead, they devoted their hours to household duties. So, too, did Kathleen Mark. All had small children and household help was hard to come by. "The men didn't have time to do anything," one British Mission wife recalled, "anything at all."

The wives who did not work outside the home, however, entered wholeheartedly into what was termed "community affairs" or "Mesa Business." Mesa Business soon assumed serious proportions. Genia Peierls, for example, devoted much of her time and energy to this aspect of life. Boisterous and opinionated, Genia Peierls also manifested a genuine interest in other people and their predicaments. Recognized as the life of the party, she was soon viewed as a Los Alamos "character." When she first heard about the success of D-Day, for example, she danced on the tables. On another occasion, Genia Peierls decided that the level of science teaching at Los Alamos High School, where their daughter Gaby attended classes, was not acceptable. Consequently, she organized a program whereby several laboratory scientists visited the school to give lectures on their specialties. This proved eminently successful. A third Genia Peierls story involved the library. Annoyed

---

[8]From Ferenc Szasz, *British Scientists and the Manhattan Project*, [3], pages 37 and 38.

that so many books were overdrawn, she decided to remedy the situation personally. So, she borrowed Rose Bethe's baby-buggy, went from house to house (nobody locked their doors) and simply borrowed back all the library books.

Genia Peierls played a prominent role in Los Alamos social life. Her broad frame, resonant voice, and distinctive accent (she was fluent in English but usually skipped all articles, e.g. "We are Peierls") often made her the center of conversations. Rumors also flew behind her back that she had once been in the Russian Army, with a rank that varied from private to captain. Her flamboyance provided a marked contrast to her quiet, soft-spoken husband. Genia Peierls once boasted that all the worthwhile physicists in the British Isles had spent at least one evening in their home. She also claimed that all international visitors of note had stopped there, too. Nobody could ever forget Genia Peierls, recalled John Manley.

Figure 8.6 Somewhere in New Mexico.

*****

More from F. Szasz[9]

By consensus, the social event that best personified this Anglo-American spirit of cooperation at Los Alamos was held on Saturday, September 22, 1945. That night the British Mission members hosted a party to celebrate "the birth of the Atomic Era." With $500 supplied by both the British Mission on the Hill and the Embassy in Washington,[10] the Los Alamos group pooled their ration points to shop with élan in nearby Santa Fe. Klaus Fuchs purchased the alcohol, and people worried when he was late in arriving.

The celebration that evening reflected a distinct "British flavor." Invitations were engraved. Guests arrived in "formal" attire, many of the women in white gloves. A "footman" announced the arrival of each guest. Dinner began promptly at eight. British Mission members all had their own tables, and they invited their own guests. The Mission wives had worked for weeks at their "most secret" (British "top secret") preparation. Genia Peierls made a thick pea soup in pails, while Els Placzek[11] and Hanni Bretscher contributed turkey, boiled ham and English potato salad. Winifred Moon's dessert of trifle[12] became an object of considerable interest to the Americans, most of whom had never seen it before. Several hid theirs in the long table drawers to be discovered much later. Rudolf Peierls recalled carving roast beef for over 100 people.

The best port wine was then liberally dispensed for a round of ceremonial toasts. The group raised glasses to the King [George IV], the President [Truman], and, especially, to the health of the Grand Alliance. The hall was absolutely packed.

After the meal came entertainment. All members of the British Mission on the Hill (except the Chadwicks who did not attend) helped stage a "British style pantomime" where a narrator told a story that was acted out by silent performers. Titterton accompanied the action on the piano to the tune of "Atcheson, Topeka and Santa Fe." Entitled "Babes in the Woods," the play told the saga of the British Mission in Los Alamos and how "Good Uncle Franklin's forces" outwitted "Bad

---

[9]See *British Scientists and the Manhattan Project*, [3], page 43.
[10]Appriximately $7000 in USD-2018.
[11]Née Andriesse, George Placzek's wife, see [2].
[12]Trifle in English cuisine is a dessert made with fruit, a thin layer of sponge fingers soaked in sherry or another fortified wine, and custard.

Uncles Adolf and Benito." The dialogue contained a number of double entendres about the project that probably eluded several of the American wives.

Through several short sketches, the members of the Mission lampooned life in wartime Los Alamos. Censorship was parodied when an actor dropped a letter into a slot — only to have it fly out back at him. Otto Frisch, dressed as an Indian maid, solved the problem of house cleaning.

*****

Many interesting details about the everyday life of physicists and their families in Los Alamos were collected by Katrina Mason and published in her book [4]. It also contains reminiscences of the "children of Los Alamos." Below I assemble a few small fragments from this book.

Figure 8.7 The Peierlses stayed 18 months in a four-family apartment house, such as the above, which were prevalent in Los Alamos. The Fermi family occupied an apartment just above them. Gaby Peierls' room was exactly under that of Enrico Fermi's daughter Nella. The children discovered that they could talk to each other through the heating pipes and medicine cabinets. That was how the friendship between Gaby and Nella formed; it lasted for half a century. From [4].

Figure 8.8 In the Los Alamos school, physical education lessons were often conducted by the Women's Army Corps (WAC). These classes were looked upon as something to be endured. They also reminded students that Los Alamos was run by the army. Courtesy of Gaby Gross (née Gaby Peierls). From [4].

Gaby Peierls was two years younger than Nella [Fermi] but so advanced academically that she took classes with the high school group. Nella and Gaby found that they had a lot in common: each was the older child of a prominent European-born physicist, and both had gone through a number of geographic and cultural upheavals. Nella had moved from Italy to New York to New Jersey to Chicago and then to Los Alamos. Gaby had moved from England to Canada to New York to Los Alamos. But while Nella had remained with her family, Gaby had been thrust out on her own — she was seven when she and Ronnie were sent to Canada (in 1940) to escape England during the bombings. There they lived with different — but related — families. Gaby attended local schools and then a Canadian boarding school before joining her parents in New York City in December 1942. Arriving in Los Alamos in August 1944, Gaby, then 11, was two years ahead of her peers in school, still getting reacquainted with her parents and disoriented.

Although the Los Alamos school was public, it could not be part of the New Mexico school system because of the secrecy of the town.

Figure 8.9 When Gaby Peierls (right) arrived at Los Alamos she was 11. Courtesy of Gaby Gross (née Peierls). From [4].

So, the University of California agreed to act as paymaster, with the federal government ultimately footing the bills.

Hiring good teachers and setting up the academic curriculum at this school turned out to be a problem. The age distribution of pupils was unusual. With an average age at Los Alamos of 24 for adults, the children were young.[13] The largest group was infants and preschoolers. There were a hefty number of elementary school children, but very few teenagers. Of the 178 children enrolled in the schools in 1943–44, only about 40 were in the high school, eight of them the children of scientists. The senior class numbered two.

---

[13]In June 1944, medical corps officer Stafford Warren wrote a memorandum to General Groves recommending expansion of the hospital to include 16 cribs for pediatrics. He based the recommendation on the observation that "approximately one-fifth of the married women are now in some stage of pregnancy" and that "approximately one-sixth of the population are children, one-third of whom are under two years of age." A widely told story has Groves asking Laboratory director Robert Oppenheimer to do something to curtail the high birth rate. Oppenheimer's reply, if any, is not known.

The parents had strong views, they all wanted the school to have everything. They wanted tennis and horseback riding, and a lot of languages in the high school, even if there were only three students. The parents would say, "We want our kids to have German," and they got a German class.

The secrecy extended to names and occupations. For instance, Enrico Fermi was known as Mr. Farmer. Fearing that people might guess the nature of the work at Los Alamos if they knew the professions of its residents, the army discouraged words like physicist and chemist. The wives quickly formed their own code names for their husbands' occupations — physicists were "fizzlers" and chemists were "stinkers." The place where the fizzlers and stinkers worked — and in a very real sense the center of town — was the Laboratory.

In a conversation with Katrina Mason, Nella Fermi [Weiner] said:

> Gaby had the room right under mine. So we used to talk to each other late at night through the pipes that were supposed to bring up the heat. That was great fun. All the houses were alike. I remember one time when Gaby and I came home from school, we kind of wandered into somebody else's house. The streets had no names, and there were no street numbers. We walked in, and a strange lady came out. She was quite nice about it, and we apologized mightily. We assumed that no street names or numbers was done deliberately because of spies.
>
> We used to go hiking on Sundays, and in the winter we used to go skiing. Either Jane and I or Gaby and I would go on long walks. It was the kind of place that invited walking.

According to Jay Flanders,[14] her (Jay's) and Nella Fermi's IQ tests were "about the same, but Gaby was much smarter. Gaby's mother [Genia] was a terror. She used to tell Laura [Fermi] what to do. She was always giving Laura [Fermi] orders about how to handle me and Nella."

Gaby's mother directed considerable energies towards making sure that her daughter would get the education she would need to compete in English schools. What Gaby remembers most about the year and a

---

[14] Donald Flanders' daughter. Donald Flanders was an American mathematician who worked on the Manhattan Project at Los Alamos. Flanders was a group leader in the Theoretical Division headed by Hans Bethe.

half [in Los Alamos] is wanting just to be left alone. She also remembers feeling a strong need to keep separate her two worlds — the world of her parents, European-born scientists, and the new world of her American-born teenage classmates. Gaby recalled (see [4])

> It wasn't a bad time for me, but it was very mixed. I had had the feeling [that I was foreign and different] since I was seven. I was just very oppositional. I just wanted to stay home, be alone, and read books. In school, I felt there was a cultural norm, and I was not part of it. For me that norm was the fire chief's daughter. She was the center of a group and went to church on Sunday, which we did not do.
>
> I was two years younger [than anyone else in my class]. That was hard in itself. My mother was always pushing. She kept worrying what would happen when we go back to England. So I did eighth-grade arithmetic by myself, then took ninth-grade algebra when I was in the eighth grade. I did Shakespeare with the tenth, eleventh, and twelfth grades. I felt more comfortable with the tenth, eleventh, and twelfth grades because they were more secure.
>
> I remember coming back from ice skating with my parents and Hans Bethe in a car. Hans was driving, and we passed some boys in my class walking. He said, "Shall we pick them up?" and I said, "No." I said it so immediately and loudly that they all criticized me.
>
> One thing we used to do a lot of was babysit. I had never done that before. When the women had babies, they would stay in the hospital for about a week back then. So we would stay with their other children. I remember going to bed at night when my parents were having dinner parties, and I would fall asleep to the adults' conversation. Hans Bethe had a characteristic laugh. My mother had a very characteristic laugh. It was a wonderful time for my parents because there were so many people whom they had known in Europe. But they were also very serious about what they were doing.

Figure 8.10 Rudolf Peierls in 1946. From Sabine Lee, *Biogr. Mems. Fell. R. Soc.*, **53**, 265 (2007). Photo by Walter Stoneman, with kind permission from Godfrey Argent Studio.

Genia Peierls' Letter to the Magazine *New Scientist*, 1961

Birmingham,
July 16, 1961

Mr. Raison,
New Scientist

Dear Mr. Raison,

I have just read your profile of my husband. I know how difficult it is to write about a man after a talk of two hours, and how few people can protect themselves. I am sorry that most of my husband's lecturers and collaborators are away on holiday or at conferences, and that his former secretary has just left to get married!

Perhaps, if I try to write my "profile" of my husband it might be of some help to you to look at, although usually the collaborators and secretaries are more amusing and detached than a mere wife.

I met my husband at a large conference in Russia. And I think I fell in love with him partly because of his "easy-goingness." To compare to others he was a treat to look after. He ate everything, slept under any conditions. Insects did not sting him, germs did not attack him. He was game to try anything, to organize anything, take part in anything. It is true he was the youngest of the visitors, but even so the order of magnitude of his "un-fussiness" was quite different.

He is still very much the same, and does not mind discomfort, noise, children playing, phones ringing, bad beds, unusual hours, unusual meals, etc. Just as well as he goes around so much and usually in a great hurry.

He is a very typical untidy professor. His pockets bulge with correspondence, theater programs, and railway and airline time tables, for the last six months. He immediately covers with a thick layer of paper all available surfaces around him. His secretaries usually resign themselves to this state of affairs, and I take a chance and tidy his study when he is traveling. He brings back so many papers that the place immediately looks the same, and he is quite happy.

On the other hand he is a very untypical professor in the way in which he enjoys and pursues any line of activity or enquiry which comes his way. He as willingly will cook, camp, paint ceilings, write funny verse (on domestic topics or science or scientists) as work on one of his many problems. If you idly ask him an idle question, he is immediately fired with the urge to find the answer and will go endless trouble to find out what is the world's production of chewing gum or the price of a ticket to Timbuktu.

He is very much what I call a "tennis player." This means he enjoys and thrives on the questions, problems, challenges which come his way from the contacts with his pupils and collaborators, from seminars, conferences and visitors. (The alternative type of scientist being the "golfer," who works alone, and is only upset by external pressures and demands. This type often does more fundamental work, but does not have many pupils and does not communicate easily. The most typical golfer is Paul Dirac.)

He is extremely critical and loves to prick the balloons of beautiful but unsound theories (here the training by Pauli probably played some part). His colleagues say that in the departmental seminar he often talks considerably more than the speaker. They complain that he has very definite dislikes and if he is against an approach (too formal or too "messy") it takes a long time and arduous work to persuade him that they have a case. They say ruefully that in the process they are just as likely to change their approach as he his distrust.

He is called "Prof" by everybody in his department, and, I think, very much liked. I am not so sure about the university as a whole. He does not pay sufficient attention to people's pet aversions, hates fuss, Empire building; he does not understand intrigues and has no time for gossip, so probably misses many local complications. After working in his youth with such intellectual giants and such great personalities as Bohr and Rutherford he is impressed by very few people, and certainly no titles, rewards or positions. He gets more "kick" out of a conversation with a bright research student than from one with an eminent but somewhat stale man.

You mention the "international community" of physicists in the late twenties and early forties. It was indeed very international and quite extraordinarily closely linked personally. It was very much a family. Conferences were like family gatherings. They all were very young. Pauli, in his early thirties, seemed an oldish man, and Bohr at 40 a patriarch! In Copenhagen, there was a great tradition of putting on plays, operas and mock discussions. We were of this generation the first to be married (namely, the generation of Bethe, F. Bloch,

Weisskopf, Teller, Landau, Gamow, Rosenfeld, Casimir) and when our first daughter was born, the announcement was pinned on the main notice board in Copenhagen.

Now there are too many people and too many centers for this. The conferences are too large to be so personal, and one has no chance of personal contacts with half the people one wants to see. But in many places the departments of theoretical physics are still extremely friendly. I know the one in Birmingham best, and since we came to this country, there stayed in our house, for months or even years, Bethe, Frisch, Fröhlich, Fuchs, G.E. Brown, Dyson, Salpeter amongst others. At Christmas our living room is covered from ceiling to floor with Christmas cards and photographs from all over the world with news of new jobs, new children and new successes.

You write that my husband is very attached to Birmingham. He is very happy in the university, which he considers one of the best-organized seats of learning and is extremely fond of his department. So much so that he refused chairs in Cambridge, Oxford, London, Edinburgh, Manchester etc., to mention only British universities. So I am resigned to die in Brum!

For twelve years my husband existed in a medium-size ex-army hut and for one year a bright young American and a very enthusiastic Italian shared as office a trailer parked outside the hut, since inside there was no room for another desk or another piece of paper. The surprise of many eminent visitors to find the largest department of theoretical physics in England and one of the largest in Europe in these grim surroundings was great, but everybody was happy. When they finally got a new building many were even worried that the old spirit would die. But it survived alright, and the difficult, often nervous, and sometimes very "rough" boys from all over the world, from Brazil to Poland, from China to Sweden, settled down in an unbelievably friendly community. The new ones were looked after by the older hands. They are taken to the shops and restaurants, helped with digs, language etc. And in a year's time, they are quite ready to welcome a new flood. They come of all ages, of all backgrounds, of all stages of development. Some come for a year (or less), some for three. Some who come for a year stay for 10 and end as professors, like G.E. Brown. This year there will be 40 or more. In the middle of all this my husband is turning around and around like a top, dictating hundreds of letters, lecturing, helping with troubles in many problems, interviewing, running to and from meetings. This is an ideal arrangement for a tennis player and he thrives on it.

It also means he has little time, and many excuses for not starting outside projects such as large textbooks, or a book on education, which he could do very well. His two books were written under great pressure. One as a result of lectures at a summer school which had to be written up, the other, "The Laws of Nature,"[15] as a *tour de force*, to help pay the college bills of our children, whose courses at Oxford and Cambridge overlapped for one year and nearly broke the family finances. It proved to be a great pot-boiler and has been translated into I don't know how many languages (I believe nine, including Armenian) and still brings in some very welcome royalties.

I don't know how long he will be able to stay this present pace, but being very easy-going and not a perfectionist he manages so far.

I think you will now have a more lively picture than after one luncheon. What else? Yes, he loves puns (anathema to me!), good thrillers (the house is littered with hundreds of them), films, good theater (for which in Birmingham we are fortunately placed), and classical music. Goes to sleep every week night with the Times crossword puzzle, and on Sunday with the Observer's Everyman. He reads all the law reports in the Times, and collects time-tables of every description, which is very useful, as with so many guests passing through from all parts of the world, he can usually [solve] the most unusual traveling problems on Sunday night.

He does not like to play chess and cannot even play bridge. He dislikes all card games, monopoly, etc. He does not play ball games, but learned to water ski at 51. Likes skiing and sailing, but never gets time for either. Likes camping, motoring, color photography. Is very good with the children and very patient with the cat. Hates new clothes, mauve color, flowery hats and "learning-by-doing."

*****

---

[15] See [5] in the list of references.

From Rudolf Peierls' AIP Interview in 1969:[16]

[Gaby and Ronald] were born in 1933 and 1935. We are now talking about 1940, so they would be 7 and 5. They were not at home at that time. Their school was evacuated to the country. Then in the middle of 1940, the possibility came up of sending them to Canada, on the invitation of the University of Toronto who invited the staffs of several universities in England to send their families.[17] We jumped at that opportunity — not because we worried too much about bombing and things like that — but, of course, in the summer of 1940 the possibility of invasion looked very real; and with our both being Jewish, my wife of Russian background, and I German, we did not think very much of our chances of getting along and thought to have the children out of the way at that point was the right thing to do.[18]

I liked England very much. In fact, I developed a strong liking for it already on my first visit in 1928, although I did not really see much of the scientific community then, but of life in general. And so I do not think I ever had any doubt that if I could manage to hang on, I wanted to stay; but, of course, I didn't feel a very firm assurance that I would be able to. I was not convinced that it was going to be my permanent home. In fact, I still remember when our first daughter was born in 1933, in choosing a name for her, we were very conscious that it should be pronounceable in any language of any country where we might eventually live. The name chosen was Gaby [...].

[...] Later on when I had many invitations and offers of jobs in America, of course I considered it. At that later time I came to the view that while I liked visiting America, on the whole I liked the life in Europe and particularly in England better.

---

[16] Charles Weiner, Session I, II, and III, August 11, 12, and 13, 1969, https://www.aip.org/history-programs/niels-bohr-library/oral-histories/4816-1
[17] See Genia's letters to Bethe, pages 176 and 179; her letters on pages 177, 182; and the letter from Toronto on page 184.
[18] See also in Verblovsky's memoir on page 130.

In Part II of this interview Charles Weiner asked Rudolf Peierls about the spread of the communist ideas in England in the mid-1930s. In particular, publications and lectures of John Desmond Bernal and James Gerald Crowther were mentioned.[19] "There was a movement developing in England, the Science and Society movement, where a number of people in physics and biology and so forth had specific, outspoken views on the relationship of science to society and what the role of the government should be in science. Some of it was influenced by their view of what the Soviet planning for science was..."

In reply, Rudolf Peierls said:

> I was aware of the attitudes of people like Bernal and Crowther and did not have much sympathy for their views [...]. I may have heard Bernal talk about this. He was in Cambridge at the time. I don't think I talked much with him privately, but he may have given some general talks on this, and I saw some of his books and articles and so on. This to me seemed just fairly regular party-line stuff and not at all convincing.
>
> There were a number of people with rather Marxist views. And of course you were willing to listen and to see that occasionally they had interesting points. But in general you treated them like some strange religious sect whose views you were not particularly interested in.

*****

---

[19] John Desmond Bernal (1901–1971) was an outstanding biologist who pioneered the use of X-ray crystallography in molecular biology. In addition, Bernal was a political supporter of communism, the USSR, and Joseph Stalin. He maintained sympathy for all of the above from the early 1930s through 1950s and beyond. James Gerald Crowther was a prolific author in the area of sociology of science. Both are charcterized as "Marxist historians," see e.g. Morris Berman, *Journal of Social History*, Vol. 8, No. 2, (Winter, 1975), pages 30–50.

From a Personal Security Questionnaire Filled Out by Eugenia Peierls
in Los Alamos, New Mexico, on July 11, 1944:[20]

◇ Born on July 25, 1908, in Leningrad, USSR

◇ Student at Leningrad University from 1926 to 1929

◇ Employment: 1930–1931, Leningrad Geophysical Laboratory;
1939–1941, a nurse at a Birmingham hospital;
1941–1943, Planning Engineer with General Electric, Birmingham, UK

◇ British citizenship since March 1940

While the above part of the Peierls' dossier is certainly correct, the subsequent paragraph is half realistic and half ridiculous and laughable. I find it a great demonstration of the preposterous nature of gossip propagation. That's how legends are born.

> The FBI File File №̱ 100-344156 states that during investigation at Los Alamos, information was received that Mrs. Peierls was considered to be a "character" and that she was loud and outspoken. It was stated that there were many rumors that she was a Russian and formerly been everything from a private to a captain in the Russian Army.[21]

I will discuss later another laughable gossip in Peierls' dossier which, unfortunately, was taken seriously and led to dire consequences.

*****

---

[20] FBI File №̱ 100-344156.
[21] See also page 191.

Protest Action in Copenhagen, 1979[22]

The film "A Man Called Intrepid" directed by Peter Carter was released in 1979[23] and was full of caricature distortions. It was based on the "idiotic" (Peierls' characterization -MS) book *Intrepid* by William Stevenson. The film treated Niels Bohr even in a more fantastic way than the book. He was shown working on heavy water for the Germans in Norway, in a lab guarded by German security, and realized what he was doing only after receiving a letter from Einstein.

The screening of the film in Copenhagen outraged Genia and led to her standing outside the theater, with a poster in hand, protesting about the factual errors and distortions in presentation of Niels Bohr for three days. She would stop every passerby and explain: "I personally knew Niels Bohr for years. There was nothing like that!"

In the 1981 Annual Letter (see page 364), Rudolf Peierls wrote: "The winter of 1979–1980 was spent partly in Copenhagen. I was working there on the Nuclear Physics Volume of Niels Bohr's Collected Works. There appeared the film "A Man Called Intrepid" which is full of libelous insinuations about Niels Bohr, and, as nobody else was doing anything about it, Genia demonstrated for three days outside the cinema with an appropriate placard. This, of course, attracted a lot of publicity, which was the object.

---

[22] Based on the conversation with Joanna Hookway of April 22, 2017, Rudolf Peierls' letter to J. Rotblat of January 8, 1979 (Sabine Lee, Vol. 2, page 794), and the 1981 Annual Letter from Rudi and Genia Peierls to friends, page 363.

[23] It was also broadcast as a TV-miniseries (in three episodes) in the summer of 1979 in 20 countries under different titles.

Memories of Genia

From Maria Verblovskaya Memoir:[24]

As for Genia, since the beginning of the war, she had worked at a factory. To protect their children — the daughter and the son — from nightly Luftwaffe air raids they sent them first to the north of England, away from bombs, and then to Canada. Genia said that *she always carried poison with her, in case the Nazis landed in England.*

Nina visited England a few times, and Genia started coming to Leningrad too. The last time she came was around 1985, when my mother died. After Nina's death in 1982, she and my mother examined the Mandelshtams' archive and Genia wept very much, reading the letters from Maria Abramovna.

*****

From Peierls' Annual Letter dated November 1983

In October Genia went for a week to Moscow and Leningrad. She had intended to go earlier, but all tours for spring and summer were booked up, and a tour is the only practical way to get there. But now there was the ban on Soviet planes, and we did not know until the last minute whether her flight would go. It did but not without trouble: There was "industrial action" at Gatwick, so the plane was delayed for two hours, the passengers had to carry their luggage to the tarmac, and there was no food on board.

Genia's main object was to look through the papers left by her sister. This proved quite a shattering experience for her, since these papers included all the letters from her mother and stepfather written when they were banished, and the stepfather spent some time in prison. He was a wonderful man, and the letters from this period showed this particularly.

*****

---

[24] M. Verblovskaya, *Flow of Time*, 2014, unpublished.

THE GENYA[25]

It is obvious to all of us that the genia should be the international unit of something or other, like the joule and the newton. But of what?

We must hold an international conference to define the genia. It will be a rowdy and contentious affair. Scholars on all continents have had experiences measured in genias. Some will say that it is the unit of loudness, some that it is the unit of big-heartedness, some that it is the unit of self-confidence, some that it is the unit of loving concern, some that it is the unit of bossiness, some that it is the unit of generosity, some that it is the unit of stubbornness, some that it is the unit of unbreakable English, some that it is the unit of compassion, some that it is the unit of bad verse, some that it is the unit of fantasy, some that it is the unit of hypnotic gaze, some that it is the unit of irresistible kindness... The conference will surely conclude that the genia is a complex, multi-dimensional, unit whose aspect changes with the axis along which one approaches it, the unit in terms of which to measure all those qualities of humaneness and humanity that cannot be measured but that are so precious.

But whatever it is that it measures, the genia itself is clearly larger than life: the milli-genia should be perfectly adequate for normal purposes.

With love, Denys and Helen Wilkinson

*****

We all know that Genia is seldom at a loss for words and that her *mots justes* are usually easily audible. One typical example of many [is...] The occasion was a party in an Oxford college to view the newly painted portrait of the Warden. It was not an easy occasion; the Warden had not liked the artist who had been chosen and the portrait was not a success. As the Fellows arrived they mumbled a few platitudes and rapidly retired behind their sherry. But not Genia — her comment was immediate and startled the subdued company. "Ah, Warden, in twenty years you will look like that."[26]

*****

---

[25] G. Peierls, *Reminiscences Collected on the Occasion of Genia Peierls' 70th Birthday*, July 1978. Copy in Peierls Papers, Suppl A.119.

[26] G. Peierls, *Reminiscences Collected on the Occasion of Genia Peierls' 70th Birthday*, July 1978. Copy in Peierls Papers, Suppl A.119.

Notes of a Former Student

Freeman Dyson

Institute for Advanced Study, Princeton

These letters[27] are a record of the thoughts of a remarkable man and his remarkable friends and family, living through a historical period of exceptional violence and danger. They illuminate the larger scene of the Hitler years as well as the intimate scene of the Peierls family. They show us how quiet courage and sanity could bring a family safely through the storm and build a solid future upon the ruins of the past.

The two chief characters in the story are the German physicist Rudi Peierls and his Russian wife Genia. The two most dramatic incidents are the meeting of Rudi and Genia in Odessa in 1930 and the first calculation of the critical mass of a nuclear bomb by Rudi and Otto Frisch in 1940. Luckily for us, Rudi and Genia were separated for a year after their first meeting, so that we have an authentic record of their epistolary courtship in the magnificent series of letters exchanged between Leningrad and Zurich in 1930–1931. The Peierls-Frisch calculation of the critical mass was the decisive event that started large-scale projects to develop nuclear weapons both in Britain and America. The correspondence of 1940–1943 gives us an inside view of the fumbling efforts to get these projects organized in both countries, before they were finally and efficiently combined in 1943 at Los Alamos.

I had the good fortune to be a boarder in the Peierls home in Birmingham for the academic year 1949–1950 while I was working as a research student in Rudi's department. Thanks to Rudi's skills as a teacher and administrator, his department of theoretical physics was the best in Britain, having left Oxford and Cambridge far behind. Unmarried research students were welcome to stay in the Peierls home, where we were well fed and entertained by Genia. The home was a warm and wonderful chaos of teenagers and babies. The four Peierls children were Gaby aged 16, Ronnie, 14, Kitty, two, and Joanna, six months, no two of them alike, each having individual needs and schedules. Genia managed them all with her unique combination of loving heart and loud voice. Food was then still rationed in England, and gas-fires were regulated by shilling-in-the-slot meters which required an infinite supply of shillings. House-keeping for a big family was

---
[27]Letters collected and published by Sabine Lee in a monumental two-volume edition, S. Lee, *Sir Rudolf Peierls: Selected Private and Scientific Correspondence*, (World Scientific, Singapore, 2009), Vols. 1 and 2.

complicated. But Genia made light of all difficulties. She always had time left over from her family to take care of personal problems of mine and other student boarders. When one of us complained about anything, she would say, "What you have to complain, compared with Jew under Hitler or prisoner in Russia?" When I expressed surprise that she had another baby so soon after Kitty, she said, "There's never convenient time for having baby". When one of the students made an unsuitable marriage or bought an unsuitable motorcycle, she said, "Worst thing for health is to be an idiot." Since the definite article does not exist in Russian, she considered that it was also superfluous in English. Having myself been raised in a proper well-ordered British home without much noise or excitement, I fell in love with the uninhibited Russian-Jewish way of life that Genia brought to Birmingham.

From time to time, Rudi and Genia would throw big parties at the house for their numerous friends and students. Massive quantities of food and drink would appear and the boarders would be put to work peeling potatoes and apples. A frequent visitor on such occasions was Klaus Fuchs, who had been a close friend and collaborator of Rudi's in the bomb project. Klaus was fond of the Peierls children, and Genia claimed he had been their favorite baby-sitter when they were together at Los Alamos. He sat quietly at the parties and spoke only when spoken to. I talked with him mostly about Harwell, the British Atomic Energy Research Establishment, where he was head of the Theoretical Division. He was enthusiastic about Harwell and the prospects for peaceful development of nuclear energy. In February 1950 we were flabbergasted to learn that Klaus had been arrested and had confessed to being a Soviet spy. At first Genia did not believe it, but Rudi went to visit Klaus in jail and confirmed that it was true. For Genia, with her long experience of living in fear of the Soviet police, the key to survival was to have friends that one could trust, and the unforgivable sin was the betrayal of that trust. For once, she was speechless with anger. After Klaus was convicted, she too visited him in jail and gave him a piece of her mind.

While Genia was welding the families of students and staff into a tight community of friends, Rudi was building his Department of Theoretical Physics into a first-class international center of research. He attracted bright young people from all over the world. During my time in Birmingham I shared ideas with Gerald Brown from America, Frank Nabarro from South Africa, Jens Lindhard from Denmark,

Stuart Buter from Australia, Wladek Swiatecki from Poland and Tony Skyrme from England, besides many others whose names I have forgotten. The atmosphere of youthful exuberance was similar to the atmosphere at the Institute for Advanced Study at Princeton where I had spent the previous year. But in many ways Rudi was a better leader than Oppenheimer for a group of young people. Rudi was a platoon-leader while Oppenheimer was a general. Rudi was friendlier and more personally involved with the troops. He was actively engaged in research as well as teaching. During my two years in Rudi's group, the most important discovery was made by Rudi himself, who found a new and elegant method of deriving quantization rules for any classical field theory. Rudi's method gave for the first time a clear and intuitive meaning to the uncertainty principle in quantum field theory.

In summer 1951 I bade farewell to Rudi and Genia and went to start a new life at Cornell. Richard Feynman had moved from Cornell to Caltech and I was invited to take Feynman's place, an offer I could not refuse. But I left Birmingham with a heavy heart. Having once been a member of Rudi's team and Genia's incomparable household, I knew that I had reached childhood's end. For the rest of my life I would be making my own way in the world, without Rudi's wisdom and Genia's warmth to fall back on. Many years later, when my own children were growing up, we spent a summer in Seattle together with the Peierls family, and my children got to know Rudi and Genia. By that time, Rudi had become Sir Rudolf and Genia was Lady Peierls. But they had not changed. Genia still gave us good advice in her uniquely Russian version of English, and my children called her "The Loud Lady." In these letters I can still hear her voice.

About Genia Peierls[28]

Gerald E. Brown

Department of Physics and Astronomy, State University of New York, Stony Brook, New York, 11794

I was settled in the most stimulating theoretical group in the world. I didn't even have the burden of making decisions about personal problems. Genia Peierls did that for everyone in the group. I gratefully accepted her advice. I could do research the way I wanted to learn how to do it, and I had brilliant colleagues. [...]

Descriptions of the Peierls' Birmingham school of physics during the 1950s, written by those who participated in it, stress the stimulation in research and the family feelings created by the Peierls, especially by Genia Peierls, who was a dynamic phenomenon indescribable by mere words. After I had spent 1950 shivering in British "digs" with Jens Lindhard, the Peierls took me into their house as a boarder. The house was a gigantic 12-room building that the Peierls could only have bought because of the British leasehold, i.e. it reverted back to the trust that owned it in $\sim$40 years. My room was diagonally across from the dining room and upstairs. Genia with her loud voice would simply call me for breakfast across the entire house, causing the walls to rattle. Genia loved parties, keeping the entire group of theoretical physicists telling their stories or playing charades until 2:00 or 3:00 in the morning, long after buses stopped. As her team guessed each correct word, helped by an elaborate set of gestures from their member operating from the "enemy camp," Genia would erupt with deafening screams of delight.

When Rudi came home from the University at night she questioned him about every detail of every conversation he had had with anyone during the day. When such a conversation had concerned a personal problem, he would usually get back to the relevant person the next day and say, "I've thought more about the matter we discussed yesterday..." and then convey Mrs. Peierls' advice. On the weekend Genia would phone the wives of group members and advise them as to what they should do to cope with the many problems they encountered in Birmingham. We were all poor, and shivering in the inadequate heating, but we were happy and the group dynamics produced remarkable results.

When I went to Birmingham in 1950, Freeman Dyson was, as I said

---

[28] An excerpt from Annu. Rev. Nucl. Part. Sci., **51**, pages 1–22, (2001).

earlier, a postdoc there. Dick Dalitz, Wladek Swiatecki, and Geoff Ravenhall had just finished their PhDs. The entire group, including graduate students, was only a dozen people. Many of the Europeans went back and became professors in their home countries. Among the Birmingham PhDs prominent in America are Jim Langer, Elliott Lieb, and Stanley Mandelstam.

Figure 8.11   Rudolf Peierls, Gerald Brown, and Victor Weisskopf at Peierls' retirement celebration, Oxford, 1979. Source: see footnote 21 on page 212.

Those of us on the staff did undergraduate teaching, but the strength of the Birmingham teachings were the courses for graduate students, postdocs, and staff. Lectures would be given throughout the year, often by Peierls, but initially by Dyson when I arrived in Birmingham, covering different subfields of physics. They were designed so that in the three years a British student spent in graduate school he would be exposed to all the subfields of physics, except for general relativity, which was not popular in Birmingham. The result was that the "products" of the Peierls school were well versed in most of theoretical physics and were easily able to change subfields. Very few of them are now working in the field in which they were trained, but they are almost universally successful. Rudi Peierls (Fig. 8.11) was actually so informal and shy, letting Genia convey most of their feelings, that, although impressed by his intellect, I did not properly credit his contributions to physics. But I kept encountering them in my life following Birmingham.

In 1995 Ronnie Peierls phoned me the night after his father died. He asked me to phone Hans Bethe. Rose Bethe answered, and I told her, but I was too overcome to continue. A bit later Hans called me back and we talked about Rudi for some time. In closing, Hans said sadly, "A big part of my life is gone — and yours too." Rudi Peierls not only taught us how to do physics, but he and Genia taught us how to interact and deal with people creatively and positively. [...]

From "Hans Bethe and His Physics" by Gerald Edward Brown and Chang-Hwan Lee, Eds.[29]

Further information about Mrs. Peierls surfaces. She is Russian, "Eugenia" by first name — "Genia" with a soft g, to friends (that solves the mystery about the terrible accent that Gerry always uses when imitating Mrs. Peierls). Gerry[30] describes the extremely nosy and prying nature of Genia Peierls, how she was cornering everybody until she had extracted the most private and intimate secrets, then proceeding to offer her advice. Her advice, Gerry concedes, was almost always sound, and opened up avenues the advisees often hadn't even thought about. He offers the example of his son, who was "ordered" by Mrs. Peierls to go into the hotel business after she had analyzed his personality. He is now running several large hotels, one of them, I was told, in Santenay, France. Gerry asks Hans [Bethe] whether the inquisition staged by Mrs. Peierls had annoyed him.

"No," he answers, "I rather enjoyed that. It was her loud voice that sometimes annoyed me." Gerry asks about how Rose [Bethe] felt, and we learn that she couldn't stand her. After all Genia Peierls had the habit of telling the professors' wives that they should stay at home and care for their great husbands. (This was also observed by Judith Goodstein some days ago, who also was offered this unsolicited "advice.") Hans [Bethe] and Gerry [Brown] agree however that Genia Peierls was the most exceptional woman that they had ever met. Peierls had met her at a conference in Odessa, and married her very soon after. Gerry recounts how Genia decided to marry Rudolf.[31] The latter could sleep in any situation. Once, they traveled to Kiev by train, and Rudolf Peierls, being fairly short, crawls into one of the overhead luggage nets, and falls asleep. The hyperactive Genia sees this and decides on the spot: "This very stable man. This is man for me!" They didn't

---
[29] World Scientific, Singapore, 2006, pages 78–79.
[30] Gerry, was how friends addressed Gerald Brown.
[31] See page 6.

spend a lot of time in Russia. But apparently Peierls was very good at languages: the Peierls conversed in Russian with each other.

Sir Rudolf Peierls and Genia — Memories
Issachar Unna
Department of Physics, The Hebrew University of Jerusalem, Israel

I spent the academic year 1969/70 at the Department of Theoretical Physics, Oxford University. This was Sir Rudy's department.

When I arrived at the department (with my wife and two children, 8 and 3), direct from the railway station, I was immediately invited to his office and received with a warm welcome. Peierls was not a person to show his emotions, still I felt very welcome. One of his first questions was: "Do you have enough cash money for the first days? I can easily lend you any amount you need. I am fluent..."

After settling in the beautiful house they prepared for us and arranging school and kindergarten, I started to work on nuclear physics. Genia offered to show my late wife, Malka, around Oxford. Malka told her that our eight year old son, Ohad, played piano. After listening to his performance, Genia decided that with his talents he had to continue learning and practicing. She found for Ohad a wonderful teacher. The bond of the teacher with Ohad became a real love affair. At the end of the year, she gave Ohad, as a farewell present, large bound volumes with the whole set of the 32 Beethoven Sonatas. Genia also invited Ohad to a piano concert played by Daniel Barenboim.

Back to the physics, there was, of course, the weekly Colloquium. Everyone who prepared a talk knew that Peierls' criticism was dangerous. If you made a mistake, or if you were sloppy or unclear he would catch you.

Years later (in 1985), at a conference celebrating 50 years of the Einstein-Podolsky-Rosen paper in Joensu, Finland, I was late for one of the sessions. Approaching the lecture hall I met Peierls pacing back and forth outside the hall. I asked him why he did not attend the lecture. "I have my reasons," he said. When I entered I understood. It was a crazy talk with which it was impossible to agree.

Every Thursday we all had lunch together in the department's seminar room. A cold meal, ordered from a delicatessen shop, and wine were served on the dishes and cutlery that belonged to the department. Usually, one of us gave a short informal talk on his or her work or something interesting found in the literature. Peierls' critical remarks never failed to show up. Immediately, after we finished, Sir Rudi would put on an apron and start washing and rinsing the dishes. Certainly, not a tradition he learned from his German professors-teachers... We all felt, of course, obliged to help tidy up the room.

Once, I happened to ask him: "How can you read this newspaper which conveys such revolting ideas?"

"I never read newspapers to get ideas. I am completely able to form my own ideas!" was Rudy's answer.

From time to time Malka and I were invited to the Peierls' home for dinner. Usually, together with several couples, sometimes in honor of some important guest (e.g. Hans Bethe). Genia directed forcefully the sitting arrangements around the table. We were seated alternatively, men and women, and never husband and wife next to each other. She said this order prevented talking "shop" (i.e. physics) or stay isolated during the meal.

A good friend (probably, a former student) of Peierls was Professor Tony Lane. He was a true Englishman but his wife was originally from Israel. When they went to Israel together for the first time, Genia (and Rudi?) took them to the airport. When they were already through the departure gate Genia shouted in front of all the crowd:

"Take care not to get pregnant!"

My best PhD student was invited by Peierls to spend his postdoc at his department. After two successful years in Oxford he was looking for an academic position in Israel. At the Hebrew University we considered him as a serious candidate. There was one problem, he had a small speech defect. He was often stuttering. In his praising reference letter, Peierls also referred to this problem: "...I believe that his stutter is due to the fact that he is such a quick thinker that his spoken words don't catch up with his thoughts...".

Six years later (1976/7) we were going to spend again my Sabbatical leave in England. This time at Sussex University, Brighton, with three children (14, 10, 5). A few weeks after we arrived Malka was diagnosed with cancer. Her first telephone call was to tell Genia. We returned immediately to Israel. Since then on, Genia and Malka carried long conversations in letters and on phone which were very helpful to Malka, encouraging her in so many ways. These continued for the next two years until Malka's untimely passing.

Sir Rudolf Peierls: My Reminiscenses

João da Providência

Department of Physics, University of Coimbra,
3030-790 Coimbra, Portugal

I was most lucky to have been supervised by Sir Rudolf Peierls at the University of Birmingham Mathematical Physics Department. This was, for me, a great honor. My Ph.D. thesis was written under Peierls's guidance. Its subject was the ground state energy of Oxygen-16 computed up to second order perturbation theory. It was submitted in 1959, under the title "Perturbation Theory in Finite Nuclei." It is not easy to describe in words the insurmountable care Professor Peierls took in regards to the success of his students and the researchers who sought the inspiring atmosphere of his Department. He paid extremely generous attention, not only to scientific aspects, but also, with the help of Lady Peierls, to the many details of our personal lives.

Those who had the privilege of learning physics from Professor Peierls could count on his kind dedication, incomparable interest, and wise advice. He was kind enough to keep a fruitful relationship with the Department of Physics at the University of Coimbra, expressing a continuous, deep interest in its research activities and projects, and extended his generous hospitality to several Physics Department members. Peierls's inspiration, enlightening suggestions, and encouragements were crucial to the constitution and development of the new Many-Body Theory Group at Coimbra, where regular research activity was pursued on short range correlations in nuclei, rotational motion in nuclei and collective motion in fermionic and magnetic systems, projection methods, etc.

The University of Coimbra eagerly received Peierls's prized visits from the 1970s to the 1990s and later expressed its gratitude to Peierls by conferring an honorary doctorate upon him.[32]

---

[32] See pages 377, 378, 380, 387, 412, 418.

1989–1993: Selected Correspondence

Rudolf Peierls to L.I. Volodarskaya

Flat B, 2 Northmoor Road
OXFORD OX2 6UP
Oxford (0865) 56497

27 September 1989

Dear Ms. Volodarskaya,

I did know Isai Benediktovich very well, though I was not a direct relative. I was married to his stepdaughter, Evgenia Nikolaevna Kannegiser. Her father, a gynecologist and surgeon, died when she was less than two years old, and her sister, Nina, not yet born. Isai Benediktovich married the mother, Maria Abramovna, a wonderful person. He looked after the education of the two girls, who adored him, and Genia always said that she owed much of her understanding, much of her judgment and taste, to him.

I met Genia at a physics congress in Odessa in 1930, and during that same visit to the USSR also briefly visited Leningrad and met her parents. When I came back to Leningrad in the spring of 1931, we married, and stayed in the parents' flat on Mokhovaya until I had to leave, and again in the summer when finally Genia was able to come with me to Switzerland, where I was then working.

We visited a few more times in the early thirties, but in [19]35 or [19]36 Isai Benediktovich and Maria Abramovna were exiled from Leningrad, first to Ufa and later to Melekess (or perhaps in the opposite order). They spent the war years there, and later were moved to Alma Ata, where he spent some time in prison. Of course he had not committed any offense, but that was just how it was in the Stalin days.

I do not know much about his early years, except that he was an engineer by training. Part of his translating therefore was technical but much of if was literary, and as you say he was an extremely talented translator. He also wrote poetry, though I do not think I ever saw any of it. I speak Russian quite fluently, but I cannot really appreciate poetry in Russian. I do not have any of his writings, not even any letters. I do have many photographs, all amateur snapshots, and if you want I could search out a number and get them copied for you.

When he and Maria Abramovna died in Alma Ata in the early fifties, their daughter Nina was with them. She took charge of any papers and writings that were left. Nina died (while visiting us here) in 1982, and her papers went to a historian, whose name I regret I do not remember. When he died a few years later, his archives went to some library. Genia died in 1986 — she could have given you much more information. Now about the only relative left, a second cousin, is Masha Verblovskaya, Kamenoostrovskii 5, apt. 20, 19732 Leningrad P13; telephone 234-1481. Another relation is Natasha Belousov,[33] Sokolnicheskaya Sloboda 14-18, apt. 25, Moscow; telephone 264-0745.

Isai Benediktovich was a man of great charm, a subtle wit and absolute integrity. I would be very happy to help you more, but this is about all the information I have.

Yours sincerely,

Rudolf Peierls

---

[33] Natalia Alexandrovna Belousova-Gurvich (1905–2007) was the daughter of Alexander Gurvich, see page 125. She was an acclaimed art critic.

Rudolf Peierls to Tom Sharpe

>Flat B, 2 Northmoor Road
>OXFORD
>OX2 6UP
>Phone: Oxford (0865) 56497
>September 3, 1993

Mr. Tom Sharpe
Journal Publishing Company
328 Galisteo
Santa Fe, NM 87501
U.S.A.

Dear Mr. Sharpe,

I am afraid I have no idea where Fuchs met Harry Gold on any occasion. The occasion you mention would have been the day of the British party,[34] when members of the Brith group at Los Alamos were entertaining their American hosts. This evidently required a lot of drink, which was unobtainable in Los Alamos. As it was an army base, no drink was sold there, but there was no ban on bringing it in. Fuchs was the only member of the British group who had a car, and it was therefore natural that he should go on this errand. I do not believe there were any comments about this. In other words, he did not leave a party to go and buy liquor, as you say, but he went to Santa Fe to buy liquor for the party.

As regards his life style, he worked hard and long hours. In his spare time he often visited people's houses for more or less formal dinner parties. He was fond of dancing and was good at it. At parties he drank a lot and could hold his drink well. He was proud of his car and liked to go for drives. He lived in a house for single people in a room next to Feynman's, and they became quite friendly. Feynman's wife was then very ill with T.B. in, a hospital in Albuquerque. Feynman had no car, so Fuchs drove him to Albuquerque to visit his wife on some occasions. I do not know how many times this was, and I don't know whether he had other things to do in Albuquerque or was doing it only out of kindness. He liked climbing, and occasionally practiced on

---

[34]September 22, 1945; see page 192.

some small rocks, but as far as I know did not do any serious climbing in Los Alamos. He would occasionally baby-sit for his friends. He did not ride, or make music, though he enjoyed music. I don't remember whether he went skiing, or whether he went at the movies.

One episode might interest you: After the war ended, we had plenty of leave to our credit, and we went on a two-weeks' trip to Mexico City. This was Klaus Fuchs, Mici Teller (the wife of Edward Teller) my wife and I. We drove in Fuchs's car. (This is described in my book "Bird of Passage.")

As regards his work, he was a member of my group, which was concerned with the hydrodynamics of the implosion, and he solved many problems in that area, but I cannot remember what were the actual pieces of work he did.

The best book about Fuchs in general is the biography by Norman Moss, but it does not deal much with Los Alamos. Thats about all I can tell you.

Yours sincerely,

Rudolf Peierls

∞

Commentary

If the *rendezvous* between Klaus Fuchs and his courier Harry Gold took place in Santa Fe before the British Party, as suggested in the above letter by Rudolf Peierls, it could not have been the first one in New Mexico, which, as is now known, occurred on June 4, 1945. The date of the British Party can be found e.g. in the book *British Scientists and the Manhatten Project* by Ferenc Szasz,[35] from which I quote:

> [...] The social event that best personified this Anglo-American spirit of cooperation at Los Alamos came on Saturday, September 22, 1945. That night the British Mission members put on a party to celebrate "the birth of the Atomic Era." With $500 supplied by both the British Mission on the Hill and the Embassy in Washington, the Los Alamos group pooled their ration points to shop with

---
[35] St. Martin's Press, New York, 1992, page 43.

élan[36] in nearby Santa Fe. Klaus Fuchs purchased the alcohol, and people worried when he was late in arriving.

The celebration that evening reflected a distinct "British flavor." Invitations were engraved. Guests arrived in "formal" attire, many of the women in white gloves. A "footman" announced the arrival of each guest. Dinner began promptly at eight. British Mission members all had their own tables, and they invited their own guests. The Mission wives had worked for weeks at their "most secret" (British "top secret") preparation. Genia Peierls made a thick pea soup in pails, while Els Placzek and Hanni Bretscher contributed turkey, boiled ham and English potato salad. Winifred Moon's dessert of trifle became an object of considerable interest to the Americans, most of whom had never seen it before. Several hid theirs in the long table drawers to be discovered much later. Rudolf Peierls recalled carving roast beef for over 100 people.

The best port wine was then liberally dispensed for a round of ceremonial toasts. The group raised glasses to the King, the President, and, especially, to the health of the Grand Alliance. The hall was absolutely packed.

References

[1] Rudolf Peierls, *Bird of Passage*, (Princeton University Press, 1985).
[2] A. Gottvald and M. Shifman, *George Placzek: A Nuclear Physicist's Odyssey*, (World Scientific, 2018), page 335–341.
[3] Ferenc Szasz, *British Scientists and the Manhattan Project: The Los Alamos Years*, (St. Martin's Press, New York, 1992; Palgrave Macmillan UK, 1992).
[4] Katrina Mason, *Children of Los Alamos*, (Twayne Publishers, 1995).
[5] Rudolf Peierls, *The Laws Of Nature*, (Charles Scribner's Sons, New York, 1956), for a modern edition see e.g. (Andesite Press, 2017).

---

[36] Momentum, zest, in French.

Figure 8.12 Paul Dirac, Wolfgang Pauli, and Rudolf Peierls in front of the Science Museum in London, 1953. Courtesy of the Peierls family.

Figure 8.13 William Penney, Otto Frisch, Rudolf Peierls, and John Cockcroft, 1946. Source: LANL.

Figure 8.14 At the table, left to right: Bryce Seligman (DeWitt), Mrs. Blackett, Rudolf Peierls, and Bernard Peters, circa 1947. Courtesy of Chris DeWitt.

Figure 8.15 Genia Peierls, circa 1956. Photograph by Lotte Meintner-Graf. Courtesy of the Peierls family.

Figure 8.16   Rudolf Peierls and C.N. Yang, 1969. Courtesy of the Peierls family.

Chapter 9

Betrayal

One of the difficult pages in the life of the Peierls family was the betrayal by Klaus Fuchs, their friend and Rudolf's colleague, who turned out to be a Soviet spy and who had passed on to Moscow crucial information on the Anglo-American nuclear weapon problem. Fuchs' espionage activities were investigated by the British counterintelligence services in 1950. Since it was Rudolf Peierls who brought him in for nuclear research in the UK, and later into the Manhattan project, Peierls also fell under suspicion and eventually lost his clearance, although there was no proof of Peierls' involvement whatsoever. Repercussions lasted well into the 1970s.

This chapter is *not* a biography of Klaus Fuchs, nor is it a systematic description of his activities as a spy. Moreover, today we know that Fuchs was not the only westerner working for the NKVD,[1] which at that time was a criminal organization by any standards, responsible for the extermination of millions of fellow citizens. I want to focus here on moral "imperatives" which compelled Fuchs and his "comrades" to willfully and passionately collaborate with them.

There are a number of serious books devoted to Fuchs' life and deeds. Rudolf Peierls mentioned in his letter to Tom Sharpe of September 3, 1993, that his choice would be Norman Moss' volume [1]. Indeed, this book seems illuminating. A few more recent publications appeared in the 1990s, more or less following the lines of [1]. A significant change occurred after the year 2000 due to the fact that accounts and recollections of the Soviet agents involved in this large scale espionage operation were published. First and foremost let me mention the memoirs of Aleksander Feklisov, who was Fuchs' "handler" (the case officer)

---

[1] In fact, Fuchs was a GRU agent from 1941 to 1943 and worked for the NKGB (formerly, NKVD), from 1944 to 1950. *GRU* is the Russian abbreviation for the Red Army Intelligence Service, see also footnote on page 116.

in post-war England. Feklisov arrived in England in August 1947 as part of Soviet *Rezidentura* under the diplomatic cover of Second Secretary of the Soviet Embassy in London. The first edition of his book was entitled *Overseas and on the Isle*, 1988, and presented a rather short brochure for limited distribution full of Soviet ideological clichés. In the second Russian edition [2] of 1994, the factual side was significantly expanded. Finally, the third edition (in English) was released in 2001 [3]. Although the general canvas of the 1994 edition remained intact, the English version contains a rather complete view of the operation "Enorez" (that's how it was called in Moscow), with a number of details, such as dates, names, etc. Most of these details were certainly unknown to Rudolf Peierls.

Another Russian source is the book *Atom Intelligence and KB-11* [4], written by L. Kochankov who at that time was Head of the Security Service of the Russian Federal Nuclear Center (ВНИИЭФ, Sarov) and thus had access to all relevant archives. This book is well-documented. I should also mention the book by Vladinir Lota [5]. The most recent publication on Fuchs has been just released [6].

In addition, an hour long documentary film *The First Soviet Atom Bomb* was released by the Russian Federal Nuclear Center-VNIIEF in 2013. It can be found in the public domain. Another revealing TV documentary is *Fundamental Intelligence. Leonid Kvasnikov* which narrates the story of the *Enorez* operation beginning in 1940.

In the above-mentioned publications the reader will find a relatively full description of data on the nuclear bomb admittedly obtained by the Soviet Intelligence — both conceptual design and technology. The amount of the data and their depth is mind boggling.[2]

In addition to making the previously classified NKVD documents public, the Russian documentaries and books produced a significant psychological effect. Indeed, prior to *perestroika*, the Soviet Union never recognized that its first atomic bomb was an exact replica of the American plutonium bomb "Fat Man" dropped on Nagasaki. Fuchs' role was not only not admitted, but his very name was taboo in the media.[3]

---

[2] See Appendix on page 251.

[3] For the first time, the name of Klaus Fuchs was spelled out in Soviet media in July 1988, when the film "RISK-II" was shown on Central Television. The Soviet public was told that Fuchs voluntarily handed over to the USSR highly classified materials about the American A-bomb, with no further details.

I first heard about him in 1985, under the following circumstances. A friend of mine gave me a wonderful gift: a photocopy of Feynman's book *Surely You're Joking, Mr. Feynman* [7]. It was so fascinating that I could not keep it to myself. I badly wanted to share my fascination with others. Upon reflection, I decided that the only way for me to do so was to translate it into Russian and

Assuming that not every reader of this book knows the story of Fuchs' betrayal and its consequences (which will be important in what follows), I will briefly outline it below.

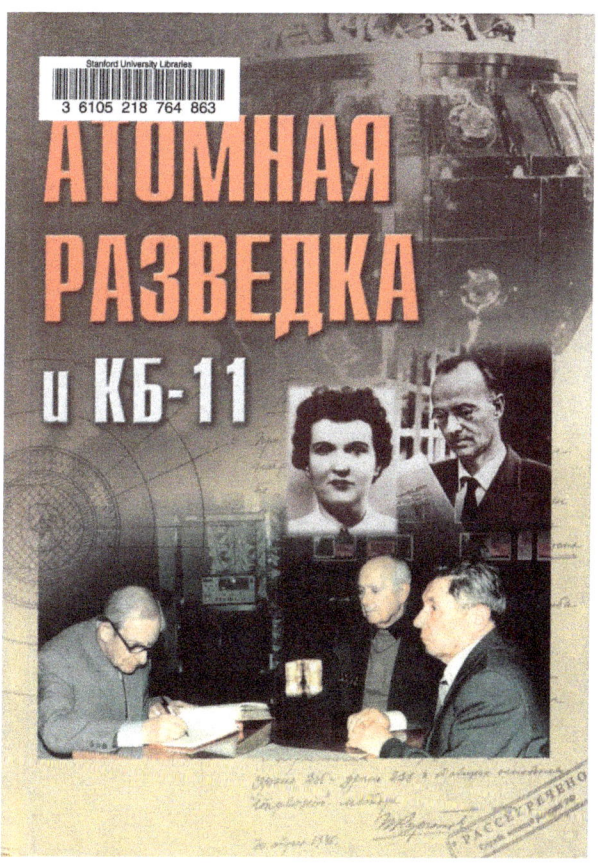

Figure 9.1 Leonid Kochankov, the title page of *Atom Intelligence and KB-11*.

---

try to publish the translation.

I called a someone who was in charge of one of the departments of the popular magazine *Nauka i Zhizn* (Science and Life). He met the idea with enthusiasm and was very supportive. He told me that I could go ahead and translate from a quarter to a third of Feynman's book, at my choice. He would push it through the board and take care of the copyright issues. "Just make sure you stay away from chapters with political connotations, and passages where he might mention our spy at Los Alamos, Klaus Fuchs," he added.

When I came to the Institute, I asked my former scientific supervisor B.L. Ioffe, who in his youth was involved with the Soviet nuclear project, whether he had ever heard about Fuchs.

"Of course," said Ioffe, "he gave our intelligence the data on 'Fat Man,' which were in turn forwarded to Kurchatov. We referred to such data as 'iks-perimental,' in contrast to our own experimental data, which appeared later." The word "experiment" sounds the same both in Russian and English. 'Iks-perimental' was supposed to show that the data were not honestly measured, but stolen.

*****

On December 29, 1940, Klaus Fuchs invited a few of his friends in London to a birthday party. That day he turned 29. Only four days previously he returned from an internment camp in Canada. He was penniless and without a job, making him feel miserable. One of those invited was Jürgen Kuczynski, formerly a professor of political economy at Berlin University and an active member of the German Communist Party. He was highly esteemed in the community of the German refugees in England. Very few knew that Jürgen Kuczynski was also a GRU agent (his codename was Karo).

Jürgen offered Klaus moral support and encouragement. After carefully phrased questions on Fuchs' political persuasions, and after learning that Fuchs was a professional theoretical physicist, Jürgen asked Klaus whether he might be interested in sharing his knowledge with the Soviets. Fuchs immediately answered in the affirmative. In a couple of days, Kuczynski arranged a *rendezvous* between Fuchs and a GRU officer, Simon Davydovich Kremer (codename Barch), who worked under cover as an assistant to the military attaché of the Soviet Embassy in London [5]. Kremer introduced himself as Johnson. Mr. "Johnson" asked Fuchs what he knew about possible uses of atomic energy. "Not much", answered Fuchs, "but I can prepare a brief review based on the literature."

I should emphasize that in January of 1941, the Soviet Union was a political and economic ally of Nazi Germany. The Molotov-Ribbentrop Pact which had established their cooperation was denounced only after the German invasion in the USSR on June 22, 1941.

We will never know whether or not this first *rendezvous* would have had a continuation, were it not for an unexpected turn of events.

Shorty after the party, Fuchs found a temporary job with Max Born at the University of Edinburg.

On May 10, 1941, Rudolf Peierls, who was in desperate need of an assistant for his MAUD-related work, wrote a letter to Fuchs offering him a position at Birmingham University with the salary £275 per year — much higher than Fuchs had ever earned before. By late summer Fuchs knew everything about the Frisch-Peierls Memorandum and all subsequent results of the MAUD group.

On August 8, 1941, Klaus Fuchs met Simon Kremer for the second time. "I have something very important to say, Mr. Johnson!" Fuchs had prepared a 6-page report on the nuclear bomb research in the UK and passed it on to Kremer. That same day a coded radiogram went from the Soviet Embassy in London to Moscow:

> To the Director.
> Urgent. Extremely important.
>
> Barch had a contact with the German physicist Otto [Fuchs' code name] who informed Barch that Otto found employment as a member of a group working on the Uranium bomb at Birmingham University. A group of other physicists work on the same problem at Oxford University. An industrial production is planned to be organized in Canada. Otto passed to Barch a review on basic principles of the Uranium chain reaction. If at least 1% of the energy is released in the explosion of the 10 kilogram Uranium bomb, the explosion would be equivalent to that 1,000 tons of TNT. The original document will follow.
>
> Barch

In early September, the "original document" was on the desk of General Alexei Panfilov at GRU headquarters.

The next time Fuchs and Kremer met was in March 1942. Kremer received from Fuchs 155 pages of documents related to Tube Alloys program. In his turn, Kremer passed on to Fuchs several specific questions which had been formulated in Moscow by Igor Kurchatov who soon after became the Head of the Soviet Nuclear Bomb program.

In the summer of 1942, Kremer was recalled to Moscow. Fuchs, who lost his only contact with the Soviet intelligence, was unaware of Kremer's departure and addressed Jürgen Kuczynski asking him to help resume communications with the GRU.

On one Sunday in October of 1942, Fuchs arrived in a small town near Birmingham to see Kremer and — surprise — it was a young elegant lady who approached him and, starting a conversation, spelled out the necessary code words. "Sonya," she introduced herself.

Sonya was Ursula Kuczynski, Jürgen's sister and a GRU officer. Also known as Ruth Werner, Ursula Beurton and Ursula Hamburger, she had worked for the Soviet intelligence since 1932, first in China, then in Poland and Switzerland. In 1941, she arrived in England.

The radiogram to Moscow sent by Sonya after her first *rendezvous* with Fuchs, contained information on the gas-diffusion method of separating uranium isotopes. Persistent contact between Klaus Fuchs and

Sonya continued till he left for the US in November of 1943. They would meet in small towns near Birmingham every three months or so. Fuchs handed to Sonya over 370 pages of top secret documents on Tube Alloys work. All were transmitted to Moscow.

Before Fuchs' departure Sonya provided him with detailed instructions as to how he could contact Soviet agents in America.

*****

Klaus Fuchs was born in 1911 into the family of a Lutheran pastor, a conscientious pacifist who preached love for one's neighbor and tolerance. Klaus chose another way, however, and became an ardent — I would even say, fanatic — communist. In 1929 he enrolled at the University of Leipzig, but the academic studies were abandoned once Klaus discovered the writings of Karl Marx and Vladimir Lenin and joined a leftist paramilitary organization, the *Reichsbanner*. In 1931, the Fuchs family moved to Kiel where Klaus Fuchs formally joined the KPD (the Communist Party of Germany). That's where he and his brothers were nicknamed "the red foxes" by their comrades.[4]

In 1933 he had to flee Germany, through Paris, to England. There he received higher education. Below I will quote a few paragraphs from Alexander Feklisov [3] which will help us to compose a psychological portrait of this person. On page 191 Feklisov writes:

> [Neville] Mott[5] made an important observation confirming the strong Communist ideals Klaus Fuchs believed in. The professor and his assistant would go to the local section of the Association for Cultural Relations with the USSR. [...] They would organize reenactments of the great purge trials taking place in Russia at the time. Fuchs always wanted to play the role of Andrei Vyshinsky,[6] the chief prosecutor, and as Mott remembered, "He played the role of the prosecutor accusing the suspects

---

[4] Fuchs means fox in German.

[5] Sir Nevill Francis Mott (1905–1996) was an English physicist who won the 1977 Nobel Prize for Physics for his work on the electronic structure of magnetic and disordered systems. In the mid-1930s Nevill Mott held professorship at the University of Bristol and employed Klaus Fuchs as his assistant.

[6] Andrei Vyshinsky (1883–1954) became Procurator General of the USSR in 1935. He was the legal mastermind of the Great Purge, encouraging investigators to procure confessions from the accused by torture. In many cases, he prepared the indictments before the "investigations" were actually concluded. Vyshinsky became known for his directives that an actual committing of a crime was not required for conviction: people could have been convicted for being perceived as bourgeois ("class responsibility") or simply if that was considered to be beneficial for the Communist Party.

in such a cold poisonous way that I never would have suspected in such a quiet and calm young person."[7]

After the outbreak of the Second World War in Europe on September 1, 1939, Fuchs, as a German citizen, was interned first on the Isle of Man, and then sent to Quebec, Canada. In December of 1940 he was allowed to return to England. At that time England alone waged war with fascist Germany which enslaved almost all of Western Europe (with a few exceptions, such as Switzerland or Ireland). The situation was difficult: nightly air raids by the Luftwaffe, strict food rationing, gloomy mood and the expectation of the inevitable invasion of German troops. Naturally, the anti-German sentiments in the country were strong. Before his internment Fuchs was Max Born's assistant at the University of Edinburgh. He continued collaborating with him during the internment and then, after his return to the UK, from January till May 1941. Max Born provided Fuchs with a good recommendation. Apparently, this played a role in Rudolf Peierls' decision (Peierls needed an assistant) to offer this position to Klaus Fuchs (on May 10, 1941). Not only did he take Fuchs into his group, but Rudolf and his wife Genia (the same Genia Kannegiser, who was once a faithful friend of the "three musketeers," Landau, Gamow, and Ivanenko) also offered him a home in their house. Fuchs let Mrs. Peierls handle his ration coupons and buy clothes for him.

Klaus Fuchs arrived in the US on December 3, 1943, and already on February 4, 1944, the first meeting with Harry Gold (code name Raymond), Fuchs' new courier, took place in New York [3]:

> On Saturday, February 4, 1944, as planned, Klaus Fuchs was standing on the sidewalk in front of the Henry Street Settlement on the Lower East Side in Manhattan. He was holding a tennis ball in his hand, which seemed odd for such a cold winter day, but after all, this rather slight young man could also be attempting to exercise the muscles in his forearms. Soon a rather short, stocky little man with a sad-looking face and a slight double chin, walked up to Klaus from the crowded street corner. He was holding a pair of gloves in one hand and a green book in the other. After an exchange of conventional passwords, Fuchs was sure that this was indeed Raymond.

---

[7] Frankly, I cannot understand how the British Intelligence Service could allow Fuchs to pass the vetting process and join the Peierls group in 1941, given the above characterization by Mott.

After Fuchs moved to Los Alamos, the heart of the Manhattan project, contacts between Fuchs and Soviet agents and the transfer of top secret information (see Appendix on page 251) continued on a regular basis. I won't go into further details, referring the reader to [1; 3] and other books. I will only mention that the first meeting of Fuchs and Gold in New Mexico, some 30 miles from Los Alamos, occurred on June 4, 1945,[8] on a bridge near the imposing hotel La Fonda in downtown Santa Fe. This time Fuchs passed to his courier documents containing information about the implosion mechanism of the plutonium bomb. The regular meetings with secret "handouts" on the bridge considered as a "hot" spot in Santa Fe went unnoticed despite the fact that during the war, many of the "employees" of the hotel and its occupants worked for the Los Alamos Security Department. Their task was to ensure the secrecy of the work on the American nuclear bomb, which was in the making on the "Hill" (that was how physicists working there called Los Alamos). As we see, they turned out to be not up to the task.

La Fonda (in Spanish simply "hotel") was a favorite evening hangout of the physicists and their wives — those who ventured down from the Hill to taste the "fruits of civilization." Fearing that the wine could untie their tongues, security agents looked after scientists from Los Alamos when they "relaxed" at La Fonda. Moreover, Oppenheimer, who worried that the inhabitants of Santa Fe could eventually guess what they were doing on the Hill, once sent the leading physicists Bob Serber and John Manley to Santa Fe with their wives to intentionally spread the rumor that Los Alamos was designing electric rockets.

There is an ironic nuance which must be mentioned. In the letter dated July 14, 1944, James Chadwick writes to Peierls,[9] "Dear Peierls, I have now had talks with both Kearton and Fuchs about the future of the New York section and in particular about their own positions. [...] The position of Skyrme is quite clear. Bethe or Oppenheimer should write to Groves asking for his services in [Los Alamos]. Groves has provisionally agreed and there should be little delay over his transfer. Fuchs's future is not so clear. I gave you the gist of a cable from Akers in my letter of July 11. I have now had a talk with Fuchs himself. He feels that he has a special contribution to make in England, whereas in [Los Alamos] he would be one of a number and can make no really significant difference to the work..."

---

[8] The date is given according to Aleksander Feklisov. Some other sources give June 2, 1945.
[9] See Sabine Lee, Volume 1, page 819.

It was a year *before* the first meeting with Harry Gold near Los Alamos which happened in June of 1945. Could Fuchs believe that Los Alamos would be such an unimportant place that it would not be worthwhile going there to spy? To my mind, it is unlikely. It seems to me, that it was rather a diversion maneuver on his side. What would happen in an imaginary world in which Klaus Fuchs returned to England in June or July of 1944? At this stage Fuchs was unaware of the implosion design nor of the weapons potential of plutonium. If so, he would not have stolen as much knowledge as he actually did from Los Alamos. The damage done would have been less significant. Why Fuchs was sent to Los Alamos when the British Mission wanted him to return to England and so did Fuchs himself is unclear.

Klaus Fuchs returned to England in July of 1946. He was eager to pass to Moscow information on new developments he had gathered. Alexander Feklisov's narrative on how Fuchs managed to contact the Soviet intelligence in London becomes quite captivating at this point. One could make a good Hollywood movie based on it. On pages 208–212 of [3] we read:

> The first meeting between Fuchs and his new case officer — myself — was scheduled for Saturday, September 27, 1947, at 8 p.m., in a remote part of London. Visual contact would take place at the Nags Head, a pub near the Green Wood underground station. [...] My mission was to organize the handling of [Fuchs] and satisfy all his requests, but also be completely responsible for his security; I was to make contact with him in person only if I was absolutely certain that neither of us was being watched or followed. The West was now paranoid about spies everywhere and, beyond the fact that an arrest would be in and of itself a serious setback, any kind of problem could be potentially disastrous for Soviet scientific research. [...]
>
> It was a foggy evening when I emerged from the Green Wood underground station. The Nags Head was across the street and I opened a newspaper, pretending to be waiting for a bus. Everything appeared quiet and clear. At ten minutes to eight a man came around the corner and went into the pub. He was tall and thin and held his head high as he walked. He had a pleasant face, wore

glasses and looked like an intellectual. I knew it was Fuchs even before he entered the Nags Head. I waited a few minutes to make sure no one was following him and then walked up to the pub myself. A little bell chimed every time someone opened the door. I went in.

The man I had seen enter was sitting on a high stool at the counter nursing a beer. He was reading the *Tribune* as we had agreed, which was actually not such a good idea because it was too left-wing for scientists who had been given the highest security clearance. He had blond hair and a high, receding hairline. We exchanged a brief glance and he had certainly spotted me as well, since I was holding a red book as my sign of recognition. I sat a bit farther away at the bar and ordered a draft beer. [...]

Since I only knew Fuchs from a single photograph and he knew nothing about the way I looked, some passwords had to be exchanged besides the signs we used. [...]

Klaus had obviously noticed my book because he got off his stool and walked over to a corner where there were some photos of British boxers. I slowly wandered there myself. Without looking at me, Fuchs said the conventional phrase:

"I think the best British heavyweight of all time is Bruce Woodcock."[10]

I immediately answered:

"Oh no, Tommy Farr is certainly the best!"

Fuchs finished his beer, said goodbye to the bartender and left. I watched him from the window to check which way he was walking. No one was following him. For added security I waited one more minute before leaving the pub. Fuchs was walking slowly and I had no trouble catching up.

"Klaus? Hello! My name is Eugene, I'm happy to see you." "Hello, Eugene! I'm happy as well. I thought you had forgotten me."

---

[10] According to some KGB archival documents, they debated whether lager or stout was the best beer.

We shook hands as we kept on walking. Contact had been made. [...]

On his own initiative during our first meeting, Fuchs handed me some very important documents he had been unable to obtain in the United States on plutonium production techniques. I also had a list of technical questions I had memorized and that he answered straight away. I gave him some cigarette paper, which could easily be swallowed, with a list of information we were interested in for our next meeting. Klaus read it carefully and handed it back to me:

"No problem. You'll get all this next time."

I would have preferred to see him take some notes in code in his daily reminder, just to make sure he wouldn't forget anything, but if he trusted his memory that was all right too. His Lubyanka[11] file had stressed his phenomenal memory.

We made plans for our next *rendezvous* and a backup meeting should problems arise. It was a lot of detailed information and as a precaution I asked him to make some coded notes. Klaus smiled.

"Don't worry, I remember everything."

"But you don't know the layout of the location!"

"I'll go there right away to make sure."

I became insistent:

"It's not that I don't trust you, but it is quite important."

"All right," he smiled, and then proceeded to repeat, point by point, every detail of our coming meetings. [...]

Now it was time to split up because there were too many risks if we kept on chatting away. "I'm very happy to be with you again," said Klaus looking at me straight in the eyes. After a moment he added, "I hope your baby is born soon!"

I didn't understand.

"Which baby?"

---

[11]Lyubyanka is the location of the KGB-FSB Headquarters in Moscow.

"Your bomb. From the questions you're asking me I estimate it will take you one to two years.[12] The Americans and our own research scientists think it'll take you seven or eight years. They're very wrong and I'm delighted."

Before parting he gave me a notebook with about forty-odd pages covered in tiny but legible handwriting. Fuchs had wanted to give it to me at the beginning of our meeting but I took the documents at the very last moment. Should our meeting have been broken up by counterespionage, the source would have a better chance of explaining things than if the documents were in the pockets of a foreigner.

*****

A chain of FBI investigations which started in 1945 after the defection of Igor Guzenko, a cipher clerk at the Soviet Embassy in Ottawa, culminated in September 1949, when a formal dispatch was sent from the Soviet Section of FBI to MI5 Headquarters in London informing the British Secret Service that there was a leak of confidential information on the Manhattan project from the British delegation and that the most probable source of the leak was Klaus Fuchs. The British Prime Minister Clement Attlee ordered his immediate investigation.

By this time (i.e. from the end of 1947 to May 1949), Fuchs had handed over to Alexander Feklisov the basic scheme of the hydrogen bomb, the results of tests of uranium and plutonium bombs on the Enewetak Atoll,[13] and basic data on the separation of uranium-235.

The subsequent actions of the MI5 are characterized by Alexander Feklisov as gentlemanly. He notes that the events would have developed in a totally different way were Fuchs arrested in the US, or extradited to the US by the British Government.

The first informal meeting of Klaus Fuchs with the British MI5 intelligence officer William Skardon who was in charge of his investigation, took place in Fuchs' office at Harwell on December 21, 1949. Fuchs indignantly denied everything. After this conversation, he was sent home.

On January 23, 1950, in a conversation with the Harwell Head of

---

[12]The first Soviet plutonium bomb was tested on August 29, 1949.

[13]Enewetak Atoll (or Eniwetok Atoll, sometimes also spelled Eniewetok) is a large coral atoll of 40 islands in the Pacific Ocean. With its 850 inhabitants it forms a legislative district of the Marshall Islands.

Security (who was Fuchs' old friend) Fuchs admitted that since 1942 he had been working for the Soviet intelligence.

In a document dated January 31, William Skardon detailed the full story of the meeting he had with Fuchs on January 24, 1950, and the circumstances under which Fuchs confessed.

Skardon wrote:

> I suggested that he should unburden his mind and clear his conscience by telling me the full story. He [Fuchs] said: "I will never be persuaded by you to talk." At this stage we went to lunch. During the meal he seemed to be resolving the matter and to be considerably abstracted... he suggested that we should hurry back to his house. On arrival he said that he had decided it would be in his best interests to answer my questions... I then put certain questions to him and in reply he told me that he was engaged in espionage from mid-1942 until about a year ago. He said there was a continuous passing of information relating to atomic energy at irregular but frequent meetings...

At a meeting with the Head of the Atomic Energy Department, Skardon was asked whether Fuchs was likely to try to escape from the country. "Mr. Skardon stated that, in his opinion, Fuchs still had confidence in the likelihood of remaining at Harwell, or, as a possible second best, obtaining a senior university post," a minute of the meeting said. "He recalled that the interrogation had shown that Fuchs regarded himself as a 'linchpin' of the Harwell organization; in his present state of mind it is almost inconceivable to Fuchs that he might be removed." [8]

A few days before his arrest, on January 27, 1950, Klaus Fuchs signed his confession statement [9], from which I quote (pages 5 and 6):

> In the course of this work I began naturally to form bonds of personal friendship and I had, concerning them, my inner thoughts. I used my Marxist philosophy to establish in my mind two separate compartments. One compartment in which I allowed myself to make friendships, to have personal relations, to help people and to be in all personal ways the kind of man I wanted to be and the kind of man which, in personal ways, I had been before with my friends in or near the Communist Party. I could

be free and easy and happy with other people without fear of disclosing myself because I knew that the other compartment would step in if I approached the danger point. I could forget the other compartment and still rely on it. It appeared to me at the time that I had become a "free man" because I had succeeded in the other compartment to establish myself completely independent of the surrounding forces of society. Looking back now, the best way of expressing it seems to be to call it a "controlled schizophrenia."

In the post war period I began again to have my doubts about Russian policy. It is impossible to give definite incidents because now the control mechanism acted against me also in keeping away from me facts which I could not look in the face but they did penetrate and eventually I came to a point where I knew that I disapproved of many actions of the Russian Government and of the Communist Party but I still believed that they would build a new world and that one day I would take part in it and that one day I would also have to stand up and say to them that there are things which they are doing wrong.

Fuchs was arrested only on February 2, 1950. Six weeks elapsed since his first conversation with William Skardon, three weeks since having been charged with espionage. During this time Klaus Fuchs was free and could have made a run, but he had not. Was this an example of a visionary stance of the British counterintelligence?

*****

Communist ideology is as hateful as the German National Socialism. If the latter is based on the postulate of the superiority of the German nation and inferiority of all others, according to the communist doctrine the carriers of progress, freedom, social justice, and enlightenment are workers while all others are aliens (or unnecessary ballast at best). The German National Socialism led to extermination of millions of European Jews, of which Fuchs certainly knew. That the communist ideology was responsible for the extermination of millions of innocent people who happened to be born into a "wrong social class," all over the world — from the USSR to China, from Cuba to Cambodia —

could have been inferred. But the mind of Klaus Fuchs was twisted and distorted to the extent that he preferred to close his eyes to the obvious. I hasten to add, though, that he was not alone — this disease struck a significant part of the Western *intelligentsia*. A terrible choice between Hitler and Stalin — the most blood-thirsty dictators of the 20th century... this was the tragedy of that generation. I will return to it in Chapter 10.

From the documents and quotations above I might conclude that not only was Fuchs delusional in inventing a disease, "controlled schizophrenia," unknown to medical science, but he may also have been close to megalomania: he honestly believed that he was such a great scientist that his absence from Harwell would hinder or even stop their programs. Moreover, he was certain that "one day" he would stand up and say to [the Russian government] "and Communist party that there are things which they are doing wrong," and which have to be changed.

At this point I honestly did not not know what to do: to laugh or to cry... Upon reflection and after discussing with my friends I came to the conclusion that Klaus Fuchs probably needed psychiatric treatment.

*****

On February 3, 1950, on the next day after Fuchs' arrest, Rudolf Peierls visited him in detention. Probably, he desperately hoped to hear from Fuchs that the whole story was a misunderstanding.[14] Well, that did not happen. Fuchs confirmed — face to face — that he had been spying for the NKVD-NKGB for eight years. Upon returning home, Peierls shared his impressions and feelings with Genia.

Genia wrote a letter to Fuchs which I perceive as a powerful document even now, 67 years later. He responded two days later. I will quote below both letters before continuing my narrative. To my mind, they show the best and the worst one can find in the human nature.

---

[14]Katrina Mason testifies [10]: "For Ron Peierls, the day the story [of Fuchs arrest] appeared in the papers was etched in his memory. He and Gaby were walking down the street in Birmingham and saw the headlines. Not too long before, the Peierlses had been skiing in Switzerland, along with Fuchs. Ron had fallen behind, his parents had spoken sharply, urging him to hurry along. Fuchs had quietly stayed with him. Upon seeing the headlines, Gaby and Ron ran home to ask their parents if the accusations could possibly be true. 'No,' said their stunned mother, Genia, firmly. 'They could not be true'."

Genia Peierls to Klaus Fuchs (in prison)[15]

February 4, 1950[16]

Dear Klaus,

Rudi just came home from London and I am writing to you in front of our sitting room fire where we so often talked about so many things. This is a hard letter to write, perhaps even harder one to read, but you know me well enough not to expect me to mince my words.

I am taking it all much easier than everybody else, because my Russian childhood and youth thought me not to trust anybody, and to expect anyone and everyone to be a communist agent. Twenty years of freedom in England softened me somewhat and I learned to like and trust people, or at any rate some of them. But early attitudes are deeper, and after the first half hour I feel I can take it. I certainly did trust you. Even more, I considered you the most decent man I knew. I do that even now. This is the reason why I am writing to you.

I understand that you have now changed your views and want the best of our civilization to go on. The best is trust in human beings, friendship, this bit of freedom and fresh air which is still lingering here and there in the world and makes life worth living, and bringing up children a joy. Your actions have tremendously endangered just these things. And this in two ways: one is directly as was your intention, and one cannot do much about that now. But one can and one must do something about the other.

Do you realize what would be the effect of your trial on scientists here and in America? Especially in America where many of them are in difficulties already. Do you realize that they will be suspected not only by officials but by their own friends, because if you could, why not they.

For your cause you did not <u>have</u> to be on such warm personal relations with them, to play with their children, and dance and drink and talk.[17] You are such a quiet man that you could have kept yourself much more aloof. You're enjoying the best of the world you were trying to destroy. It is not honest.

---

[15] Sabine Lee, Volume 2, page 209.
[16] This letter is undated, but one can recover the date from the context.
[17] Gaby Peierls, Genia's and Rudi's daughter, remembers, "He would come to our house every Sunday and we would go hiking or do something. He was very nice to us, perhaps one of the few grown-ups who didn't talk down to us. He always gave very appropriate presents. Just before he was apprehended he had given us a Christmas present of records for dancing." (See [10]).

In a way I am glad that you failed in this, because these [...] taught you the value of humanity, of warmth, of freedom. What did you do to them, Klaus? Not only their value in decency and humanity is shaken, but for years to come they will be suspected of being involved in this with you. Perhaps you did not think about it at the time but you must think now.

[...] to say who were your real connections. It is awfully hard, perhaps the hardest thing of all to do. But you went all the way in one direction, don't stop half-way now. You are not soft, and not one for the easy way out. You are a mathematician. This problem has no rigorous solution. Try to find the best approximation.

Rudi told me that you don't want to give the impression that you want to ease your own position. Klaus, don't be a child! This [is] a schoolboy code of honor. Impressions don't matter, you personally do not matter. The issues are too important for that, and you know it, otherwise you would have taken the only easy way out for you personally — to take your life. Thank you for not doing that. You could not leave all this terrible mess for the other to sort out. This is your job, Klaus. And you never shirked.

Not even washing up!!

Oh, Klaus, my tears are washing up my ink. I was so very fond of you, and I so much wanted you to be happy, and now you never will be.

I still think that you are an honest man, it means that you do what you think right, whatever the cost. Do the right thing. Try to save as much as you can of this decent and warm and tolerant, this free community of international science, which gave you so much these last ten years.

This letter is just a sea of ink, I am asking Rudi to copy it. Would you like me to come to see you? You are now going through the hardest time a man can go through, you have burned your god.

God help you! *Genia*

Figure 9.2 Klaus Fuchs. From police file. Circa 1950.

Klaus Fuchs to Genia Peierls[18]

February 6, 1950

In replying to this letter, please, write on the envelope Number 994, name Fuchs, K, to Prison

Written on prison notepaper

Dear Genia,

It was wonderful of Rudi to visit me on Saturday [February 4], although I could not do anything to cheer him up. On the contrary, it is up to you.

---
[18] Bodleian, D53; Sabine Lee, Volume 2, page 212.

Do you mind if I talk of other things? Sometimes I shall try and describe to you what went on in my mind. But you will to be very patient. I have been sitting here for an hour trying to think what to write next, when your letter arrived. I have told myself almost every word you say but it is good that you should say it again. I know what I have done to them and that is why I am here. You ask "Perhaps you did not think about it at the time." Genia, I did'n, and that is the greatest horror I had to face when I looked at myself. You don't know what I had done to my own mind. I thought I knew what I was doing, and there was this simple thing, obvious to the simplest decent creature, and I didn't think of it.

I had used my god to make myself into this, and that was the point where it finally crashed down.

Controlled schizophrenia is the the nearest description I can give to it, but I couldn't control the control, it controlled me.

I know that it is my job to try and clean up the mess I have made. I am afraid I did shirk it at first, and that made the mess even worse. They give me a much easier way out: I could have left Harwell to go to a university free from everything, free from friends, with no faith left to start a new life. I could even have stayed at Harwell, if I admitted just one little thing and kept quiet about everything else. I bungled the "take your life" stage; yes, I went through that too, before the arrest. The elaborate precautions taken after my arrest, I am glad to say, were quite unnecessary, though a trifle inconvenient. I was only afraid they might discover the safety pin which held my pants together. In that case my appearance in court the following morning might have been somewhat undignified.

I suppose you would almost enjoy the kind of thing I am learning about here. All these people in their way are kind and decent. Even the chap who apparently made prison his home, with occasional excursions to pick up a few hundred pounds and have a few riotous weeks on them. He grew quite sympathetic when I admitted that I hadn't made any money out of it. Nothing could shake him from the belief that I had been double crossed.

Many many thanks for your letter. Funny that women see such things much clearer than men. And that they are much kinder by saying hard words straight out.

Klaus

Sorry, I haven't got anybody to type this for me. I hope you can read it. And don't worry, if you don't see the tears. I have learned to cry again. And to love again.

K.

☙

*****

I close my eyes and I see Genia with tears dripping down her face onto the letter. I admire her courage and wisdom and decency. I see the bewilderment and insincerity of Klaus Fuchs. He tries to fence off by little and irrelevant things — a pin in the pants, a thief who liked him. He invents a controlled schizophrenia as an excuse. He is not too scrupulous in referring to women. He tries to play a heroic note ("I could have stayed in Harwell if..."). And the culmination of it all — "what I had done to my own mind..." I see a self-indulgent person who could not care less about the Peierls, about his other friends, in dismay after he was caught by the hand. As we will see below, he did not name any of his NKVD-KGB helpers, although the evidence he provided indirectly incriminated Harry Gold — the same Harry Gold with whom he met in Santa Fe (see page 234).[19] Most importantly, Fuchs deliberately misled MI5 investigators with regards to Alexander Feklisov, his key link in the atomic espionage in post-war Britain. This follows from the declassified (British) National Archives file № KV-2/1263, as well as Feklisov's recollections [3], pages 228–264. A few months after Fuchs' trial, Feklisov returned from London to Moscow with no obstacles.

*****

On February 28, 1950, trial hearings began at the Central Criminal Court, the Old Bailey. The UK Attorney General, Sir Hartley Shawcross, prosecuted Fuchs, who was defended by Derek Curtis-Bennett and his legal team. Before the trial started Curtis-Bennett

---

[19] I believe that Fuchs gave Harry Gold up because Gold was the person who led FBI to the conclusion that the Los Alamos spy was Fuchs. A map of Santa Fe found in Gold's residence carried Fuchs' fingerprints on it and it was presented to Fuchs during an MI5 interrogation which led Klaus to believe that at that time Gold had already been arrested.

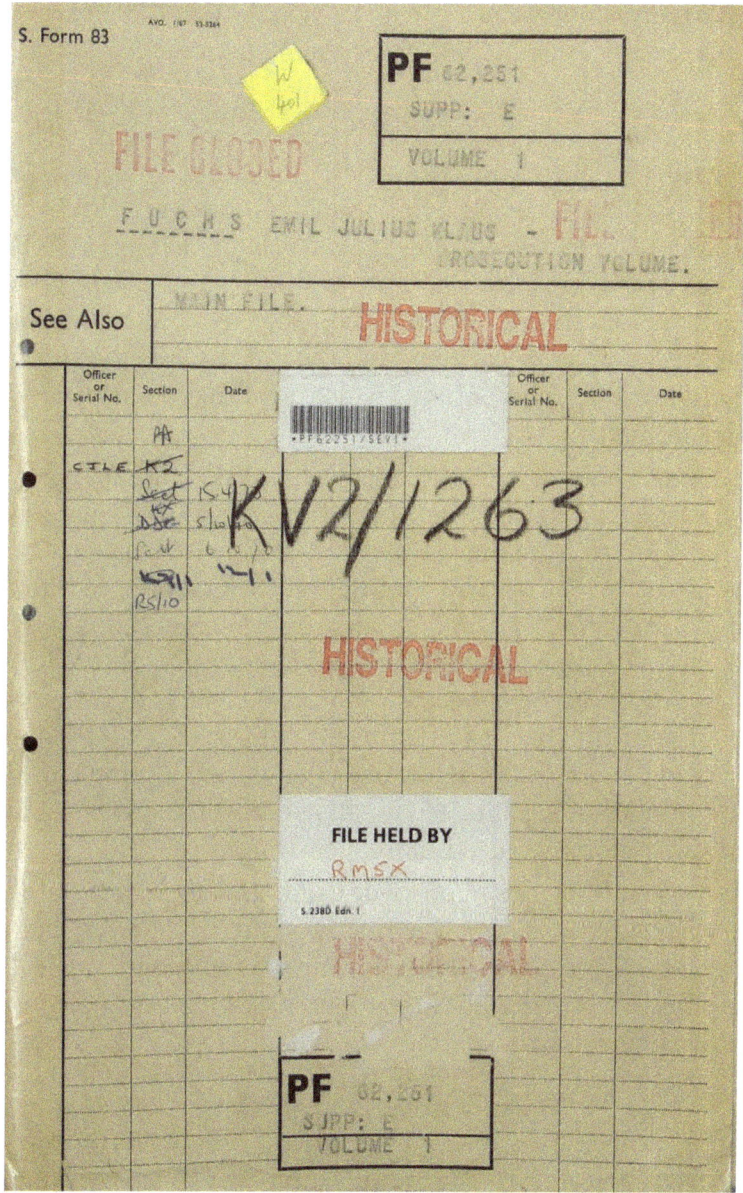

Figure 9.3 The cover page of the TNA Klaus Fuchs file KV-2/1263.

conferred with Fuchs. He told him he would do his best to minimize the sentence, but warned him that the maximum penalty was likely.

– Do you know what that is?
– Yes, I know, it's death.

– No, you bloody fool, it is 14 years! You did not give secrets to an enemy, you gave them to an ally.

Now I will quote a few paragraphs from [1] (pages 159, 184, 188–189).

> Sir Hartley then said that strictly speaking, there was no need for him to go into the prisoner's motives. However, he went on, in the statement by Fuchs which formed the basis of the prosecution, the questions of motive were so inextricably mixed with questions of fact that in fairness to him, and as a warning to others, it was right to say some word about motives, which would explain some of the facts.
>
> "The prisoner is a Communist," he said, "and that is at once the explanation and indeed the tragedy of this case. Quite apart from the great harm that the prisoner has done to the country he adopted and which adopted him, it is a tragedy that one of such high intellectual attainments as the prisoner possesses should have allowed his mental processes to become so warped by his devotion to Communism that, as he himself expresses it, he became a kind of controlled schizophrenic, the dominant half of his mind leading him to do things which the other part of his mind recognized quite clearly were wrong.
>
> In this country the number of Communists is fortunately very few, and it may be that a great number of those people who support the Communist movement believe, as the prisoner at one time apparently believed; misguidedly if sincerely, that that movement is seeking to build a new world. What they don't realize is that it is to be a world dominated by a single power and that the supporters of Communism, indoctrinated with the Communist belief, must become traitors to their own country in the interests — or what they are told is the interests — of the international Communist movement."
>
> Russia acquired the atomic bomb sooner than it would have otherwise. A Federal judge in New York City, Judge Irving Saypol, sentenced the Rosenbergs to death because, he said, North Korea would not have started the Korean War if Russia did not have the atomic bomb, and

so the Rosenbergs were partly responsible for the deaths of the Americans who were killed in Korea. [...]

Certainly if Russia's acquisition of the atomic bomb was linked to the outbreak of the Korean War, then Fuchs was the person in the West who was principally responsible. There was one more blow in store for Fuchs. In December 1950 [he was stripped of the] the British Citizenship [...] Sir Hartley Shawcross gave evidence before the committee, and said that the only consideration in the question was "whether it is in the public interest that Fuchs should continue as a British subject."

It may be important that he [Fuchs] never worked out his regret for what he had done and his distaste for the state of mind that had inspired him in political terms. The nearest he got was a rejection [for a short while] of all political ideology: "Blame me, and if you can't do that, blame Hitler and Karl Marx and Stalin and all their blasted company," he had written to [his friend] Arnold.[20]

Under British prison regulations, a prisoner who does not get into any trouble in prison or offend against the regulations is entitled to remission of one-third of his sentence for good conduct. Fuchs satisfied these conditions, and so he served only nine years and four months.

Shortly before his sentence came to an end, Peierls wrote to him offering to help him find a job in England, but he received no reply. Arnold visited him in Wakefield, at the request of MI5, and offered to help him sort out any financial affairs. Fuchs told him *how hurt he was at being deprived of his British citizenship.*

[Fuchs left Wakefield Prison on June 23, 1959. A police car drove Fuchs straight from prison to Heathrow airport, where he boarded a Polish airliner for East Berlin. In DDR he applied for membership in the Communist Party, and was accepted. He was elected to the Academy of Sciences and somewhat later appointed deputy director of the Institute for Nuclear Research in Rossendorf.]

---

[20] Henry Arnold, the head of security at Harwell.

The Director of this Institute was Heinz Barwich, a German physicist with pro-Soviet sympathies who had gone to the Soviet Union after the war and worked on the Soviet atomic bomb program, and earned a Stalin Prize. At the Institute, Barwich had his difficulties with the Communist Party. A party bureaucrat was installed at the Institute and interfered with the way it was run, until Barwich found the situation intolerable and demanded his removal; after a struggle, he got it. When he learned that Fuchs was to join the Institute as his deputy, he looked forward to having at his side a man who, as he saw it, had already shown great moral courage and independence. He assumed that Fuchs would be an ally in any future struggles with the bureaucracy.

But he was disappointed; he found that Fuchs followed the Party line on everything dogmatically, and never differed from its dictates or its officials.

I do not know whether Klaus Fuchs had indeed found the radiant future of all the mankind in the DDR or not. Did he ever regret the consequences of his previous life? What would he say now, when DDR is no more?

An excellent topic for a psychological drama.

Appendix (see page 228)

Excerpts from Leonid Kochankov, [4]:

Needless to say, the Soviet Intelligence provided our scientists not only with the drawings and descriptions of the American A bomb, tested in 1945, it also acquired a huge set of data on the development of the nuclear industry in the US. For instance, we managed to obtain data on special features of the construction of the gas-diffusion cascade which produced Uranium-235, including sectional design of the industrial facility and technology of membrane production, information on the 100,000 kWt nuclear reactor generating Plutonium in the amount 100 g/day, the top secret report on the secondary neutron spectra and multiplication coefficients in the bulk pure metallic Uranium, survival doses of radioactivity, behavior of subcritical masses, production technology for extracting Uranium from the ore, which was so unique that an emergency project was launched to build an appropriate plant within a year. In April 1945 we obtained detailed information on the Fermi design of the American experimental reactor, which became a prototype for our own reactor at the Laboratory №2 launched in December 1946 (page 6). Only in the *Rosatom* Archive one can find 13,500 pages of data obtained by our agents and 1200 drawings; and this is only in the declassified section of the Archive. How many documents still remain classified? (page 10). We do not know.

Excerpts from Feklisov, [3]:

The United States would later accuse the USSR of having stolen their atomic secrets. This is true. I was in a position to know that Soviet nuclear weapons were very closely based on American prototypes. In my country we have asserted that only the first bomb built by Kurchatov[21] was a replica of the one made by Oppenheimer. This is not altogether the case; the second and the third bombs were also replicas. The Soviet hydrogen bomb generally attributed to Andrei Sakharov was, just like Teller's H-bomb, as big as a hotel room. The fact is that, because of the information provided by Klaus Fuchs, we were able to reduce its size and easily deliver it to the target (page 201).

This confirms my earlier statement that Klaus had only admitted to the very minimum and was far from volunteering to confess everything to the British authorities (page 243).

Skardon had nevertheless tried hard to get him to speak of his liaison

---

[21] Igor Kurchatov (1903–1960) was the director of the Soviet atomic bomb project.

agent in London, but Fuchs only gave a very sketchy portrait.
"How old is he?" asked Skardon.
"He's neither young nor old. Let's say thirty, thirty-five."
"Is he tall?"
"Well, yes, rather tall."
"How well does he speak English?"
"Very well. He's very fluent and has an excellent vocabulary."
"But does he have an accent?"
"Oh yes! He does have an accent."
"A Russian accent? Or Slavic?"
"I wouldn't know. Possibly Slavic."

We had enough sources within British intelligence at the time to find out all these details later on. Since Fuchs had been given a local liaison agent (Harry Gold) in the United States, MI5 reached the conclusion that in London his case officer was also a covert agent — an "illegal" in Russian terminology — perhaps a Czech or a Pole who had been living in London for a long time and was probably a naturalized citizen.

Only amateurs would be satisfied with such evasive answers and I wouldn't dare insult British counterespionage that way at all. MI5 had the photos of all the employees of the Soviet offices in London. In the one I had furnished for my file I was wearing diplomatic-style clothes with a tie. Naturally I always wore business suits in town. Klaus and I had met six times, and it would have been impossible for him not to recognize me! Yet he said nothing (pages 248, 249).

One week after the trial TASS issued a statement which, I subsequently found out, had been written by Foreign Minister Vyshinsky himself:

> *Reuters* has published news of the trial of British atomic physicist Fuchs that took place in London this week. Fuchs was sentenced to 14 years in prison for having violated State Secrets. The public prosecutor of Great Britain, Shawcross, who prosecuted the case, stated that Fuchs had transmitted atomic secrets to "agents of the Soviet government." The TASS agency is authorized to say that this declaration is a gross fabrication since Fuchs is unknown to the Soviet government and no "agent" of the Soviet government has had any contact with him. (page 247).

References

[1] Norman Moss, *Klaus Fuchs: The Man Who Stole the Atom Bomb*, (St. Martin's Press, 1987).
[2] Alexander Feklisov, *Za Okeanom i na Ostrove*, (Moscow, 1994), in Russian.
[3] Alexander Feklisov, (with assistance of Sergei Kostin), *The Man Behind the Rosenbergs*, (Enigma Books, 2001).
[4] Leonid Kochankov, *Atomnaya Razvedka i KB-11*, (Sarov, RFYATs-VNIIEF, 2011), in Russian.
[5] Vladimir Lota, *GRU and the Atomic Bomb*, (OLMA Press, Moscow, 2002), in Russian.
[6] Frank Close, *Trinity*, (Penguin, 2019).
[7] R.P. Feynman, *Surely You're Joking, Mr. Feynman! Adventures of a Curious Character*, (Norton & Co Inc, Scranton, Pennsylvania, 1984).
[8] Jamie Wilson, *Guardian*, May, 22, 2003, https://www.theguardian.com/uk/2003/may/22/nuclear.russia
[9] *Statement of Emil Julius Klaus Fuchs* of January 27, 1950, available at The National Archives (Kew), KV 2/1263; reprinted in R.C. Williams, *Klaus Fuchs Atom Spy*, (Harvard University Press, 1987), page 140.
[10] Katrina Mason, *Children of Los Alamos*, (Twayne Publishers, 1995).

Chapter 10

Tragedy of That Generation

In this Chapter I address not only my readers but — in the first place — myself, in an attempt to understand the mindset of Klaus Fuchs and many other people who made up their minds (and continue to do so at present) to forgo moral principles for the sake of a certain Idea which, ironically, had been originally designed to make social life just and fair for everyone, being based on the most perfect moral principles.

Compassion and empathy are basic human emotions. The dream of universal justice is built into us simultaneously with the consciousness we acquire in early childhood. When we look around, we see an imperfect world in which there is grief, poverty, resentment, death of loved ones, betrayal, blood and sweat. Life is generally unfair, in principle. For example, someone may be born in Moscow, into a family of loving parents, but someone else in the middle of nowhere into a family of alcoholics. Someone may have had wonderful teachers who managed to instill in the person love for knowledge and skills of systematic work in school, but someone else had to settle for indifferent and ignorant teachers — there were no others around.

Thousands of accidents make our starting conditions unequal, and the maximum of what can and should be striven for is the equality of all before the law and the opportunity for everyone to choose his or her path in life from a wide range of different roads.

The mirage of the leftist *intelligentsia* is the equality of all people in the "mechanical" sense of the word: the same thoughts, the same interests, the same incomes, the same expenses, the same assessments of all events — and all this under the control of "society." This is my experience of 40 years in the USSR. The thought that this nebulous utopia is unrealizable and, moreover, terminal for humanity as a whole, does not cross their mind. They do not want to understand that the "control of society" almost immediately brings to power a "Big Brother," and a few of those close to the Big Brother will become more equal than others.

Human society is a very complicated system, and simple radical solutions to improve it suggested so far not only do not work — they lead to the misery of millions, and to social and human tragedies of unprecedented scale. To think that the communist ideology can cure social ills is the same as asking a butcher to perform surgery on a sick patient. I am a physicist and believe only in experimental data. There were many experiments with establishing communist or extreme leftist regimes, starting from the USSR to East Europe, China, Cambodia, North Korea and currently Venezuela. All these regimes gave rise to blood-thirsty dictators, who dismantled the elements (or traces) of democracy which had existed before them and suppressed social life and independent thought almost completely. They failed not only in not achieving social justice, but they also exterminated the notion of human dignity, making everyday life for ordinary people hellish. Thus, as a physicist I must conclude that the original Idea was grossly wrong and is not implementable.

Klaus Fuchs was a physicist too. This means that he should have thought logically rather than emotionally. He should have talked to eye witnesses. Even before the WWII he could have obtained reliable information on what was going on in the Soviet Union at that time. Already in the early 1938 Niels Bohr and his circle in Copenhagen — the Mecca of nuclear physics at that time — were well aware of mass arrests and tortures unleashed by Stalin [1]. Moreover, both Bohr and his team, and a number of physicists in Paris, were heavily involved in the campaign to save a number of outstanding mathematicians and physicists (including Landau) sent to Gulag. In England, Fuchs could have discussed the situation with David Shoenberg, professor at the Mond Laboratory at Cambridge, who spent a year in Moscow (from September 1937 to September 1938) and had witnessed the arrest of Landau and hundreds of other innocent scientists and the onset of the Great Terror. Also, he could have spoken with George Placzek, who returned from Kharkov in early 1937; before his departure for the US in 1938 he stayed for some time in Copenhagen, London, and Paris to explain the consequences of the communist ideology to the left-leaning colleagues he was in contact with [2].

After the end of the WWII, the Red Army invaded the East-European countries and installed puppet pro-Stalin governments there.

With the Red Army came the NKVD. Intolerance, political repressions, personality cults became palpable in Eastern Europe since 1946. In 1945 Raoul Wallenberg was abducted by a Red Army patrol in Budapest and vanished without a trace. Political assassinations became rather common since 1947. Suffice it to mention Jan Masaryk's assassination in Prague in 1948. (Jan Masaryk was the Foreign Minister of Czechoslovakia who was not ready to blindly follow instructions from Moscow.) A wave of political trials swept Eastern Europe in 1949. Charged and tried (Soviet style) were the veterans of the Spanish Civil War, communists who returned to East Germany from exile in the Western countries, East-Europeans who fought the Nazis on the Anglo-American side, and others (see e.g. the first footnote on page 281). In 1948 the so-called anti-cosmopolitan campaign was launched in the USSR. It was a vicious (albeit slightly camouflaged) anti-Semitic attack instigated by Stalin. Simultaneously, Soviet scientists in the areas of biology, cybernetics, history and linguistics were accused of bourgeois-idealistic methods of research and repressed. These events were widely covered in the West-European and American press.

Being a physicist, and a highly educated and cultured man, Fuchs could not have been unaware of all these horrors unless he deliberately made a choice not to notice.[1] His mind poisoned by ideology turned him into a delusional person. As we will see below, unfortunately, he was by far not the only one in this cohort. What upsets me even more is that such mindset is not the phenomenon of the past; some of our contemporaries are eager to follow this road even now, after all the lessons history taught us in the 20th century. A famous Soviet dissident and cultural theorist Grigory Pomerants once wrote:

> The devil begins with a foam on the lips of an angel,
> who fights for the holy righteous cause.
> Everything turns into dust — both people and systems.
> But the spirit of hatred in the struggle
> for a just cause is eternal.
> And because of this, evil has no end.

*****

Now I would like to pose and answer a question which is trivial to everyone who has ever lived or spent considerable time under the

---
[1] A direct proof that he was aware of the crimes committed by the Soviet regime, at least after the end of WWII, is discussed on page 266.

communist or radical leftist regimes. Most of the Western readers, especially from the younger generations, are ignorant in this respect, and tend to idealize Marxism, radical socialism and similar teachings. I want to briefly outline where this road ends in actuality. Although I will speak of the USSR, this particular example is general and can be easily extended to any other society or country in which a hateful ideology — the Idea to be implemented at any cost — prevailed (or prevails) over moral principles.

Whom Did Klaus Fuchs Work For?

Less than a year after the *Bolshevik coup d'état* in 1917, on September 5, 1918, the Executive Order "On the Red Terror" was enacted. This order created the basis for the repressive policy of the communist regime: the creation of concentration camps to isolate "class enemies" and physical elimination of opposition "involved in conspiracies and riots." The Cheka was empowered to take hostages, pass extrajudicial sentences and enforce them.

In that same month in the first wave of the Red Terror, over 500 people were executed only in the capital. Thousands of people were executed throughout Soviet Russia, some of whom, on charges of belonging to "counter-revolutionary" classes and social movements. Among other victims were landlords, clergy, Imperial Army officers, and members of non-socialist parties.

In the subsequent two to three years about two million people were killed in the Civil War, or executed by local *Bolshevik* committees, including the Russian Emperor and all his family: men, women, children... About two million of the White Army officers, dissenting *intelligentsia*, entrepreneurial classes, industrialists, clergy and members of all parties except the *Bolshevik Party* were pushed out of the country and found refuge mostly in Western Europe. By 1921 the country was exhausted to the extreme: economy in shambles, widespread food shortages and even famine in some areas, resulting in riots by peasants in rural districts. The *Bolshevik* Party was forced into a temporary tactical retreat from the communist doctrine. The so-called New Economic Policy (NEP) was introduced, allowing small-scale private businesses to resume operations, along with some minimal steps in social liberalization. Nevertheless arrests and persecution of "alien elements" continued unabated. The general population by and large was indifferent to these processes once they did not affect them personally and were implemented within the Marxist doctrine. Bourgeoisie (in the broad sense of the word) was considered by many a valid target for persecution.

I need to explain who were considered to be *bourgeois* by the late 1920s when the NEP was folded (in 1928). My maternal grandfather was a self-employed tailor. His only property was a Singer sewing machine which provided a means of living for the whole family: his wife and four children. This was enough to put him on the list. Twice he had to flee overnight to escape imminent next-morning arrest as a "bourgeois" element: the first time in the 1920s (from Belorussia to Siberia) and the second time in the 1930s (from Siberia to Moscow). In Moscow he did not repeat his mistake. He hid his Singer sewing machine and found employment at a factory. This saved the family.

In 1929–1930 the Communist Party inflicted a fatal blow to Russian peasantry — once and for all. It never recovered. The private farmers were accused of being the last carriers of the bourgeois (anti-Marxist) ideology. The so-called *collectivization* was implemented: complete confiscation of private land, livestock and agricultural implements, with the subsequent forced labor of former farmers in the so-called *kolkhozes*. Millions of the most hard-working and productive farmers were exiled to Siberia, were they were left to die *en mass* during the Siberian winter. This resulted in disaster — the massive famine of 1932–1933 which struck Ukraine and some other areas of the USSR. The affected areas were cordoned off by the Red Army patrols so as not to allow the starving people to reach big cities. Till present the exact number of victims of this man-made tragedy — *Holodomor* — is unknown; indirect data provide evidence of two to three million victims of starvation.

A phase transition in the communist slaughter machine occurred in 1937. Before 1937 the majority of people residing in big industrial cities were not personally involved in the repressions (and could even pretend that nothing out of the ordinary was going on), however, the situation dramatically changed in the summer of 1937. At that time, arrests, executions, and Gulag came to every family. No family was spared. On page 260 you will see a copy of Stalin's top-secret telegram dated July 3, 1937, sent to local party leaders ordering the immediate start of mass executions of "anti-Soviet elements." On July 5 the Political Bureau of the Central Committee of the Communist Party issued an order "On the members of the families of convicted traitors to the homeland," according to which "the wives of the unmasked traitors to the homeland

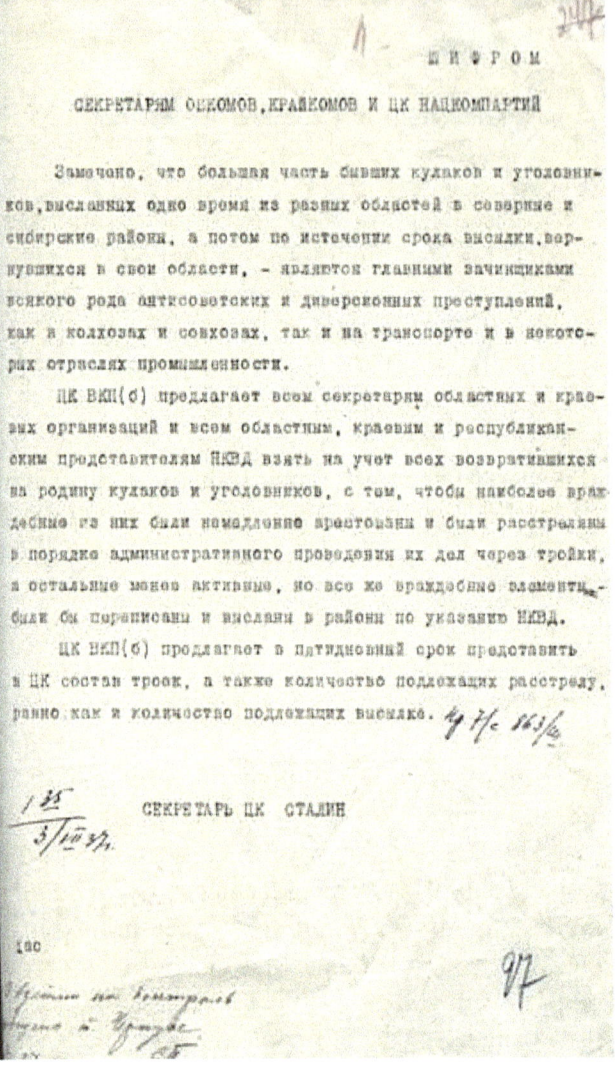

Figure 10.1 Copy of top-secret Stalin's telegram dated July 3, 1937, sent to local party leaders ordering immediate start of mass executions of "alien elements." Lists of such individuals were supposed to be delivered to the Central Committee of the Communist Party within five days (!). Within a year, 700,000 "accused" were shot and a large number sent to Gulag. To deal with the huge number of the "anti-Soviet elements" an extra-judicial mechanism — the infamous *troikas* — was established by Stalin. This unleashed the Great Terror.

and the Trotskyite spies were to be imprisoned in the Gulag camps for no less than 5–8 years," while their children were to be sent to special orphanages (with their names changed).

On July 16 Stalin held a meeting with the NKVD leadership to discuss details of the upcoming operation. According to the memoirs of the participants, the NKVD Commissar said: "If an extra thousand people are shot during this operation, there is no harm in that. Act boldly, arrest more — we will sort it out later."

The executive orders of July 20 and 24 authorized arrests of all Germans working in the defense industry or servicing railways. Finally, on July 31, in the aftermath of Stalin's telegram of July 3, the NKVD issued the Executive Order №00447 detailing groups of people subject to repressions: former wealthy farmers (the so-called *kulaks*) settled in the cities, former members of the socialist parties other than *Bolsheviks*, the priesthood, former "White" officers, and anti-Soviet elements in general. The "operation" was to be carried out also in prisons, Gulag camps, and labor settlements. The executive order established generous quotas for the "first category" (i.e. those subject to immediate execution) and the "second category" (imprisonment in the Gulag camps) for each region of the USSR. It also specified the composition of the *troikas* which were put in charge of processing the cases through a simplified procedure and handing over the verdicts. The sentences were taken in *absentia*, i.e. without summoning the accused, and without the participation of defense. The verdicts were not subject to appeal. The executions were to be carried out in secret locations with no delays.

This was the onset of the Great Terror.

In 1937–1938 the slaughter grew to genuinely "industrial" proportions. The lists of those to be arrested and their sentences were composed *a priori* and signed by authorities (in some instances, by Stalin), and then formally processed by *troikas*. From August 1937 to November 1938, in four months, the NKVD arrested 1,548,366 people, more than 680,000 of them were shot. In addition to 680,000 shot, about 115,000 people "died under investigation" — in other words, under torture.

In the Soviet archives 388 execution lists were found signed personally by Stalin and his top lieutenants. They were forwarded to the Military Collegium of the Supreme Court for rubber-stamping. The total number of innocent people executed by Stalin's direct orders is around 45,000, i.e. around 130 per list. On one day, December 12, 1937, Stalin and Molotov[2] sent to death 3,167 people.

---

[2]Vyacheslav Molotov (1890–1986), was an *Old Bolshevik*, and a leading figure in the Soviet government from the 1920s, when he rose to power as a protégé of Joseph Stalin. Molotov served as Chairman of the Council of People's Commissars from 1930 to 1941, and as Minister of Foreign Affairs from 1939 to 1949 and from 1953 to 1956.

> С 8 августа 1937 года
> по 19 октября 1938 года
> на Бутовском Полигоне
> было расстреляно
> 20 762 человека

Figure 10.2 20,762 people were shot by the NKVD firing squads from August 8, 1937, to October 19, 1938, at the Butovo Firing Range. The memorial plaque at Butovo-Kommunarka, Moscow (opened to the public in 2017).

The Decree of the Presidium of the Supreme Soviet of June 26, 1940, "On the Prohibition of Unauthorized Absenteeism of Workers and Employees of Enterprises and Institutions" qualified being late to work by more than 20 minutes as an offense punishable by the Gulag imprisonment. The collective farmers (*kolkhozniks*) were deprived of their identification documents. This automatically implied that they could not change their place of residence.

Thus the majority of the USSR citizens were turned into serfs.[3]

Stalin had a sharp mind. Already at this early stage he understood that in the absence of positive motivations the socialist economy can function only in the atmosphere of terror, in which the workers and collective farmers worked not for profit or personal benefits but out of fear of the Gulag imprisonment. The decay of the Soviet economy in the 1970s and 1980s which led to the collapse of the USSR demonstrated that the Leader of all Progressive Humankind Comrade Stalin[4] was right. With great sadness I observe that this simple thought does not immediately come to mind for the present-day social experimenters.

---

[3] All the data and documents presented above can be found — in a concise form — at http://bessmertnybarak.ru

[4] That was how he was referred to in newspapers.

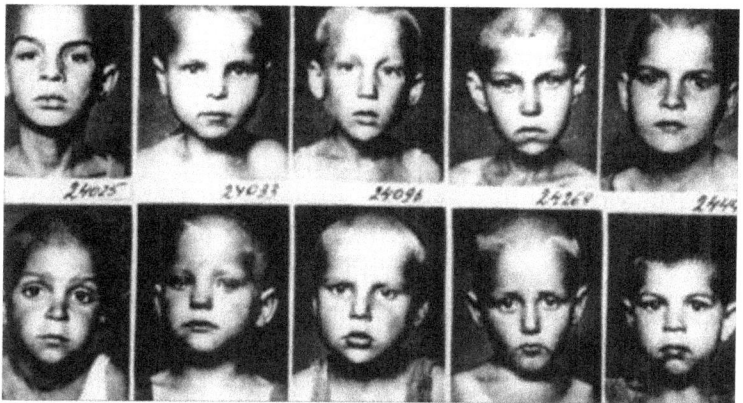

Figure 10.3 Children of the executed "enemies of the people" sent to special orphanages.

*****

Now I will move to Western Europe.

Against the backdrop of a complete economic meltdown in Western Europe (especially in the countries that lost the First World War), the influence of national-socialist ideas was growing. Populist slogans by Hitler were widely supported and brought the National Socialist Workers Party of Germany into the political mainstream. Horrified by such a development of events, the European *intelligentsia* found solace and hope in communism — the radiant future of all mankind, the construction of which in the USSR was in full swing. The communist idea was an epidemic which swept the continent: from Spain to Germany. Once, in an English film, I saw a scene in which a wealthy German refugee explained to an English official: "My only alternatives were either to join the Nazis or the Communists."

In 1937, Leon Feuchtwanger, a world-famous novelist, came to Moscow for two months. He was received by Stalin and attended the second Moscow show trial — the so-called "Process of a parallel anti-Soviet Trotskyite center" — in January of 1937. Of 17 accused top echelon party leaders, 13 were sentenced to death and shot immediately. The remaining four were executed by firing squad a few years later. Returning to the West, Feuchtwanger published in Amsterdam the infamous book *Moscow 1937* presenting his impressions of Stalin, life in the Soviet Union, show trials, etc. Summarizing in a nutshell, he was absolutely delighted. The Idea of "the radiant future of all mankind" blinded him to the extent that he did not notice the monstrous falsity of everything that was shown to him. He cited official

Soviet statistics indicating that Soviet people ate better than Italians or Germans. It did not occur to him to doubt the accuracy of this statistic — to him, if the State were based on the Idea, then, of course, it should not fall into distortion of facts. It was 1937, the beginning of the Great Terror, a time when from 300 to 1,000 people were shot every day at the Butovo-Kommunarka range near Moscow, without trial, by the so-called "sentences" of the *troikas*. The artificial famine caused by forced collectivization, millions of peasants expelled to Siberia and dying on the way, and the mass eviction from Leningrad after the murder of Kirov, all these tragic events escaped Feuchwanger's eyes — his vision was blurred by the Idea.

Fleeing from one monster — Adolf Hitler — the Western intellectuals, chose to serve another no less blood-thirsty dictator, Iosif Vissarionovich Stalin. When I write "chose to serve," this statement must be understood literally. All the European Communist Parties at that time unquestioningly obeyed the instructions from the Comintern in Moscow.[5] Thinking people, entering the party, became "soldiers of the party," and as any other soldiers, were supposed to obey orders without asking questions. From that moment on, they were led through life by the Idea of the Radiant Communist Future. "Ends justify means if the Idea is at stake," was their slogan.[6]

Figure 10.4 Feuchtwanger (left) and Stalin (center). Moscow, 1937.

It is possible that these soldiers of the party contributed to Hitler's rise to power in Germany in 1933. In the elections that year his party received 33% of the vote. If the German communists joined in a block

---

[5]The Communist International, abbreviated as Comintern (1919–1943), was a a communist organization with headquarters in Moscow. It united almost all world communist parties. The Comintern program included the following clause: fight "by all available means, including armed force, for the overthrow of the international bourgeoisie and for the creation of an international Soviet republic as a transition stage to the complete abolition of the State."

[6]In the early 1990s, when I had just arrived in the USA, I learned that one of the professors in our Department, Erwin Marquit, was a lifelong communist. Once at a faculty dinner he tried to explain to me that the Soviet way of life was way superior to that in the US in each and every respect because of the guiding Idea. He would not listen to my objections. Erwin Marquit recently died, but he left recollections *Memoirs of a Lifelong Communist*, https://www.academia.edu/38348295/_Memoirs_of_a_Lifelong_Communist_by_Erwin_Marquit_PART_I and https://www.academia.edu/38348677/_Memoirs_of_a_Lifelong_Communist_by_Erwin_Marquit_PART_II. If you want to understand, at least approximately, the mindset of such people and the degree of ideological indoctrination, I advise you to read this manuscript.

with the Socialists, they would have a majority in the Reichstag (37% of the vote) and they could, in principle, receive a mandate to form government. But Stalin imposed a ban on a unified front of communists and socialists, whom (socialists) he regarded as traitors to the Idea. And in the "workers and peasants" state, Stalin's word was tantamount to a law for everyone, including the Comintern. More precisely, it was above any law.

The way to hell is paved by good ideas.

Working on my previous book, *Physics in a Mad World* [1], I looked through a notable number of files from the archive of the German and Austrian sections of the Comintern. This archive is now kept in the Russian State Archive of Social and Political History (RGASPI) in the public domain. I was amazed by the number of German and Austrian communists who were agents of the Comintern in Western Europe and carried out the order of Comrade Stalin with an iron fist. In many dossiers there is a note "performed special assignments." Special assignment is an euphemism that could mean anything: from espionage to discrediting opponents among Russian emigrés, from eliminating disobedient agents, to assassinating defectors from the "socialist paradise," Trotskyists (and Trotsky himself),[7] and other "undesirable elements".

In 1934–36, many of the Comintern agents fled or were recalled to Moscow, and almost all disappeared in 1937–38: they were either sent to Gulag, or were executed immediately after their arrest by the NKVD. A tragic story of one of the most active Comintern agents, Fritz Burde (party name Edgar), is narrated in Alexander Weissberg's book [3] *The Accused*.

A few survivors of the Great Terror were transferred to the Gestapo in 1940 under a secret clause in the Molotov-Ribbentrop pact. As they say, they planted and then shook.

*****

A few days after Rudolf's conversation with Fuchs on February 3, 1950, the Peierls went to Brixton prison together to see Klaus Fuchs. They decided beforehand that there were three specific questions to which they wanted answers, and they dropped these into the

---
[7]Leon Trotsky (born Lev Davidovich Bronstein; 1879–1940) was a Russian revolutionary and Soviet politician. Ideologically a Marxist-Leninist, he later developed his own extreme version of Marxism, known as Trotskyism. On August 20, 1940, Trotsky was assassinated in Mexico by a Ramón Mercader, an NKVD agent, on Stalin's order.

conversation.[8] The first of Rudolf's questions to Fuchs was: "Why had you brought back from America Kravchenko's book *I Chose Freedom* as a present for Mrs. Peierls?"

To my mind, this was an act of utmost mockery by Fuchs. Let me say a few words about this book. Victor Kravchenko (1905–1966) was a high-ranking official in the Communist Party of the Soviet Union who defected to the United States during World War II. He wrote a book *I Chose Freedom*, published in the US in 1946 and in France in 1947, based on his own experiences. This was the first detailed exposé of the atrocities committed by Stalin and his henchmen in the USSR available to the Western reader. He depicted mass arrests and executions by the NKVD, the Gulag, artificial famine in Ukraine caused by forced collectivization, the atmosphere of total fear... This currently forgotten book was a best-seller in the US and Europe. At the same time it was met with ferocious attacks from the European Communist parties. Of course, now, after Solzhenitsyn [6] and the archival data that became known in the 1990s, we know that Kravchenko in no way exaggerated the situation — in fact the reality had been much worse than Victor Kravchenko could have imagined in 1946.[9]

An especially vicious attack on Kravchenko's character by the French communist weekly, *Les Lettres Françaises* made him sue *Les Lettres Françaises* for libel in a French court [7]. The trial which lasted from January till March 1949, featuring hundreds of witnesses was dubbed "The Trial of the Century."

The communist weekly was defended by counsel Joe Nordmann and his team (including Maître Leo Matarasso) who summoned numerous witnesses supposed to confirm that Kravchenko was lying; from the Soviet side: General Sergei Rudenko, famous writer Ilya Ehrenburg, directors of the Magnitogorsk, Dnepropetrovsk and Nikopol factories, Kravchenko's ex-wife,[10] and from the French side: Nobel laureate Frédéric Joliot-Curie, MP Fernand Grenier, journalist Pierre Kurtada, famous writer Simone de Beauvoir, and many others. Their testimonies

---

[8] Norman Moss, [4], page 153. Corroborated in the letter by Rudolf Peierls to Hans Bethe of February 15, 1950, [5], page 353.

[9] The very fact that Fuchs presented this book to Genia almost certainly meant that he had read it and knew its contents. At least after the war he had a clear-cut idea whom he was working for. After I became aware of this episode I thought: "What a disgusting and perverted mind!"

[10] Joe Nordmann stated that at the request of the defense team Kravchenko's ex-wife Zinaida Gorlova expressed her desire to travel from the USSR to Paris to testify against Victor Kravchenko. ... Expressed her desire ... Everyone who has ever lived in the USSR, especially in 1949, would infinitely laugh at the unimaginable, monstrous falsity of this statement.

were backed with false documents written by journalist André Ulmann (who worked for Soviet intelligence). The left-wing press focused on one thing: "... the indignation of all honest people of our country [France] who have not forgotten the services rendered to France by our Soviet ally in the fight against Nazi barbarism, and 17 million dead Soviet people will also appear in court as accusers of the traitor Kravchenko."

Between two evils: the tragedy of that generation.

Kravchenko's legal team enlisted witnesses mostly from the survivors of Soviet prisons and camps. Among them was the Gulag survivor Margarete Buber-Neumann (widow of the German Communist leader Heinz Neumann, who perished in the hands of the NKVD[11]) who came from Stockholm. Kravchenko had received around 5,000 letters from displaced persons all over Europe — former Soviet farmers, workers, and engineers — Gulag prisoners in the 1930s later deported by the Nazis from occupied Ukraine to German labor camps. They all wanted to testify that Kravchenko's book was truthful. Kravchenko selected 20 of them as witnesses. I should also mention André-Remy Moynet, a Deputy in the French National Assembly and a former air captain who had received numerous decorations from both the French and Soviet governments for missions flown against the Nazis in Soviet territory.

Their testimonies fully corroborated Kravchenko's allegations concerning the essential similarities between Stalin's Communism and Hitler's National Socialism. When I read the reports of the trial (see [7; 8]) I was astonished by the humiliations that Buber-Neumann, and other witnesses on Kravchenko's side had to go through just to say what they knew was true.

To finish with this digression I should add that in 1968 *Les Lettres Françaises* tried to criticize the Soviet invasion in Czechoslovakia. This was the end of it: the Soviet government decided to withdraw its subsidies, and so did the French Communist Party. *Les Lettres Françaises* being stripped of its financial lifeline ceased publication in a couple of years. In the 1990s Boris Nosik interviewed Maître Matarasso:

---

[11] Heinz Neumann (1902–1937) was a German communist politician and a journalist. He was a member of the Comintern, editor-in-chief of the party newspaper *Die Rote Fahne* and a member of the Reichstag. In 1927, Heinz Neumann was an adviser to the Communist Party of China and participated in the uprising in Canton. In 1932–1934 he was an "instructor" of the Comintern in Spain. At the peak of the Great Purge he was arrested in Moscow on April 27, 1937, by the NKVD and sentenced to death on November 26, 1937. His wife was sent to Gulag.

He was a nice man, he had a Law Office on Rue Tournon. When I came to him he felt unwell, but then he got a little more comfortable and told me: "Yes, yes, we all said a lot of stupid things. Many stupid things were said at that time. But you noticed, Boris, I did not ask Buber-Neumann a single question..."

Figure 10.5 First page of a 1953 issue of *Les Lettres Françaises*. The headline reads "That's what we owe to Stalin." Articles in this issue (or should I say odes) were written by Louis Aragon, Frédéric Joliot-Curie, Pablo Picasso, and French writers and journalists Henri Bassis, Pierre Curtade, Pierre Daix, and Georges Sadoul. In the American press a similar ode "To You My Beloved Comrade," was penned by Paul Robeson: "Through his deep humanity, by his wise understanding, [Stalin] leaves us a rich and monumental heritage."

*****

I am not sure whether this Chapter needs conclusions. Although new generations usually are reluctant to learn from mistakes of the past, still I will risk to give an advice: if you believe in a social theory or idea, try to find at least some experimental evidence of its validity. There is a wonderful principle in the medical ethics, "Do not hurt," which, as they say, can be traced back to Hippocrates of Kos who lived around 2,400 years ago. A basic principle of physics is as follows: "Even beautiful theories must be abandoned if experimental data do not support them." Social systems are subject to much more complicated regularities than physical laws. Simplistic approaches or quick fixes do not work.

If there are even slight reasons to doubt the validity of your sacred Idea, do not try to force it on other people. Try first on yourself or volunteers. Facades are often deceptive, go deeper and see what is really going on there with your own eyes. Fashionable trends about social engineering have so far led nowhere. Finally, no idea of a universal happiness in the nebulous future deserves to be implemented if *en route*, today, it requires lies, suppression of freedom, and persecution of opponents and dissidents. Even if it were only 10 innocent people that have to be executed *en route* for the sake of the "radiant future," this is not the future in which you would want to live.

References

[1] *Physics in a Mad World*, Ed. M. Shifman, (World Scientific, Singapore, 2016).
[2] M. Shifman, *George Placzek: A Nuclear Physics Odyssey*, (World Scientific, 2018).
[3] Alexander Weissberg-Cybulski, *Hexensabbat*, (Frankfurt, Frankfurter Hefte, 1951), in German, English translation: Alexander Weissberg, *The Accused*, (New York, Simon and Schuster, 1951) and *Conspiracy of Silence*, (Hamish Hamilton, London, 1952). A German edition was released recently in Austria under the title *Im Verhor: Ein Überlebender der Stalinistischen Sauberungen Berichtet*, (Europa Verlag, 1993).
[4] Norman Moss, *Klaus Fuchs: The Man Who Stole the Atom Bomb*, (St. Martins Press, 1987).
[5] Sabine Lee, *The Bethe-Peierls Correspondence*, (World Scientific, 2007).
[6] A. Solzhenitsyn, *The Gulag Archipelago, 1918–1956*, (Perennial Library, 1973).
[7] Gary Kern, *The Kravchenko's Case: One Man War on Stalin*, (Enigma Books, 2007).
[8] Nina Berberova, see in *The Last and the First. The case of Kravchenko*, (Sabashnikov Publishers, 2000), in Russian.

Chapter 11

On the Other Side of the Iron Curtain

By late 1942, the stream of data on the Anglo-American nuclear bomb project obtained by the Soviet intelligence had reached a critical mass. On September 28, 1942, the State Defense Committee issued an executive order №2352cc "On Organization of Work on Uranium." In October 1942, L. Beria,[1] the Head of the NKVD, briefed Stalin. On April 12, 1943, the so-called Laboratory №2 was launched, the Soviet analog of the Manhattan Project.[2]

On July 24, 1945, the US President Truman told Stalin in Potsdam that the United States "now has a weapon of extraordinary destructive power." According to Churchill's memoirs, Stalin smiled, but did not inquire into the details, from which Churchill concluded that he did not understand anything and was not aware of the events. That same evening, Stalin instructed Molotov to order Kurchatov to speed up the work on the atomic project. Shortly after, Molotov was replaced by L. Beria who thus became responsible for the entire complex of problems associated with the nuclear bomb development and production. In September 1945, a Special Committee headed by Beria was created. It included leading members of the Party and Government (and also two scientists — Kurchatov and Kapitza).

On April 1, 1946, the town of *Sarov* was "erased" from all maps of the USSR, since it was chosen as the location of the first Soviet nuclear center, the Soviet analog of Los Alamos, which became known by its code name, Arzamas-16.

---

[1] Lavrentiy Pavlovich Beria (1899–1953) was the chief of the Soviet secret police (NKVD) since 1938; since 1941 also deputy premier. He was in charge of the Soviet atomic bomb project. After WWII, Beria organized the communist takeover of the state institutions of Central and Eastern Europe and political repressions in these countries. On June 26, 1953, Beria was arrested on Khrushchev's order. His case was considered by a special committee, with no defense counsel and no right of appeal. On December 23, 1953, he was found guilty of high treason, sentenced to death and executed.

[2] Currently, the National Research Center Kurchatov Institute. Igor Kurchatov was the Director of the Laboratory №2.

As a rule, construction work at atomic sites as well as mining of the uranium ore were carried out by Gulag prisoners. In the latter case no safety measures were taken — the prisoners were dying in large numbers but were immediately replaced by new prisoners.

An elucidating paragraph from B.L. Ioffe's recollections [1][3] which I present below hardly requires any comments:

> I will tell a story which was told as an entirely true story: about the way in which the International Joint Institute for Nuclear Research in Dubna was organized. At that time, it was called the Hydrotechnical Laboratory (HTL) — presumably for the reason that it stood on the Volga river. But it had nothing at all to do with hydrotechnology. This institute was created at the suggestion of I.V. Kurchatov to do research in nuclear and elementary particle physics. In actual fact, the research undertaken in Dubna was in no way related to nuclear weapons. (Though for many years the powers-that-be were persuaded that the opposite was true.) At the time of making the decision to create the Institute, the natural question of its siting was discussed. A special commission was set up to study this question. Beria called a meeting at which the commission presented its recommendations and suggested three possible sites for the future institute. Beria listened, then asked for a map, pointed a finger at a place for the future Dubna (this was not one of those recommended by the commission) and said:
>
> – We'll build it here.
>
> – But — somebody timidly objected — this is an area of swamps, the ground conditions are not suitable for accelerators.
>
> – We'll drain them.
>
> – But there are no roads.
>
> – We'll build them.
>
> – But there are few villages there, it will be difficult to find laborers.
>
> – We'll find them, said Beria.

---

[3] The episode below refers not to Sarov, but to a research laboratory in Dubna built a few years later, but the principle was the same.

And he was proved right. This area was surrounded by Gulag camps, this is why Beria had chosen it. As late as 1955, when I first visited Dubna, the camps stretched along the road, there were guards everywhere and one had to tell them "We are going to see Mikhail Grigorievich." (This was M.G. Meshcheriakov, director of HTL.)

*****

The remainder of this Chapter contains a part of the recollections of Olga Shiryaeva entitled "The Story of One Life" (see [2]; also pages 154–159 in [3]). Olga Shiryaeva herself was a Gulag prisoner (albeit quite privileged because of her profession), and for a number of years worked at the Sarov site. I hasten to add that at the very end, her story had a relatively happy end — an extremely rare occurrence at that time. I wish the reader could read her memoir in its entirety. Unfortunately, for the time being this is impossible, since it was written in Russian and is not yet translated in full. However, even the part presented below speaks volumes about the abyss between the civilization based on moral principles and that based on ideology.

*****

The Story of One Life[4]

Olga Shiryaeva

A few days after my return to Riga, Germany invaded the Soviet Union. The 22nd of June was a Sunday, but in spite of that we were working when we heard Molotov's speech from a street loudspeaker. At 3 p.m. the first aerial bombardments of Riga began. For another three days, we stayed in the town, burnt documents and prepared for evacuation. Then we were told that women must leave immediately. Together with my friend Anna Fridbur we boarded the transport train directly from work. On the way some trains were bombed, but we were lucky — our train passed through. On stations, we saw people with luggage, crying children. It was terrible!

In Moscow, we found no such horrors, but all the same people were gloomy and weighed down. Military units were marching, leaving for the front. Women, children and old people were being evacuated. My husband, Basov, was already mobilized for the front. His parents had left Moscow for their dacha and were staying there with my son Seriozha. They did not plan to return to Moscow but did not want to leave for safety even though I insisted on it because I myself had firmly decided to defend our country. On the 22nd of July, the Germans started bombing Moscow. I was just returning from the Basovs' *dacha* and entered the Metro station Komsomolskaya when the air raid alarm was sounded. We went down into the tunnel and stayed there until daybreak. Up in the street I could not get to my house: everything was sealed off because a bomb had dropped nearby.

Then I understood that I could not leave my child in Moscow and began preparations to join the evacuation together with a group from the Union of Architects. The city was bombed every day at the same time. In the beginning of August, we — my son Seriozha, his nanny and I — boarded the transport train going to the town of Buzuluk.[5] We were all sure that we would return home soon.

First, I went to the village Labaza near Buzuluk. I thought that the situation with food would be better in a village. There was a terrible heatwave, the houses were dirty, and my son very soon fell ill with dysentery. I was required to join the work of the kolkhoz. I sent the nanny and stayed with the sick child. After a few days, some local

---
[4]Courtesy of Alina Chertilina. Translated from Russian by Wladimir von Schlippe.
[5]A small town in the Orenburg Region, approximately 1,200 km south-east of Moscow.

women came to my window and shouted: "Shiryaeva, take off your silken dresses and go to work at the kolkhoz." As soon as I had nursed my son back to health, I left him with the nanny and returned to Buzuluk.

There I rented a room in the house of "uncle" Peter on the outskirts of the town and arranged to work as a caretaker and technician at the railway station. Then I returned to Labaza to fetch my son and nanny. In the meantime, transports of evacuees from Moscow kept arriving. My friend Marina Dvorez with her son arrived in one of these. She was in a distressed state, and I took her in with me. She was not a member of the Union of Architects, and therefore had a ration of just 800 grams of bread for the two of them. Therefore, I shared with her everything I had. I did not work long at the railway. Soon there was a reduction of staff and I was made redundant.

The end of 1941 was a very troubled time. The Germans were approaching Moscow. Nobody believed that Moscow would be taken by the Germans, but the anxiety was depressing. In Buzuluk a small community of architects had formed. We were supporting each other with food supplies but also morally. We were jubilant when the Germans were driven back from Moscow by two hundred kilometers at first, then four hundred. Naturally, that was only the beginning and ahead was much suffering, but it was a turning-point. We wholeheartedly believed in our victory.

Soon I got a job as warden at the barrack of the Buzuluk garrison headquarters. The military personnel never stayed long with us — they arrived, stayed for a few days and moved on to the front. Sometimes international units came, mostly Poles. To tell the truth, they were rather arrogant, treated the Russians like servants. After a while they were moved to Iran under the command of general Sikorski.[6]

Marina's husband arrived from Moscow. He behaved rudely, drank. When I told him that this was my lodging, he said that, if I didn't like anything, I could get lost.

Near the end of 1941, the Czech army was beginning to be formed in Buzuluk. At first, colonel Svoboda arrived with a small group of officers. When I first met Ludvik Svoboda in the barracks, he strongly reminded me of my father. Like my father he had grey hair, the same old-time military bearing, civility in company, similar physique and

---

[6]Władysław Sikorski (1881–1943) was a Polish general and politician, Prime Minister of the Government of Poland in exile. In 1941–1942, he participated in the creation of the Polish Anders' army formed in the Buzuluk area and subsequently transferred to Iran and then to Palestine.

height as my father. He looked at me very attentively, so that I felt somewhat embarrassed. After a while, his flag lieutenant came and told me that the colonel would like to improve his Russian and asks me to meet him. It was my peculiarity to believe anything. I knew that I was not too good at Russian grammar, so I suggested that Marina's husband could do that instead of me. They had a few sessions, after which Svoboda stopped them.

The Czechs, unlike the Poles, are easy-going people. They frequently visited us, we danced and arranged simple meals. Actually, this had to be done in secret from Marina's husband, since Marina had a stormy romance with one of the officers. One of the officers, Matthias, also courted me. He proposed to me, and I accepted. Not because I loved him, but just because I understood that with Basov everything had ended, and I wanted to have a shoulder for support. Matthias was very considerate, cared for me, but soon after we had decided to get married, he was transferred to Kuibyshev.[7] As I learned later on, they had a rule: he had to ask for permission from a senior commander to get married. Matthias approached Svoboda with such a request and was immediately transferred to another unit.

Somewhat later Marina and her husband left for Saratov, Seriozha's nanny Niura got employed at a factory and was given a place in a hostel, and I started looking for accommodation nearer work, because returning from work in the evening was scary. Once I was walking in town and saw a handsome woman sitting outside a house. The house was conveniently located, close from work to walk, and I asked the woman whether she had a room available. Apparently, she too seemed to like me, and she said that she would find a room for me. We entered the *izba*: a porch, then a spacious kitchen with a Russian stove on which "batya," the father, slept, and further on a narrow passage in which four evacuated Jews lived. Nastya, as the woman was called, wrinkled her nose: "They smell of dogs." "What has this to do with dogs? — I wondered, — it is just stuffy here, it wants some airing." "You don't know, Jews always smell of dogs," and waived her hand. We carried on through the hall (which was what the large room was called), behind which there was a small room of 7 square meters.[8] "This room I can let you. Four hundred grams[9] of bread per day." As my daily ration

---

[7] Currently Samara, a large industrial city on the Volga river, approximately 1,000 km to the east of Moscow.
[8] Around 75 square feet.
[9] Around one pound.

was one kilogram of bread, and another 400 grams for my son, I could afford to rent the room and I agreed. On one side, the Russian stove protruded into the room, to which *batya* attached a cot, and we moved in.

At that time, I was convinced in the infallibility of the Soviet powers and of Stalin, who was continuing the mission of Lenin. I had been a member of the All-Union Communist Party (Bolsheviks) since 1933. My husband, after graduating, worked at the Lubyanka[10] and brought home many interesting books, among which were books about spies. I was particularly impressed by books about female spies. I would lose myself in a novel about Mata Hari, admiring her patriotism which led her to engage in such work in spite of the risk. We went to performances at the Bolshoi Theatre and saw Stalin who was sitting in his box. Sometimes Michael and I went to parties, and afterwards some of those present got arrested. At the time, I did not pay attention to this and only later found out what was going on.

I made no secret of socializing with the Czechs. Once I was called to some military person who asked me to sign a statement, according to which I undertook to report to a certain Victor Ivanovich everything to do with the attitudes of the Czechs towards us. I agreed and signed the paper. Full of patriotic enthusiasm, I decided, like Mata Hari, to arrange a salon where I would uncover conspiracies of spies. Although I was 30 years old at that time, I was very silly and naïve.

Let me say a few words about Svoboda. In 1939, after the occupation of Czechoslovakia by the Germans, Svoboda moved to Poland, and then to the Soviet Union. Before he came to Buzuluk, he stayed in a *dacha* near Moscow. He was an energetic person, with willpower, strict but friendly; he had a solid military education and experience. In the battalion that stayed in Buzuluk, he was held in great esteem. He was not very knowledgeable in literature and the arts, was not interested in museums. When he fled from Czechoslovakia to Poland, his family stayed behind in Prague. At that time, he could not even imagine that anything could happen to them. However, the Fascists[11] took his family hostage; his son, Mirek, was tortured and killed, while his wife Irena and daughter Zoya disappeared. Ludvik loved his family very much, and he suffered greatly at the thought that he had been unable to prevent the horrors that happened to them. In the Soviet Union,

---

[10]The NKVD Headquarters were on Lubyanka Street. In colloquial language "Lubyanka" served as a synonym for the NKVD.
[11]That was how the German Nazis were referred to in the USSR.

he was totally absorbed in the task of forming the Czechoslovak army and of training them. Apart from the military life, Svoboda helped establishing the cultural life of his soldiers: a small choir was formed, and on Sundays the musicians performed Czech songs.

Ludvik was certain that the end of the war was near, and that after the war Czechs and Russians would be great friends. He campaigned for his army to join the Red Army fighting against the Germans. But the Beneš government, which was based in London at the time, was categorically opposed to this. They considered that the Czech army should stay in the rear until the beginning of the liberation of Czechoslovakia.

My relations with Ludvik began in March 1942. One evening, on my way home from work, I saw Ludvik slowly walking along the street. He approached me, greeted me and kissed my hand as is the custom of the Czechs, and said: "Tonight is such a wonderful warm evening. Would you not like to join me for a walk to the river?" I agreed. After that walk, we began to meet. Between us a real love arose, strong, passionate, all-embracing and powerful like a mountain river. It so happened that the house, where I was lodging, was across the road from Ludvik's house.

Once I heard a legend, that in early times there were no men and women on earth, and people were giants. But they enraged God, and he cut them in half. And ever since, everybody is looking for their other half. If one finds one's other half, they will both be happy. I think that Ludvik and I were the two halves of each other.

At the beginning of summer, thieves broke into Ludvik's room and stole his gabardine mackintosh and shoes. I heard of it from Bozhena Verbenski's hairdresser. The Verbenskis — husband, wife and son Vladek — lived in the same house as Ludvik. We wrote a jocular poem about this case:

> The colonel saw a wonderful dream
> But was woken by a terrible noise.
> It's morning, on the colored window
> Plays the gentle sun.
> But looking around he understood
> What considerable loss!
> Where's the coat, where the boots? O, my dream!
> How sweet was the dream,
> how bitter the awakening!
> Calamity has come to town:

> The colonel in grey? Never!
> I won't keep the secret,
> Who is marked in Buzuluk,
> That a grey coat would suit him fine.

After this incident, it was suggested to put a sentry at Ludvik's house, but he refused categorically, since then we would be unable to meet.

I did not, of course, succeed in establishing a salon to expose conspiracies of spies. Victor Ivanovich did not bother me, and I completely forgot about him. With Ludvik, we went for walks along the river, watched football matches between our and Czech soldiers, spent evenings with the Verbenskis having coffee and discussing the progress of the war, rejoicing at our victories. Only one thing was depressing me: the Czechs were eager to get to the front.

I remember Christmas 1942 when, amidst the joy, suddenly the door and windows opened and men in white camouflage armed with submachine guns burst in shouting: "Hände hoch!" pointing their guns at the officers. There was a moment of silence and consternation since it was known that the Fascists were sending commandos into our rear. But then, one of intruders removed his face mask and announced: "We wish you a merry Christmas!" We all gasped: they were Czech soldiers. "Well, some joke!" said somebody. Many of the guests were really frightened, some had even taken cover under tables.

But those were jokes. One must say that life at that time was very strained. In their spare time from military training, the Czechs did work a lot in our factories, making an effort to contribute to the struggle against the Fascists.

At the end of 1942 the Czech units were sent to the front. I remember our farewell with Ludvik. He said to me: "Oliusha, believe me, you are my only one. Soon I'll return to you with victory, and we'll be together! My sweetheart, my dear, don't forget me. You are the only one I need." He promised to write. On January the 30th 1943 the transport was formed. On the coaches hung slogans: "Death to the Invaders!," "With Svoboda for Freedom!"[12] I stood on the platform seeing off the Czechs. Suddenly Otakar Jaroš came up to me and said: "Olga, I want to say good-bye to you. I know that I must perish in the war. Allow me to kiss you!"

At his words, I felt a cold shiver down my spine; I replied: "But no,

---

[12] In Russian, this is a pun: the Russian word for freedom is *svoboda*. -VvS

why bury yourself in advance? We will dance czardas together." He smiled sadly and replied: "May God grant it, but it is unlikely." Then I forgot Jaroš. My heart was full of pain for Ludvik. The train moved away, I looked around: the streets looked the same, the same life, but everything was empty.

In March 1943, the Czechoslovak army joined the fighting on the Voronezh front. Later I learned that on the 8th of March, Otakar Jaroš had fallen in a battle near Sokolovo. He was the first foreign soldier who was awarded the order of Hero of the Soviet Union.

I must move somewhat back in time. Shortly before the departure of the transport, on my way to my Buzuluk "home," Victor Ivanovich approached me and asked me to follow him. He led me into some house, invited me to take a seat at a table and sat himself opposite me. He stared at me and started to speak. I sat as though petrified. "Instead of helping us, you have cocked up our arrangements. I hold the order for your arrest. Look, now you are going home with bread, but your son won't get it, he will be taken to an orphanage. From now on he is an orphan!" I didn't understand anything; my head was in a dense fog. He handed the paper to me and said: "Take the order and leave. And don't dare saying anything to Svoboda." I left that house as though drunk. Returned home and lay down on the bed unable to do anything. I understood one thing: I was completely incapable of falsehood, too straightforward and possibly too stupid to be a spy.

In the evening, I met Ludvik. Naturally, he noticed that something was wrong with me. He asked what the matter was, but I could not tell him. I excused myself saying that I was worried about my son: just then Seriozha had fallen ill. Ludvik tried to set my mind at rest, said that everything would be all right.

And now, on my way home from the railway station, I understood that I had to live for the sake of my son. He needed me, and therefore I had to control myself. During six months of 1943, I lived through an age of sorrow, everything around me seemed to be empty. There was one saving light at that time: I received letters from Ludvik. In one of them he described his flight. The airplane had to land on a small airfield. The attendant on the airfield did not have a green flare, and so he used a red one. But to the pilot that meant that he could not land on that airfield, and since they did not have enough fuel to reach another one, they landed in a nearby field. The airplane's tail caught on some obstacle and the plane broke up. Luckily nobody got hurt. They escaped with bruises and a fright.

In summer of 1943 Ludvik came to Buzuluk. I was at home when a soldier came to invite me to the officers' house, because Ludvik Svoboda had arrived. I rushed there like on wings. We threw ourselves at each other, I felt the warmth of his hands, his cheek — here he is, my own, close to me! Then there was a banquet. Ludvik said: "I feel as though I have come home." The evening and night were ours, and in the morning he was off to the front again. Before parting he asked me: "Oliushka, try to get back to Moscow as soon as possible, I'll be there often."

I remembered that I had an acquaintance in Moscow: a cheerful, big chap, called Solokha. Once he had told me that, if ever I needed anything, I should just write to the Kremlin to Solokha. He said, just write, don't hesitate, and I'll help. That I did, and, to my surprise, I soon got an order to move to Moscow. In August 1943, together with Seriozha we boarded the train for the Capital. Farewell Buzuluk!

In Moscow the atmosphere was tense, full of worry. Everywhere one could see the marks of the bombing and of the fortifications. But there was also a shimmer of hope: in February the Germans were routed in Stalingrad, and nobody was in doubt that our cause was right, and that we would be victorious. But how many lives would still be lost in this war, how much blood would be shed!

The Verbenskis were in Moscow, and I saw them frequently. Soon also Ludvik came. He visited me and stayed for the night, but in the morning he said that he had been imprudent: he, as a foreign general, should not stay at night in private quarters. So, afterwards we used to meet in the hotel "National" where Ludvik stayed.

In autumn Seriozha went to school, and I started working on developing the interiors of buildings that were being restored. During that period, my private life was the most important thing for me. Cultural life was re-emerging in Moscow. A few times we went to theaters, attended receptions at General Píka's Mission[13] and receptions at the Embassy. I remember, at one of the receptions, the music started playing, and dancing was to commence. Svoboda always started dancing with me. Suddenly, unexpectedly, Valentina Serova[14] approached Svoboda and invited him to dance. He declined, saying with a smile: "I am busy." I was then very proud of our love.

---

[13] General Heliodor Píka (1897–1949) was a Czechoslovak army officer. In 1941, during the WWII, Píka was appointed chief of the Czechoslovak Military Mission to the Soviet Union. Loyal to the London-based government of exiled Czechoslovak President Edvard Beneš, Píka supported his democratic policies despite Soviet opposition. In 1949 he was executed by the Communist regime in Prague after a show trial, on charges of high treason, see page 257.

[14] Possibly related to Deputy Minister of the Interior Ivan Aleksandrovich Serov (Serov's wife was called Vera). -WvS

The year 1944 arrived. Our forces advanced on all fronts. This was a year of great success and great victories. But, alas, not for me.

My friend Marina returned with her family to Moscow. Her husband had, after all, been at the front and was now greatly showing off, saying that he was battle-scarred: he had lost one finger of his left hand. Full of himself and greedy for gain, he kept saying that he'd go to Yugoslavia and bring back a car. Before long he did go there indeed and bought a car, but the airplane, in which he was returning, crashed. It was said that there had been several people with much luggage. To avoid attention, they persuaded the pilot to land not in the airport but nearby. At the landing, the plane clipped a tree and crashed — all the passengers perished.

During one of his visits, Ludvik told me that his wife and daughter had been found. For three years they had been hiding in a cellar. They had managed to get away from the Fascists, and some tender-hearted country people had given them shelter. But throughout this time they could not leave their cellar since there had been a high reward declared for their capture. Ludvik felt a mix of the joy of their deliverance with the bitterness of parting from me. I listened to his words as in a terrible nightmare, and finally said myself: "So, it is not our fate to be happy together."

General Píka invited me to meet the new year of 1945 at his Mission. We went there together with Bozhena. After all that had happened, this was a wonderful, unforgettable night. Píka courted me fulfilling all my wishes. Near morning he saw me home. Outside the lift he asked me: "Do, please, allow me to kiss you." I offered my cheek to him. He kissed it and said with disappointment: "You do love him very much indeed! But he does not deserve it." I never met Píka again.

Victory Day in Moscow I very well remember. The weather was sunny but there was a cold northerly wind blowing. Together with Seriozha I went to Red Square where enormous spontaneous meetings were held. People were singing, walking around, dancing, playing accordions. Everybody who was able to do so carried bottles with alcoholic beverages along Gorky Street offering drinks to strangers. In the evening there was the fireworks salute. Portraits of Stalin were floating in the air, lit by search lights, and shooting was heard from all sides. There were such huge crowds that one got afraid of getting squashed. With difficulty the two of us got out of the crowd and went home.

But there was no clarity any more. Doubts about the infallibility of Stalin had appeared. In one of my conversations with Marina I had called Stalin ambitious for power and cruel. Soon she got arrested, and subconsciously I began expecting my turn. I had a dream in which Marina was offering me a cup of black water. I took it and drank it. As soon as I had drunk it up I heard a loud crash. As I ran out into the vestibule I saw that a mirror had fallen down and broken.

In June, Svoboda was coming to Moscow. We saw each other sporadically. As Minister for Defense he was signing the agreement on the new borders of Czechoslovakia. On the 30th of July he left, sending me a note from the airport: "I live in a palace but have no happiness."

I was arrested on the 14th of August 1945. In the middle of the night the doorbell rang. Nadezhda Ivanovna opened. There were three of them. My first response was to jump out of the window. But one of the men went immediately to the window and, turning away, said: "Get dressed. I'll wait." Mechanically I put on my clothes. Nadezhda Ivanovna gathers some things in a bundle and hands it to me. "Take some warm things," said the MGB operative.[15] I answered: "I don't need anything." We left. The other operatives stayed behind to search my room. Thank God, my son was in a [summer] Pioneers' camp[16] at the time, so I was spared the agony of parting from him.

We were driving through the city. Familiar buildings were flashing by, and in my thoughts, I said farewell to them. We got to the Lubyanka. I got body searched in a humiliating manner, then I was put in small closet — one meter by one. There was a bench, a dark bulb, and fat bugs were crawling on the walls. The soul injured, the body dishonored, I fainted. I came to senses when a jailer shook me and held smelling salts under my nose. "All will be well — they'll sort it out and you'll go home" he said to me. I calmed down somewhat. I had not committed any crime, after all. But then I remembered the case of Charlotta. There was such a girl. She was lovely, like a china doll. She was only nineteen years old. Her parents got arrested, and she was also due to be arrested. But she was staying at her *dacha*. Her nanny came to her and asked her to flee. But instead, Charlotta lay down in the kitchen, closed all the doors and windows and poisoned herself with gas. I remember her lying in the coffin, fresh, young, even with a slight blush on her cheeks. Only now I understood how right she

---

[15] The MGB was an abbreviation for the State Security Ministry, a successor to the NKVD. Later it changed its name again, see footnote on page 116.

[16] Young Pioneers was a communist organization for children.

had been! Before my eyes passed the faces of those who had already been arrested. Had they really all been criminals? For what did they want to arrest the nineteen-year-old Charlotta?

I was moved to a cell on an intermediate level. There were five beds, one of which was free, and I was allotted to it. I lay down and sank into a mix of sleep and delirium. That was the beginning of my life in prison. Getting up in the morning — toilet, then breakfast, lunch, dinner and the ring-off for the night. During the day, we were not allowed to lie down, one could sit only facing the eyehole so that the jailer could see that one's eyes were open. For interrogation one was fetched at night. The cell was 4 to 5 meters in length and 3 in width. It held five beds, and in a corner was the slop-pail. The window was dressed with a metal shield so that one could see a little daylight only through narrow slits. At the ceiling was a bulb, lit round the clock.

In the morning I got to know the other inmates. One of them was Olga Nikitishna Mironova, who was arrested for not having denounced her sister, who had been working for Vavilov[17] and was accused of attempting to assassinate Stalin. Then there was a female historian, professor Heifets,[18] who was accused of distorting history. The women changed frequently, only Olga Nikitishna and I stayed. At one time a former KGB officer shared our cell. She had been the head of the Soviet Mission in Canberra (Australia), and she got arrested on her return home. She was semi-literate and screamed all the time: "It's not yet clear, who will put away whom! I brought back a load of things, and they put an eye on my goodies. They won't get them, they are mine, here — see this!" and she showed a finger and added a shower of expletives.

The first time I was taken for an interrogation, the soldier forced me to face a wall. And I thought: that's it, now I am going to be shot. But nobody did shoot me. My case was handled by two investigators. One of them was called Obraztsov. For three days he kept asking me about Svoboda, but did not keep any record. Then he sharply cut

---

[17] This is a reference to Nikolai Vavilov (1887–1943), the prominent Russian botanist and geneticist, Director of the National Institute of Genetics. Vavilov was arrested on August 6, 1940, on fabricated charges. He was sentenced to death in July 1941. In 1942 his sentence was commuted to twenty years' imprisonment. He died of starvation in prison in 1943. His brother, Sergei Vavilov (1891–1951), was President of the Soviet Academy of Sciences between 1945 and 1951. -WvS

[18] Fanny Aronovna Heifets was a modern history scholar. In the 1930s she was a professor at the NKVD Academy in Moscow; in 1941 she became the Head of Modern History Section at Tomsk University. She was arrested in 1945 for "anti-Soviet propaganda." Her last known publication refers to 1964.

short this talk and started shouting at me: "You anti-Soviet woman! I know that you called Stalin greedy of power and cruel, you praised the Czechoslovak officers and sang praises to the capitalist life!" At first I was denying all of this, but then I was confronted with Marina. I had indeed called Stalin greedy of power and cruel, she had confirmed it, and I signed the confession.

Then I got confronted with my former colleague Lev Dvorets. There was complete madness: as though Lev had created a clandestine anti-Soviet organization, and I was its member. I refused to sign the record, and the investigator did not insist. At the end of the interrogation I asked Lev why he had said that, but he simply waved his hand and said: "It's all the same."

For six days Obraztsov called me for interrogations, and for most of the time he drank tea, walked around the office, talked on the telephone and looked out of the window. In the early morning I was led back to my cell, and immediately the reveille sounded. After six sleepless nights I began to understand Lev, and to me too it was all the same. I decided that, whatever the investigator put in the protocol, even if it said that I was the Empress of China, I would sign it. As soon as I had signed all the records, Obraztsov turned quite kind. Before that he had shouted at me, insulted me but used no expletives. Once, in a rage, he had crumpled up a record and thrown it in my face.

One day he called me in daytime. The window was open, and music sounded from the street. "Your Pioneers are returning!" he said. I looked out of the window. There were children returning home from Pioneer camps, and my son Seriozha was among them. My heart sank and tears came streaming down my face.

Sometimes I was interrogated by investigator Karpov. He engaged in talks about life and never wrote any notes.

Then I was again interrogated by Obraztsov. He kept asking me about my acquaintances, but I gave them only positive references. Then he showed me the references which Marina and Lev had given about me. "Here, look at your friends. You shield them, but they denounce you!" But I said that that was on their conscience. Then he showed me two fat folders and said that they were my closed case. Of the five months that I had spent at the Lubyanka, one and a half had been taken up by the investigation. What saved me during this time was that there was a library in the prison, and the inmates could take books to their cells.

Professor of History Heifets got eight years, since she was not broken and refused to sign anything. Marina signed everything, agreed to cooperate and denounced all her acquaintances, after which she was discharged. At New Year of 1946 I was in the Lubyanka. In January, Olga Nikitishna and I were put in a "Black Raven"[19] and moved to the Butyrka prison. I remember marveling at the huge prison gate and at the guard, who wore a long sheepskin and had, hanging on a chain, a bunch of keys the size of an axe.

We were led into a cell with many inmates, all of whom had been sentenced under article 58.[20] Next day Olga Nikitishna was summoned, and after a while I too. There were three persons in the office, a general and two colonels. They read out the verdict: "According to the investigation, you are accused of anti-Soviet agitation and are sentenced under article 58.10 to a term of five years in the Gulag [prison camp]." As they read this out, I was under the impression that they made an effort not to look at me. They could not possibly fail to be aware of what unjust and dark deed they were carrying out.

Next day some of the women, among whom were also I and Olga Nikitishna, were called up with belongings to be transported to another place. We took our belongings, and then we were pushed into a van crammed full of common criminals. We were brought to the distribution prison in Krasnaya Presna,[21] where we were put in a large square hall referred to as "Railway Station." Outside it was freezing, the hall had no windows, it was stuffy and stinking. Lying directly on the floor were women, packed densely, some of whom were clad only in underpants and bras. Many of them had their clean-shaven heads all over daubed in green disinfectant. In a corner, on a free space of about 2 by 2 meters, sat a red-haired woman painting her eyelashes. We took out all the food we had brought, and it was immediately snatched away, leaving us only a small amount.

In the morning I realized that my pocket that contained my only warm garment, my mittens, was empty. I turned to the red-haired woman, asking her to return them to me. "They are the only thing left to me from my mother," I said to her. An hour later, the mittens, as though by themselves, appeared at my feet.

---

[19]"Black Raven" is the Russian for "Black Maria." -WvS
[20]Infamous Article 58 of the RSFSR Penal Code was applied in political cases to charge those accused of counter-revolutionary activities and crimes against the state. It included 14 clauses: from high treason to anti-Soviet agitation and propaganda.
[21]Krasnaya Presna is a district of Moscow. -WvS

The next day we were transferred to cells, where the inmates were a mix of "blatnye"[22] and those sentenced under article 58. First of all, we had to arrange line up in formation, then our cases were checked and our profession established. The *blatnye* liked the sound of my profession and from then on always called me "Architect." There was nothing to do in the cell, and I began telling fairy tales and stories, of which I knew a lot, and that earned me the image of authority.

My mother wanted to bring me warm clothes but could not come in time. Before we were sent off to the camp, we were given some torn great-coats and worn boots. That helped me out a lot since the outside temperature was minus 30 to 40 centigrade, and I was dressed only very lightly. With Olga Nikitishna we were trying to keep closely together. When we were loaded into the railway carriage, the red-haired *pakhan*[23] gave the command: "Space for the Architect!", which meant that we could establish ourselves on a higher-level bench where it was a bit warmer. During the night, we woke up from loud screams — the carriage was full of smoke, some of the women were vomiting. My first thought was of gas chamber. But it turned out that a guard, walking on the roof, thought of a practical joke: he covered the ventilation shaft. Smoke and carbon monoxide filled the carriage. Of course, there were many gaps in the walls so that there had remained some ventilation, but a few of the women had been poisoned by the monoxide. Luckily, Olga Nikitishna and I got away with no more than a headache.

We were transported to Nizhniy Tagil. As soon as we were on the territory of the camp, we had to line up in formation, hand in our things for disinfection, then line up again. Then complete check-up: case, article, surname, year of birth, profession. After this I, together with another woman, Zhanna, a Lithuanian, who was sentenced under article 1A, i.e. treason, were detailed for the medical unit. After the transport and the running around in the cold, the place in the medical unit, under the staircase, seemed like a corner of paradise.

Next day we got transferred into the common barrack on bunks. I was given the task of doing exercise with children who were starving. It was the first time I had seen anyone suffering from dystrophy. They were living skeletons. Most of all I was shocked by their complete absence of buttocks.

I persuaded Zhanna to say that she was a technical draftsman. On

---

[22] "Blatnye" is, in Russian, the slang for common criminals; the singular is "blatnoi" for a male and "blatnaya" for a female. -WvS

[23] A "pakhan" is a recognized gang leader in the criminal world. -WvS

the next day we were both moved to the projects bureau. The projects bureau was a zone within the zone; a separately enclosed compound, along the fence the same watchtowers with soldiers. The compound was divided into two halves — in one half the living quarters of the men, and in the second half the design bureau. There were few women in the bureau, and they were brought for work from the common barracks. The living conditions for the workers at the technical bureau were slightly better than for the other inmates. In addition to the usual prison soup they sometimes got a piece of omelette, fish, sugar. The beds were not wooden planks but netted. For some reason the common criminals called the workers of the bureau "Mister Boa constrictor." The Head of the project bureau was a certain Khanap Issakovich, to us just Khanap. Not a bad person, generally speaking, but terribly afraid of everything. He tried to appear strict, but did not succeed too well in this.

All women, except myself, were just copying, and they accepted Zhanna without fuss into their collective. But to me they were not so kind since I, as an architect, was permitted to move freely around the camp. All the people in the technical bureau had been sentenced under article 58.10, except Kuzma. He was a common criminal, but had a great sense of humor. He had a strip of carpet at his disposal which he would lay out when the bosses were coming. I remember it quite clearly: it was red with a green border. When it was laid out, the barrack looked pretty straight away.

Like me, in the building group, there was another architect, Zhenya Kirillov, a very talented artist. He made colored-in pencil drawings from photographical portraits for all the bosses. Naturally, he was a favorite and enjoyed additional privileges and liberties.

In this project bureau I got to know Georgy Aleksandrovich Rerikh, with whom we became friends. Also with a Pole, Cesar Turski, a zealous catholic who kept himself separate from all others.

I was straight away given the task of planning the project of a park of culture and rest for the city of Tagil. As preparation for the work, I had to visit the intended place of the park. As a convoy, I was given a very nice young soldier, who for some reason was shy of walking close to me and as soon as we were outside the camp, he kept himself at a distance from me. At that time, there were many German prisoners of war in Tagil. When they saw me through the fence, they would wave their hands and shout: "Frau, Frau!!"

Afterwards I had to go many times to the site of the park. As soon as we got out of the camp, my convoy would give me a time and location for a meeting and would leave on his own business. These few hours of freedom were to me an essential breath of fresh air. The sun, the grass, the leaves — they all felt sweet and kind. It is impossible to express in words my feelings at that time.

Zhanna worked in the project bureau only for a short time. Before long it was realized that she had been sentenced by article 1A, and she was transferred to common work. Her story was quite interesting: she spoke perfect German, and when the Germans occupied Lithuania, she managed to get work at the Headquarters of Rosenberg.[24] She stole some secret documents, and got through with them to the resistance fighters. But instead of a getting a good reception, she was accused of treason and betrayal of the motherland.

In our camp life, there were also occasional funny situations. In the summer, when it was particularly hot, we went during a break outside our barrack and lay down to sunbathe. The soldier on duty on the watchtower thought that it was scandalous that we were having a rest, joking and laughing. He raised the alarm. Before long Khanap came running and sent us back inside, saying: "Inmates are not supposed to sunbathe!"

Once we got hold of the journal "Ogoniok"[25] with a description of the Fascist legal procedures: a special council, three people reading out the verdict, camp sentences of three, five and ten years, and other characteristic features. Reading this we were discouraged.

At the end of summer, we were told that we would be moved to another camp. Zhenia Kirillov's special position cost him dearly: his end of term was coming up, but the bosses didn't want to part with him and gave him an extension of his sentence, and so he could not come with us.

In the transport were most of the workers of the technical bureau. Men and women were separated. At first we were going by horse-drawn carts, then in relatively goods trains. In the Sverdlovsk transition prison I was amazed: in the large cell there were small children running about, pale like potato shoots. They were the families of people convicted of treason and of those who had been taken prisoner and had not returned [from Western Europe]. They were being transported up north into exile. I had a small supply of vitamins, and handed them out to the children. With great joy they grabbed them!

---

[24] Following the German invasion of the USSR, Alfred Rosenberg was appointed head of the Reich Ministry for the Occupied Eastern Territories.
[25] The most popular general magazine in the USSR.

From the Sverdlovsk transition prison we were sent to Arzamas; there we were put in a large, clean and bright cell for five inmates. My mates in that cell told me that I didn't know real camp life, because I was always getting privileged conditions. Maybe. I am very grateful for this to my father, who insisted that I should acquire a profession. In Arzamas we stayed for three days, and then the transport with the men from our bureau arrived, and we moved on to Shtokov. From Shtokov we went by narrow-gauge train to our final destination — the secret town of Sarov (also called Arzamas-16), which I, like Yakov Zeldovich,[26] frequently just called "Ensk."

Chekhov said that, if there was a shotgun hanging on a wall, then that shotgun had to fire a shot. I remembered scenes familiar from childhood: hanging on the wall above my brother's bed were pictures of Sarov monastery and an icon of St. Seraphim feeding a bear. And now I was in Sarov myself.

I was struck by the seething activity on the staircase of the building where we had our barrack. Well-dressed beautiful women kept going up and down. They were mainly from the Baltic countries and from Poland. A group of nuns was staying in the same barrack. They considered that they had been struck by the wrath of God, and so they did not grumble, but they refused to work on religious holidays. For this they were put in the cooler where they would sing and pray. I think that in the end they were left in peace.

Compulsory in every camp was the roll-call in the morning and in the evening: everybody was lined up in formation, then the sentence articles and the names were read out and the prisoners carefully counted. Sometimes they would lose count, and then they started over and over again. Then the dismissal, which was a tragi-comical sight: a brass band played, columns were lined up, the foremen were running around with boards [showing daily tasks]. As the dismissal took a long time, some of the inmates would pee at the spot, as the music played. Then the command was given to make two steps to the side, a gun-shot sounded, and the columns moved to different locations for common work.

At first, we were led to the mechanical plant, where the builders had allocated us room, and we started work on the reconstruction of one of the buildings into a guest-house and the finishing of the laboratory

---

[26]Yakov Zeldovich and Andrei Sakharov were instrumental in the development of the Soviet Union's nuclear bomb project. They were in charge of the theoretical aspects of the project.

and design buildings. I worked again with Georgy Rerikh and we spent together all lunch breaks. I called him George.

To my misfortune, one "pakhan" (a kind of gang leader), called Ivan, had taken a fancy to me. I was beginning to be chased. George tried to talk to Ivan as man to man, but that didn't stop him. Fortunately, this came to the attention of the administration, and Ivan was sent away. I was saved!

My job is my happiness — I always liked it very much. After a couple of months, George came up to me and said that he had been called to the "sly house" (the house of the operatives) and was invited to cooperate. He said: "If I don't agree, then I'll get sent to another camp, and to lose you — the only life that I have — is unbearable!" I replied, that that was our fate, but to agree to be a squealer was impossible, even under threat of prison. George listened to me and refused, and soon he was sent away.

The five-storey house that I had designed was being built. I was moved to another camp site where I had an office. I was also given an orderly, an old woman from Byelorussia, who was imprisoned for having given a drink of water to a *banderovets*.[27] I had a lot of work. In these three years I created a plastics workshop, taught a team of decorators, made many wall-paintings. In Sarov there were not only prisoners, there was also an institute, referred to as "Object," where nuclear physicists were working. I decorated the cottages where they were living, and separately the house of general Zernov, the Chief of the Object,[28] which was also used by visiting bosses.

A group of project designers arrived from Leningrad; their head was Georgy Aleksandrovich Zimin. These Leningraders were friendly to me, but as an inmate I was not allowed to be with them all the time. Together we worked on the reconstruction of a church, converting it into a theater. I worked on decorating the theater and on the conversion of the refectory into a restaurant.

The refectory had a dome of about one hundred and fifty square meters, which I decided to emblazon. I remembered the hall in Pavlovsk, near Leningrad: the sky, trees descending, a balustrade showing at the side. I decided to emblazon the dome with a similar design. I painted the sky without the balustrade, because I was afraid of getting the perspective wrong. In the end I had the sky, clouds, branches hanging

---

[27]"Banderovtsy" were members of a movement of Ukrainian nationalists, in armed opposition to Soviet rule in 1940–1959. Named after their Leader, Stepan Bandera. -WvS

[28]Pavel Zernov (1905–1964), Director General of Arzamas-16 Nuclear Center since 1946.

down on three sides. Along the perimeter of the dome I made a cornice-soffit for lighting in the evening. To create the sky, I divided the dome into five sectors, mixed five color shades, and got each painter to work on his sector. Then I used a brush to blend the joins. As a result, the dome was painted in colors from sky blue to a hazy lilac. I painted the trees in three planes, making the oil paint semi-transparent, like watercolor, adding to it whitewash and paraffin. Finally, I painted swifts flying all over the sky.

When the scaffolding was taken down, I was so nervous that I ran away. I couldn't look at the painting immediately. Then somebody came running, shouting: "It is good! It is beautiful!" Pupils were led to the theater for excursions. Later on, the restaurant was converted into a concert hall, but my wall-painting on the dome was left intact.

In the camp there was a system of credits: for good work and behavior the prison term got reduced. Some time before the end of my sentence, I was joined by a radiologist from the Object. She had come by contract but she didn't like it in Sarov. Her husband and son stayed behind in Moscow, and she wanted to cancel the contract, but that was not accepted, and since she insisted, she got a prison sentence.

Even while I was still in the camp, I violated the regimen: I went to the cinema when there was a film about Czechoslovakia was shown. I wanted to see Ludvik's eyes — he was at that time Minister of Defense. Alas, although he was shown in close-up, he did not look even once at the screen. I thought that he was hiding his eyes from me, since he could not possibly be ignorant of what had happened to me.

There were some comical cases during my time in prison. The Head of the 2nd Department did not like me. She had the task of forming the transports away from Sarov, and several times she tried to include me. But all lists were seen by the Head of Construction, Anisimov, who was keen to keep me, since I was a leading specialist, and he always crossed me off the list. I had a pass but had to be present in the camp for the roll-call in the morning and in the evening, and stay in the camp overnight. But there were situations when I was kept late with work, and then my superiors phoned the zone and warned the guards about my coming late. Once in such a situation, general Zernov ordered me to be taken to the camp gate by his car. Imagine: the car of the Head of the Object arrives at the gate, the sentry comes out with military salute, and suddenly a convict gets out of the car. A minute of consternation. I walk through the gate showing my pass, and behind me I hear loud laughter — they too thought it funny.

The last Head of the camp was Morgunov — a big, sympathetic man with a kind face. He would say: "All of you are my slaves." The "slaves" worked hard: they washed him and cleaned and shaved him, mended his clothes. Admittedly, he walked by himself. But he could have demanded to be carried on a litter.

Then came the end of my time in camp. From a camp inmate I turned into a free worker with a obscure background. My liberation took place by my moving out of the camp into the house of a copyist, Varya. I had now a little spare time, and I used it to play tennis; I always liked sport. Once, when I came to the court, on the neighboring court a man was playing, who was not tall, stocky, very agile, bespectacled. When I arrived he literally stared at me. I must say that, although he was not of an impressive appearance, his eyes were very special. They sparked and shone and gave the impression that they emitted a shower of stars. I don't like being stared at, and so I quickly collected my things and left the court. But unfortunately, I forgot my balls on the bench, and they were in short supply. I had to go back. At that moment that person had already changed and came along with me. Our conversation did not go well — I was cross and he was shy. On our way we saw a motorcycle with a side car, and I said how great it would be to have a ride on a motorcycle.

In the morning, when Varya and I got up and opened the window, we found a big bunch of flowers under the window. "Your beau is putting it on!" exclaimed Varya, but I was pleased. Sometime that day a motorcycle stopped under the open window of my office, and that very man was sitting on it, Yakov Borisovich Zeldovich. At first I thought that he was a domestic worker. He offered me to mount the bike for a spin. I agreed, of course, and we went. But we did not get very far: the engine spluttered and stopped on a country lane. At that moment a herd of cattle was passing by. Yanik, as I called him, looked at his watch, hopped off the bike and ran off, leaving me alone surrounded by the cows. The herd passed, and I got off the bike and left, leaving the bike on the road. At first I thought that it was strange and hurtful, but then, at work, we all laughed at this happening.

A few days later Yanik came and gave me a letter that he had written in Moscow. Apparently, he had run off because he was late for his train. The letter was lovely, full of love and warmth. That evening we went for a walk. Yanik said that he, like Thumbelina (he was shorter than I), wants to warm the heart of a frozen swallow, such a big and beautiful bird. The words, full of warmth, worked on me like a healing balsam.

I had contact with the outside world only through my mother. My sister Tatyana was married to a party functionary who hated me. My sister Shura was the wife of the principal architect of "Metrostroi,"[29] to whom an imprisoned relation was not a credit. Basov's parents too did not write to me. From my mother I knew that my son, Seriozha, lived with them, went to school and studied well.

A while ago I had been acquainted with Orypiansky, the director of the "Proletkult" theater in Moscow. He invited me to work as an artist and I did three shows with him (decorations and costumes): "Rokovoe nasledstvo" (Fatal Inheritance), "Osobniak na naberezhnoi" (Mansion on the Waterfront) and "Bez viny vinovatye" (Guilty without Guilt), after Ostrovsky.[30] And after that play I myself became guilty without guilt.

Let me go back to life in Sarov. Yanik dearly wanted a child, but I thought that that was already impossible. And then suddenly I started having sharp pains. At first I thought that it was appendicitis, but then it turned out that I was pregnant. Yanik and I went everywhere together: to the Zababakhins,[31] the Negins, to the theatre and cinema, and almost every evening we went for walks along the river. On one occasion, when we returned home late in the evening, somebody shone a torch at us, and a bass voice said: "Ah, one of us." That was how I saved Yanik from bandits. We also went into the forest, to a beaver brook, hidden among the trees, where the beavers felled trees and built dams. On the 2nd of May 1950 we were swinging on a birch tree that was lying across the stream and fell into the water. Yanik later on wrote this up as a short poem: "On the river, whiter than white, a birch tree tricked us." It was quite warm, and we undressed and dried our clothes, and then we ran to his home at the Victory Street and drank vodka with orange juice. That was when I got pregnant.

One evening, in May 1950, I was at the theater and suddenly everybody started congratulating me. I did not understand why they did that. Then Yanik arrived and explained: he together with Khariton[32] had created the atom bomb and had been awarded the order of Heroes.

---

[29] "Metrostroi" was the organization in charge of construction of all underground railways. -WvS

[30] Alexander Ostrovsky (1823–1886) was a Russian playwright, generally considered the greatest representative of the Russian realistic period.

[31] Yevgeny Zababakhin (1917–1984) was one of the chief designers of nuclear weapons in the USSR, who worked at KB-11, see page 229. Evgeny Negin (1921–1998) was one of the chief designers at Arzamas-16.

[32] Yuli Khariton was the Chief Nuclear Weapon Designer since the inception of the atomic bomb project by Joseph Stalin in 1943. In this sense one can compare him with Robert Oppenheimer in the American project. He remained associated with the Soviet program for nearly four decades.

Yanik said that he would be given a *dacha*. He wanted it to be in the Crimea or in the Caucasus, but Stalin made it in Ilyinsk, close to Moscow, which was also not bad.

Love, interesting work, a home with Yanik, and I was beginning to spread my wings. I got permission for my son Seriozha to join us, and Yanik brought him. But then a catastrophe happened. While Yanik was on one of his trips to Moscow, I was summoned by Shutov, the Head of the MVD of the Object, and he suggested that I should cooperate with him. I refused and he said: "Think about it, you are expecting a child, and we can send you to a very faraway place, where even Makar does not send his calves.[33] What you have suffered previously is just a little foretaste."

In the evening, when I went to bed, I thought that I'd have to agree — not for my own sake but for the sake of the child. But in the morning, when I got up, I understood that I could not square it with my conscience. Shutov kept calling me several times, but then he got angry and said: "Blame yourself!"

Yanik returned. He noticed immediately that something was wrong with me. But I had signed a statement that I would not disclose what had happened, and therefore I did not say anything to Yanik, putting it down to ill health. We went for a walk. The sky was clear, we looked at the stars, and suddenly I saw a shining tree stump. I had known previously that rotting stumps shone in the dark, but what I saw was beyond expectation. The stump was glowing and around it was a blue and yellow irradiance. I filled my pockets with мању-many pieces of the shining stump and we went home. Yanik took me home, but then went to visit the Zababakhins. In the early morning I was arrested.

I was sobbing bitterly, lying on the bunk in the zone where I had been taken. The women comforted me, they all knew me. In the morning came Seriozha, brought me a bag from Yanik. It contained my blue gloves, a scarf and 5000 rubles. On Seriozha's back the shirt was torn showing a birthmark. Only one thought went through my head — who will now mend the shirt? I was hoping that Yanik wouldn't abandon Seriozha.

Whereas the first time my verdict was read out to me by a *troika* of judges, this time nobody said anything: armed people came and took me away. I was in the zone for two days, but neither Zeldovich nor anyone else of my acquaintances came to see me. They were afraid.

---

[33] The Russian idiom meaning a hellish place.

Only Seriozha came. Two days later we were loaded on a train, men and women together. Before we left, Shutov came to the carriage and, looking at me, took one of the men off the train. What pain weighed on my heart! I kneeled down, pressed my brow against back of the carriage and remained like this, sobbing, until the train departed. Nobody touched me. I was in the same coat that I had worn last night, with the pieces of the rotten wood in the pocket which had stopped shining.

God punished Shutov. Soon after I had left, his only daughter, a radiant beauty, was killed by lightning. He himself was dismissed from his job.

The transport was long and hard. At changes and stops no sentences were read out to us, only a head count was done. Life in the carriages was busy, some people were travelling in whole families, even including children. The common criminals are never bored in camp. Somewhere they got hold of a record player and kept playing records. When all the food that had been brought from home was finished, one of the women decided to tip out my bag, and my five-rouble notes fell on the floor. Amazingly, all the money was picked up and put back into my bag.

I always wanted to see the Baikal, and now we were going along its shore. We were moving on the trunk-line — the only way. It was a small embankment among endless swamps, and here and there one could see prison camps. A terrible sight. My eyes never dried up from tears. To comfort me, some women found a needle and a piece of black cloth, pulled threads of different colors from undershirts, and I tried distract myself by doing embroidery.

We were brought to Vanin Bay, and then across the Okhotsk Sea to Magadan.[34] Part of the way we moved in complete darkness, since the Japanese were shooting at Soviet ships. In Magadan we were put for a few days in a transit camp. And then by road on trucks. Some of the people were left at intermediate sites, but the most dangerous criminals were taken to the place, in Shutov's words, "where even Makar did not take his calves." Thus, I found myself at some 1000 kilometers from Magadan on the bank of the river Khan-Galas, at the gold-mine settlement Distant, district Winter. The earth was already covered in snow.

The slopes of the ravine in which the settlement was located were covered in fur trees, quietly gurgled a river. One gets the impression

---

[34]Magadan is a port town located on the Sea of Okhotsk (the most north-eastern part of Russia) serving as a gateway to the Kolyma region.

that this small corner is cut off from the whole world, forgotten and lost. In the settlement there were just over a dozen houses and an office. Apparently there were Germans living there before us: cut out on the walls and tables were words in German. The houses were block huts made of pine logs, with small single windows, but well-wrought. The doors, without a porch, give directly out onto the street.

At first we were put in a large common barrack which, apparently, had been a dining hall. From the ceiling hung colored lamps, and on the wall at the end — a portrait of Stalin. It all smacked of profanation. Next day we were distributed into the blockhouses. I was put in a room of 12 square meters together with Maria the Romanian, who was also pregnant, and a bandit Lyoshka with his wife. In the second room of our blockhouse was the bakery.

We were warned that we were not allowed to leave the settlement. We had no documents and nothing was explained to us. Soon we were detailed for work. Most of us had to wash gold. I was given the task of economist: at the end of the working day I had to receive the washed gold, weigh it, seal it in special pouches and lock it in a safe. For our work, we were paid. There was a shop in the settlement where we could buy bread, canned food, cereals and butter.

I was terribly depressed by overpowering sorrow. It seemed that there would be no return to normal life. I was frequently ill. In the settlement I met a Polish Jew, Leva, who like a friend kept my spirits up, made every effort to console and help me. Yanik sent me letters, full of love, telling me of his efforts to get me to a different place, and also support me materially.

A telegram at the start of winter told me the sad news of Mother's death. The sun did not rise in winter higher than the surrounding hills, my sadness grew. I had the feeling that my life had come to its end. Only the tiny life that I was carrying under my heart would bring me back into reality.

It was already difficult for me to work and I asked the head of the section, Kozlov, to grant me unpaid leave. He said: "All right, I allow you not to come to work." I stayed at home on the following day, but in the evening the head engineer of our section rushed in with the words: "You must immediately write a formal request for unpaid leave and take it for Kozlov to sign! In a couple of days we will be inspected by a commission, it will investigate all those who do not work and pass sentence on them. I have seen your name on the list." I wrote an official request and went to the boss. He made a face, but he did sign his agreement. I was saved again! The commission did arrive, several people who did not work were indeed sentenced and taken away.

January 1951 came. On the 11th of January my contractions began. The outside temperature was $-52°C$, the stove had to be kept going non-stop.[35] There was no hope of getting medical assistance, as cars could not run (the lubricant would freeze at that temperature) and the township had no medical centre of any sort. I had prepared a piece of silk thread in advance, keeping it in a sterilized jar. A neighbor came to help me, the wife of Lioshka the bandit, who at some stage had worked in a military field hospital. I could hear the swearing of the bakers through the thin wall, I could see the snow lying on the floor, I could feel my hair freeze to the wall. At last the baby cried and life became easier. We measured the umbilical cord using my technical ruler, we cut it with scissors and tied it securely with the silken thread I had ready. My girl was quite a large baby, her birth had caused many tears in me, but there was no-one to put this right.

I named my daughter Anna, I felt that this was a strong name and that it would bring her luck. All my thoughts and all my care henceforth centred on her. On the third day, the cord dropped off. My milk never came. I was appalled, as there was absolutely no way of getting baby food in the area. Liova came to my help — he advised me to dilute tinned condensed milk with water and feed the baby with this mixture.

After a while I was moved to a separate house, Maria asked to remain with me, so I took her as well. I was also allowed to hire help, an old German man. Soon Maria too gave birth and was allocated a room of her own, and I remained in the house with my daughter and the old man. Our workmen built a pram as a present for me, using local wood, and Annushka was taken about in it.

Summer in the Extreme North starts in June and ends in mid-August. The temperature sometimes is unbearably hot — up to $30°C$ — but the soil does not get warm, due to permafrost. The worst feature of summer is mosquitoes — they are horribly numerous and they are fierce, much worse than in a moderate area. The old man and I took turns to sit at the baby's pram and chase the mosquitoes away.

Nature is bountiful even in the Extreme North. At altitude, on the flat areas of the volcanoes, everything is covered with a carpet of

---

[35] The temperature $-72°C$ in the manuscript (rather than -52 above) seems to be a misprint.

blueberries, one cannot avoid squashing berries when walking across. Cranberries, blackcurrants, cloudberries grow there as well. Mushrooms too, but they are completely different from ours.

At the end of August I was told to get ready to move. Where to, what for — I had no explanations whatsoever. I could only hope that nothing would be worse than what we had. To start with, I had to go to the main mining base. Liova got permission to accompany me. We were brought to the river, we were rowed across it, but there was no car on the other side. The temperature had already begun to fall below freezing, I stood there, holding my baby, and did not know what to do. Luckily, Liova was with me. He walked to the road and got us a lift from a passing vehicle, so we were taken to the mining base. Liova stayed with us up to the moment when I was sent off — on a lorry, guarded by two soldiers. But this time they were not guarding me as a criminal, they were instead protecting me against criminals (bandits) who had become quite numerous about the road.

The journey covered one thousand kilometers — quite a long way. We would stay at night in barracks built for local drivers. These were sullen unshaven men. But when I would unwrap little Anna, when she would happily wave her little arms and legs, their faces always softened. One of the mining engineers — who knew the conditions prevailing at the Magadan center for sorting prisoners, gave the address of a local woman who would take us into her home. Once in Magadan, the soldiers helped me to find this address, then they handed over to me both my passport and the official pass which allowed me to go by air through Moscow to a town called Vytegra. The Far Eastern Zone was still a "closed" area and one could not travel without a special pass.

When I entered the house of the woman to whom I had been sent, she was terrified and stammered. "I cannot take you in, you have a baby! Or maybe my neighbor might do it ..." The neighbor turned out to be a large and brave woman, a former professional criminal. When I offered to pay her, she immediately agreed to take me into her home. I was aware that travel to Moscow was not cheap, but I did have the six thousand roubles that Yanik had sent me. I immediately handed this money to my landlady, asking her to hide it securely, and I also gave her some of my clothes as a present.

Next day I bought a ticket for Khabarovsk and also registered Annushka's birth at the local Registry Office. She was already eight months old, but I had not been able to register her birth earlier, simply because there had been nowhere for this to be done. It was only when I had already entered the Registry Office that I discovered that I had not brought Annushka's adoption papers which Yanik had sent — I had forgotten them in Dalny. That meant I had to register her in my name only. I left the very next day and my landlady said as we were going — "That was a good idea of yours to hand me your money for safekeeping straight away — otherwise I would have stolen it."

Among the passengers of our plane I met the surveyor of our mine and his wife — they had been volunteers for the job. On arrival to Khabarovsk we went together to the hotel. They were immediately given a room, but the reception clerk told me: "We have no right to let you in." But she saw the tears in my eyes and the baby in my arms and added: "Let us say that I cannot see you. Go to your friends, you may spend the night in their room." In the morning I bought a plane ticket for Moscow. I had no things of my own, I only had Annushka's things, her baby bath and her pram, but I had to leave even these behind at the airport, because there was no one to help me.

The flight to Moscow took six days and six nights. In those days the rules were for passengers to change planes every eight hours, because the pilots were assigned to their own machine. All the other passengers were military personnel. They looked askance at me, not one of them offered me any help. At the time there were no facilities on aeroplanes, I had to tie my daughter to me to prevent my dropping her if I dozed off.

I arrived in Moscow in early September and stayed in the flat of my sister Shura. I unwrapped Annushka — she was all floppy. I panicked and immediately called a private doctor. She understood straight away what the trouble was: a sharp change of climate. There is a legend in the North saying that one may not take children away from there, because there are no infections in that cold climate, therefore the child does not develop any immunity to them. A voyage by ordinary surface transport lasts three months on the very least, and children easily become ill in these conditions.

A policeman came on the following day and said — almost as if he were asking me to forgive him: "You must understand that I am only carrying out my duties, you are not allowed to be in Moscow, so leave town within 24 hours."

*****

A Fragment from "Memoirs" by Andrei Sakharov [4]:

One evening in the spring of 1950 on my way home from work, I caught sight of Zeldovich. The moon was out, and the bell tower casted a long shadow on the square in front of the hotel. Zeldovich was walking deep in thought, his face somehow radiant. Catching sight of me, he exclaimed: "Who would believe how much love lies hidden in this heart?"

In many respects, the Object was a big village in which nothing remained secret. I knew Zeldovich was having a love affair with Shiryaeva, one of the prisoners, an architect and artist by profession. Her husband had renounced her after her arrest on charges of anti-Soviet slander; such stories were common in those days. It was Shiryaeva who had painted the murals in the VIP dining room, in our theater, and in the homes of the Object's bosses. She had been granted trusty status, apparently as a reward for her services. A few months after our encounter in the moonlight, Zeldovich woke me in the middle of the night. Romanov, in the other bed, looked up for a moment, but then turned over and remained silent: he never asked questions. Zeldovich was agitated. Could I lend him some money? Fortunately, I had just been paid, and I gave him everything I had. A few days later I learned that Shiryaeva's term had expired, and that she was being sent to Magadan, far to the east, for "permanent resettlement." Zeldovich managed to get the money to her, and after some months I learned from him that Shiryaeva had given birth to their daughter in a building where the floor was covered in ice an inch thick.

Zeldovich managed to obtain some improvement in Shiryaeva's situation, and twenty years later, at a conference in Kiev, he introduced me to Shurochka,[36] the daughter born in Magadan. She looked amazingly like his other daughter by his wife, Varvara Pavlovna. (Zeldovich had a number of other affairs — too many — most of them strictly sexual liaisons. I don't like some of the

---

[36] Andrei Sakharov apparently meant Anna, Annushka.

stories I've heard.) He dreamed of someday bringing all his children together. I hope he succeeded. Time is a great healer, provided there's complete honesty.

*****

References

[1] Boris Ioffe, *Atom Projects, Events and People*, (World Scientific, Singapore, 2017), page 59.
[2] Olga Shiryaeva, *A Life Story*, Ed. Alina Chertilina, http://samlib.ru/c/chertilina_a_s/istoriaodnoj.shtml, in Russian.
[3] *Under the Spell of Landau*, Ed. M. Shifman, (World Scientific, 2013).
[4] Andrei Sakharov, *Memoires*, (Alfred A. Knopf, New York, 1990), page 134.

Chapter 12

Rudolf Peierls:
The Lesson of the Fuchs Case

R. Peierls to Niels Bohr[1]

Birmingham
February 14, 1950

[...] There is no doubt that this whole [Fuchs] case will have disastrous effects, quite apart from personal relations on the political atmosphere and the positions of scientists both here and particularly in America. It is, of course, quite illogical if all security clearance and investigations have missed such a case to seek a remedy in submitting people to further checks and clearances. Nevertheless this will, of course, be done. We are beginning to wonder whether the real lesson is not that it is impossible to maintain secrecy in a project involving so many people without creating the atmosphere of a totalitarian country in which everybody is ready to suspect his best friend of being an informer. Russia has found how to stop leakages very effectively. If this is the only effective solution do we want to go that way ourselves or should we not say that at that price security is not worth having. [...]

*****

---

[1] Sabine Lee, Volume 2, page 216.

Memorandum by Rudolf Peierls:
The Lesson of the Fuchs Case[2]

[Birmingham,
Around March 1950]

To all those who were associated with Dr. Fuchs during his work on the atomic energy project the disclosures at his trial have caused great distress. One could wish nothing less than to go on talking about this, particularly in public. However, the case will do such serious harm and there seems to be so much contradiction and confusion about it that I feel it necessary to write up the picture as I see it. The main point will be the conclusions to draw (or not to draw) and I shall describe the past events only as far as they have a bearing on this.

For me the story starts in 1941 when a small team was then working on atomic energy in this country. I was mainly responsible for theoretical physics and more help was needed on this side. Most people of suitable ability were then already on high priority work but when I heard that Fuchs was available I knew he was a man of the right scientific qualifications. I knew he had left Germany because of his opposition to the Nazis and I respected him for this. I knew of his connection with left-wing student organizations in Germany since at that time the communist controlled organizations were the only ones putting up any active opposition. It was natural for a young man who wanted to fight the Nazis to work with any available allies, as indeed this country did later during the war.

Approval for his appointment had, of course, to be given by the authorities. I do not know what their methods of investigations were, and what was disclosed, but I assumed that they had to weigh the value of his help (at a time of great shortage of scientific manpower) against any risk of his having retained from his early contacts in Germany (8 years earlier) a loyalty to a party that owed allegiance to a foreign power.

During all these years we saw much of him. Shy and retiring at first he made many friends and in many conversations of politics was, of course, a frequent topic. His views seemed perhaps a little to the left of ours, but he seemed to share the attitude to Communism — and to any

---
[2]Sabine Lee, Volume 2, page 219.

kind of dictatorship — of most of his friends. I remember an occasion when he talked to a young man who was in sympathy with communism and in the argument Fuchs was very scornful of the other's dogmatic views.

When I heard of his arrest I regarded it as quite incredible that anyone should have hidden his real beliefs so well. Looking back it seems that at first he shared in the life of his colleagues and pretended to share their views and attitude only in order to hide his own convictions. But gradually he must have come to believe what was at first only pretense.

There must have been a time when he shared one attitude with his colleagues and friends arid another with the agents to whom he then still transmitted information, and when he was himself in doubt which of the two was conviction and which was pretense.

I do not want to enter into speculations about the state of his mind during all this time. Some have described it as an abnormal case of a split personality, others tend to regard it as a superb piece of acting, but either way it is certainly quite exceptional.

In the past his close friends were mostly amongst people who shared his extreme views. Of course, the case for the democratic way of life must have been made to him also by many people who felt a genuine conviction for it, but apparently this had not converted him. The years spent here and in America on the project brought him more and closer association with new friends, and it is one of the most unusual features of his case that a man who was not selfish should, in spite of his position, allow these close associations to form on false pretenses. But as a result there was something new that grew on him. Nobody ever argued the case because nobody knew that he needed convincing, but he discovered that implicitly all shared principles which gave him a strength that his ideals were losing.

From his point of view this his is perhaps the most tragic: that he does not now even have the satisfaction of suffering for a cause in which he believes. But it contains a slight piece of comfort: the story has shown up a weakness in the defense of democratic countries beca[u]se the atmosphere of mutual confidence that is so essential a part of our life, makes this kind of betrayal harder to guard against. Yet, it also shows the strength of our system which in time won over such a strong supporter of a different ideology, though, in his case, only too late. Our problem must be how to reduce the risk of further cases of this kind, while yet preserving those features that make us so sure (and that ultimately convinced Fuchs) that we are right.

How, then, can we avoid further leakages: As an ordinary mortal I do not presume to know the methods of the security services but broadly speaking they can work in three ways; by "counter-espionage," i.e. by infiltrating into the espionage organization which they are trying to frustrate, by "clearance," i.e. by investigating the background of people employed or to be employed on secret work and by "supervision" of the conduct of the men on the job.

The first is obviously a good method if practicable, but one would not imagine it to be a complete safeguard in itself.

"Clearance" investigations are, of course, employed in connection with secret work. In the case of Fuchs, they would have had to probe very deeply to disclose his continued adherence to the communist cause and that would have required a depth of human insight that is very hard to achieve. Anything that could be done to raise the level of knowledge in this way would, of course, be most valuable. But, in any case, such investigations would presumably have shown that he had been a member of a left-wing organization in his youth. Should we now exclude others of whom this is found? Fuchs was German born; should one now all be suspicious of foreign born people? Fuchs was a scientist; should one mistrust all scientists? Should one mistrust all men with the initial F? The Fuchs case came as such a shock to the public that I would not blame anyone for advocating all these measures, except perhaps the last one. But we must not be under the illusion that they would bring safety. They would not even have prevented the case of Nunn May.[3] But they would have lost the country a great deal of ability. I believe that it is fair to say that if from the atomic energy teams in England and in America one would have excluded all foreign born scientists as well as those who in their youth had held extreme political views of one kind or another, the leakage of atomic energy would have been prevented by the fact that there would have been no atomic secrets. The work could not have continued effectively under such restrictions. This may sound an immodest statement for me, as a foreign born scientists, to make. But a glance at the names in the Smyth Report[4] which summarizes atomic energy work in America will make my point obvious. I am not saying that one should take no notice

---

[3] Alan Nunn May (1911–2003) had worked at the Chalk River Plant of the Manhattan Project. In 1946 he was sentenced to 10 years of hard labor for spying for passing information on the Manhattan Project in to the Soviet Union. -SL

[4] Henry De Wolf Smyth, *The Official Report On the Development of the Atomic Bomb Under the Auspices of the United States Government*, (Princeton University Press, 1945).

of the background of the people to whom one entrusts secrets. As long as there are any secrets (and all this story increases our longing for a state of the world in which they would not be necessary) it is important to judge who can be trusted with them, but one cannot insist that the precautions should be such that they would necessarily detect a second Fuchs. We are not likely to find a second person who can for years maintain the impression of being a politically inactive but generally liberal and reasonable person. But if there should be further cases of the same kind of psychology (or of equally perfect acting) they may well be people who had never openly professed communism.

Should one then rely more on supervision? The difficulty in the large number of scientists and others on secret work. To "shadow" a person day and night takes more than one investigator. Where would one find the necessary number of intelligent investigators and how does one check their reliability? Probably this method had its best chance in the atomic bomb work in Los Alamos, New Mexico, which was located at the remote spot largely just in order to reduce the risk of leakage. While the gates of this "atom city" were not actually locked, travelling by its members was discouraged and few ever travelled beyond the immediate neighborhood. We always assumed that on our rare trips we would be watched by the efficient army security services and that this applied particularly to those employed by the British rather than the American authorities.

Yet one of the charges against Fuchs relates to February 1945, a time when he was working at Los Alamos and presumably just absent to attend some meeting or collect some technical information elsewhere. If his secret *rendez-vous* could pass unnoticed in these circumstances the prospects of generally keeping all people under supervision does not look promising.

If one considers these problems objectively, one sees that as long as there are large projects employing thousands of people we cannot have absolute assurance against leakage except in one way. The governments of totalitarian countries presumably find it easy to keep their secrets and by adopting their methods we might succeed, too. If we build up an iron curtain preventing travel across the border, except in rare cases, if we suspect people who are talking to a foreigner, if we give the police the right to act on suspicion and, above all, if we build up a state of affairs in which everyone suspects his best friends of being police informers (and half of them probably are) then our military secrets might be safe, but at what price?

If this were really necessary, we would lose most of the assets of democracy including even the pleasure of convincing a man like Fuchs in the end that we are right and he was wrong because there would not be much difference.

Nobody has yet proposed such drastic measures, but the insistence that one now hears frequently on security measures without specifying them exactly and the very understandable desire for certainty that there will be no further such cases, may logically lead us in that direction.

Must we then choose between helplessly tolerating all foreign agents and becoming a police state? Fortunately things are not as black as that. Of course the authorities will continue to find out what they can about the people entrusted with important secrets and they will make the job of any future Fuchs or Nunn May as difficult as they can; they will not pretend that they are infallible. A good general knows he is bound to lose a battle occasionally.

The details of all military equipment such as tanks and aeroplanes have always been considered as important secrets. Nevertheless, no country ever succeeds to hide their main features indefinitely, but this does not even out the assets. The country with the better technical skill, the greater ability for research arid design and the greater industrial potential will still be better off because no leakage can replace the value of the right skill and knowledge of the man actually on the job. The question of the importance of atomic weapons for the future safety of this country and of the United States is a controversial one which I do not want to raise here but accepting their importance more can be gained by assuming a positive need through efficient development work and good planning than by a frustrating attempt to seal up hermetically all possible channels by which others may get to know things which, after all, they might discover for themselves.

Once fallacy that would be particularly dangerous in this context is to extend the principle of clearance to cover not merely the employment of men to be entrusted with secret work but to a wide variety of cases which it is argued that people with extreme political views might abuse their position for seditious propaganda. This is dangerous because it would lead to political discrimination and to a restriction of the freedom of expression. It is clear that in certain circumstances the spreading of extreme political opinions might be a danger, but the difference is that propaganda is something that cannot be pursued in secret. If people misuse their position to advocate their own views, this can easily be known and they can be dealt with on the basis of their actions. There is no need to suspect them in advance.

In the cases where the job is concerned with non-political matters, in particular, technical information, anybody engaged in political propaganda would, in fact, not be carrying out his duties properly and could be dealt with on that basis. In jobs concerned with the discussion of such problems as international relations of political theory or practice, it is most desirable that all views should be heard and that people should be in a position to make up their minds on a full knowledge of all arguments. This means that it is, in fact, undesirable that people should be prevented from expressing any views however extreme or unpopular, provided one takes care to balance their views by having others available who would speak for the other side. This has always been the tradition of this country and it is important that the danger of disloyal acts which, as the Fuchs case has reminded us, is serious and should not be confused with the danger of extremist propaganda which at the present time is negligible and, in any case, must be fought by argument and not by prohibition.

Chapter 13

The Peierls FBI File

After Fuchs' arrest, the Peierls family became the prime suspect. The Federal Bureau of Investigation (FBI) in cooperation with the British authorities began vetting immediately. In fact, there were two investigations — one in 1950–51, and the second in the mid-1960s. In the 1990s, under the US Freedom of Information Act, the Peierls file was declassified, and now it is publicly available.

The very fact that the Peierls family was investigated by the FBI is natural. Peierls had to be a prime suspect: he had visited Russia several times before the WWII, he hired Fuchs; he was the person who asked for clemency for Nunn May; he was on the Manhattan Project which required the highest level of clearance, and he received it. He spoke German and Russian, and, finally, his political views were somewhat left of center by American standards. It would be highly surprising if this investigation had not been conducted.

I carefully read the Peierls dossier. From my point of view the investigation was conducted rather unprofessionally. (I hasten to add, though, that I am an outsider and my perception could be biased). At certain points, I just did not know whether to laugh or cry. To my mind, the corresponding paragraphs seem like "baby talk." Below I will briefly review the Peierls file and try to explain my assessment.

Two reports on the Peierls family are collectively comprised of about 50 pages. Nowhere, in no place, can one find a slightest hint to any action on the part of Rudolf or Genia Peierls, which could be qualified as a link (let alone assistance) to foreign intelligence. The only circumstance that connected the Peierls with Fuchs was their geographical proximity.

Therefore, the thrust of the FBI inquiry revolved around two questions: was Rudolf Peierls a member of the Communist Party; was Eugenia Peierls (née Kannegiser) a member of the Communist Party. The first question was answered in the negative. Therefore, the whole burden of suspicion fell on Genia. What was the evidence?

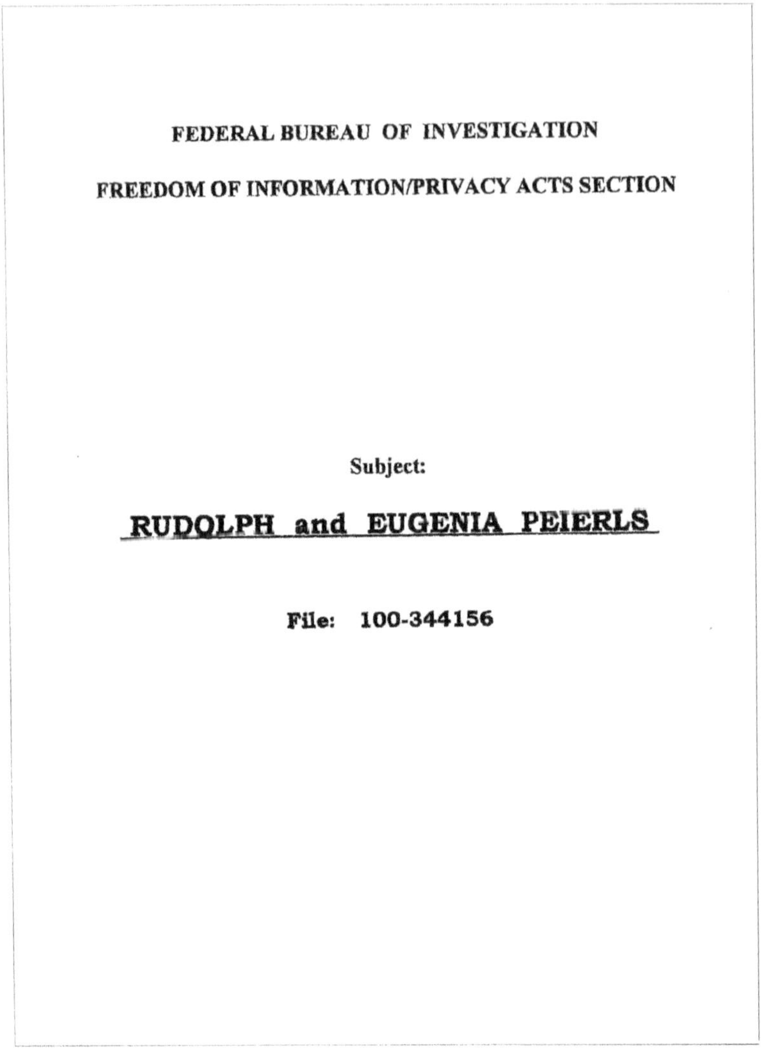

> **FEDERAL BUREAU OF INVESTIGATION**
>
> **FREEDOM OF INFORMATION/PRIVACY ACTS SECTION**
>
>
> Subject:
>
> **RUDOLPH and EUGENIA PEIERLS**
>
> **File: 100-344156**

Figure 13.1 The front page of the Peierls dossier.

Two facts caused trouble. First, in the archive of the German Communist Party, which was obtained by the US after the defeat of Germany in 1945, it was found that a certain "Comrade Kannegiesser was lecturing at a conference on Marxism in 1930,"

> Regarding Mrs. Peierls, it should be noted that captured German documents [...] consisting of original documents of the German Communist Party, listed one

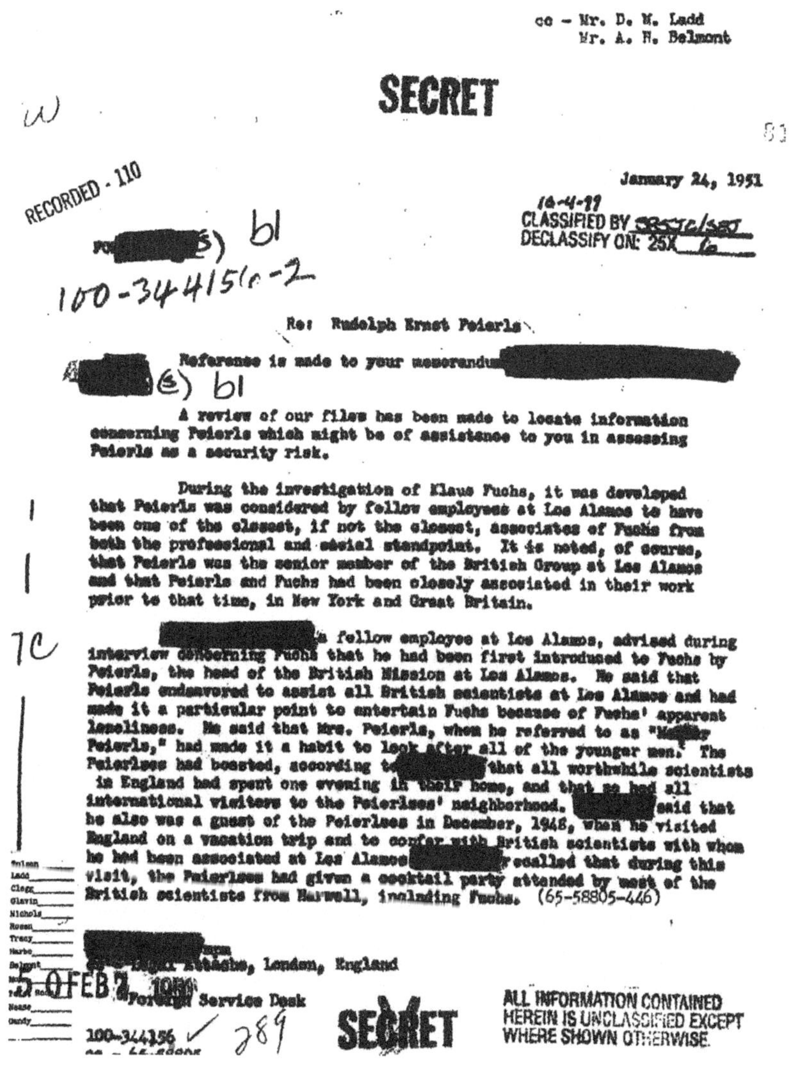

Figure 13.2 One of the pages of the FBI file. Some names are redacted.

Kannegiesser, the maiden name of Mrs. Peierls. One document, a report to the Central Committee of the Communist Party of Germany dated June 4, 1930, at Berlin, concerning the Association for Marxian Pedagogy, Berlin, 113 Wisbyerstrasse 30, reflects that on May 21, 1930, a meeting was held, pursuant to a call by the Central Committee Agitprop,[1] to consider the topic "The school —

---
[1] Department of Agitation and Propaganda.

> political situation of the present and the task of the Association for Marxian Pedagogy." This stated that Comrades [...], Kannegiesser, and [...] were present. The report was signed by [...]. It is, of course, not known whether this information has any connection to Mrs. Peierls, née Kannegiesser.
>
> The information set forth above was obtained collaterally during the investigation of Fuchs. This information is being furnished to you as of possible assistance in making your assessment of Peierls. [...] March 1951.

Neither the first name of the Comrade Kannegiesser, nor his/her gender were established. The FBI failed to establish the fact that in 1930, Genia Kannegiser was still in the USSR and therefore could not have delivered lectures on Marxism in Germany.

The second fact, which took months if not years to investigate, was that an employee of the Los Alamos Laboratory X (the name redacted) in an interview with an FBI officer told the officer that "he seemingly had heard one of his acquaintances mentioning Genia's membership in the British Communist Party."[2] Neither the circumstances of the alleged conversation, nor its timing could be recovered. X could not even remember who told him that. Later X, answering the question of the FBI officer, replied: "I think it was Y, but if not he, then probably Z."

Further investigation, outlined on many pages, was reduced to searching for Y and Z. Later still, they identified one of them (I could not understand whether it was Y or Z) and this person declared that "he never heard anything indicating that Mr. and Mrs. Peierls had pro-Communist sympathies."

As will be seen later, the FBI was not even able to establish when and under what circumstances had Rudolf and Genia gotten married and when they left the USSR. Had they tried to trace Genia's parents who stayed in the USSR, they could have found out that from 1935 to their deaths in 1953 and 1954, respectively, Genia's mother and stepfather were persecuted by the authorities, being sent from one exile to another with no intermission (see page 112 and Peierls' letter to Viscount Portal of Hungerford on page 320). Then they would probably understand that

---

[2] The primary statement of March 29, 1951, from the FBI file: "Los Alamos employee in 1944–46 told Agent [...] he knew Peierls and another member of the British Mission there, and that someone, <u>probably</u> told him Mrs. Peierls was a member of the British Communist Party.

Genia's membership in any Communist Party was highly unlikely. In this context I cannot help quoting Genia's letter to imprisoned Fuchs:

> [...] My Russian childhood and youth taught me not to trust anybody, and to expect anyone and everyone to be a communist agent. Twenty years of freedom in England softened me somewhat and I learned to like and trust people, or at any rate some of them.[3]

Of course, tracing Genia's parents in the Stalin time was unrealistic, but it was not difficult at all to read about the fate of her cousin Leonid and her uncle in the abundance of Russian press which circulated in Paris and New York. Thus, the only basis for the FBI to suspect the Peierlses was the fact that Genia was born in Russia which by itself could hardly serve as the evidence that could substantiate any suspicion.

In 1957, Rudolf Peierls' clearance was revoked and he was denied access to classified works. This was a blatant injustice towards him, the co-author of the famous Frisch-Peierls Memorandum:[4]

> The attached detailed report concerns the possibility of constructing a "super-bomb" which utilizes the energy stored in atomic nuclei as a source of energy. The energy liberated in the explosion of such a super-bomb is about the same as that produced by the explosion of 1,000 tons of dynamite. This energy is liberated in a small volume, in which it will, for an instant, produce a temperature comparable to that in the interior of the sun. The blast from such an explosion would destroy life in a wide area. The size of this area is difficult to estimate, but it will probably cover the centre of a big city.
>
> In addition, some part of the energy set free by the bomb goes to produce radioactive substances, and these will emit very powerful and dangerous radiations. The effects of these radiations is greatest immediately after the explosion, but it decays only gradually and even for days after the explosion any person entering the affected area will be killed.

---

[3] See page 242.

[4] The memorandum opens with the paragraph cited on page 19.

Some of this radioactivity will be carried along with the wind and will spread the contamination; several miles downwind this may kill people.

It was this Memorandum that prompted the British government to establish a committee to consider the possibility of making A-bombs. So, outside Germany (as we know, the German nuclear program that started earlier than the British and American one, failed miserably), they were *the first*.

Excerpts from the 1966 FBI Report

UNITED STATES GOVERNMENT
Memorandum[5]

From: Robert E. Tharp, Assistant Director of Security, Division of Security, HQ
Subject: RUDOLF ERNST PEIERLS, UK SCIENTIST
Date: March 9, 1966

[...] Information received to date has been included with information contained in Division of Security files and is summarized below.

Dr. Rudolf Ernst Peierls, Information Prior to 1951

Dr. Peierls was born June 5, 1907, in Berlin, Germany. He received his education in Germany and Switzerland and emigrated to England in 1933 because of racial prosecution [sic] in Germany. For two years he was a "Research Fellow" at Manchester University and for the next two years at the Royal Society Mond Laboratory in Cambridge, England. He was appointed to the Physics Department in Birmingham University in 1937. There is no positive information on how he met and when he married his wife. It has been alleged that he spent some time teaching in Russia sometime during 1930–39 [WRONG] and that he met and married his wife in Russia. Another source advised the FBI that he believed Dr. Peierls met his wife while attending a scientific conference in Moscow [WRONG]. In any case, based on the age of their oldest

---
[5] I indicated errors in the square brackets.

child in 1944 it appears they were married sometime in 1933–34 [WRONG]. Dr. Peierls became a naturalized citizen of Great Britain in March 1940.

In 1951, Peierls was Vice President of the Council of the Atomic Scientists Association, reportedly a Communist front organization [WRONG]. He was reported to have been a member of a delegation that visited the British Home Office on August 1, 1947, to ask clemency for Dr. Nunn May, self-admitted and convicted Soviet Agent in the Canadian-Soviet spy case who was then in prison in England.

Eugenia Kannegiesser Peierls, Spouse

Mrs. Peierls was born on July 25, 1908, in Leningrad, Russia. A Personnel Security Questionnaire filled out by Mrs. Peierls at Los Alamos, New Mexico, in 1944 reflects the birth of both her parents in Russia. Mrs. Peierls listed attendance at Leningrad University from 1925 [WRONG, should be 1926] to 1929 and indicated employment at the Leningrad Geophysical Laboratory during 1930 and 1931. From 1939 to 1941 she gave her employment as a nurse in a Birmingham hospital in England. During 1941 through 1943 she showed employment with General Electric in Birmingham as a Planning Engineer. She became a naturalized citizen of Great Britain in March 1940 at the same time as her husband.

Allegations concerning her membership in the German Communist Party [RIDICULOUS, Genia Peierls has never resided in Germany] and the Communist Party in England are discussed below. [...]

The above 1966 FBI report by and large could be viewed as the admission of the fact that the suspicions with regard to Rudolf andGenia Peierls were unsubstantiated.

The original 1951 Peierls files were declassified only after his death; he could not have seen them. But Rudolf Peierls was a true visionary and already in 1951 came to the same conclusion as to the quality of the secret services investigation as is apparent now. On April 9, 1951, he wrote to Viscount Portal of Hungerford:[6]

---
[6]See Sabine Lee, Vol. 2, pages 271–274.

Dear Lord Portal,

On thinking over our recent conversation I feel I would like to send you my comments in writing, both because this will allow me to add one or two remarks that did not occur to me on the spot, and because this will make it possible for you, if you see fit, to pass the letter on to the people who raised the matter in the first place. There are several different ideas that may have been in the minds of the security people when they drew attention to the position, and I shall comment on some of them though they may not all be relevant. [...]

The points you raised worry me because, on the face of it, the implication seems to be that of a very low standard of efficiency in the security services. It is quite correct that Professor P. is an old friend of mine, though I have not, in fact, seen him for at least eight months (outside university senate meetings, etc.). But my association with Dr. B. is much more tenuous than that. He stayed at our house when he first came here in 1939 as a very poor refugee. About 18 months ago when their child was born, my wife tried to help and advise them on their domestic difficulties. As a result of differences in outlook which became apparent then, the two ladies have not seen each other since then. My own contacts with B. do not go beyond exchanging, on rare occasions, a few words about some topical news item. I mention these thoroughly unimportant facts only because they tend to show that the information available to the security services is somewhat obsolete and not terribly significant. I would have thought there were a great many facts about me that could be made to look much worse. I would naturally assume that these were known, and regarded as innocuous, but I am now beginning to wonder whether perhaps they have not yet been reported?

Is it known, for example, that, when we were in Cambridge, we were on friendly terms with D.P. Wooster,[7] and spent a summer holiday with him and his family in

---

[7] Peter Wooster, a Cambridge crystallographer, and his wife Nora held pronounced left-wing views. -SL

1937? Or that my wife is on very friendly terms with Mrs. Betty Waddington at Cambridge, whom she has known well since about 1934, and whose views since then have shifted so far to the left that I believe she is now a member of the communist party? My wife still visits her every time she is in Cambridge, and when we go to Cambridge together I usually do as well.

Is it known that I am acquainted with Prof. P.Y. Chau of Peking, from the days when we were both research student[s], and that, when he recently visited Birmingham as a member of the official "goodwill" mission on behalf of the Chinese government, he spent an evening at our house? Is it known that, when in 1949 we arranged an exchange visit which brought a Belgian girl to our house for a few weeks, and my son later to her house in Brussels, she turned out to be the daughter of the General Secretary (or similar high official) of the Belgian Communist Party? Is it known that I am well-acquainted with Dr. Gremlin in the Physics Department here, whose name appeared on the letterhead of the committee organizing the Sheffield "Peace" congress? It is true that my social contacts with the Gremlins are not very frequent, but rather more so than with Dr. B. Is it known that I was greatly disappointed when the proposal to get Professor C.F. Powell of Bristol for our physics chair fell through, in spite of his reputation as a left-winger?

Is it known that I had recently in my department two Polish scientists who came with scholarships awarded by the Polish government,[8] and that we were on friendly terms with both of them, one even staying in our house for a few nights? I suppose it is known that my brother,[9] like myself, has a Russian-born wife (this is pure coincidence); but is it known that, before she married him in about 1930, she was a secretary in the Russian Trade Delegation in Berlin?

On the other hand, is it known that my wife is the cousin of Kannegiesser, a counter-revolutionary who

---

[8] Jerzy Rayski and Jan Rzewuski. -SL
[9] Alfred Peierls; his wife's name was Nina.

assassinated Uritsky, who was then the head of the Russian secret police? With the same, very rare surname, she was never allowed to forget this connection. Is it known that her family was banished from Leningrad in 1935, partly because of this old connection, and partly no doubt because of her marriage to a foreigner. They have not dared communicate with her for several years, and we do not know whether they are still alive. [...]

I appreciate the great importance of security checks, and I have great sympathy for the difficulty in the way of such investigations, in particular in the case of intellectuals who rarely present a case without complications. But with so many facts in my case that could be open to unfavourable interpretation, if the attention of the experts is caught by just the two men in question, one naturally wonders whether they have missed many of the other facts, or whether they have a rather curious sense of proportion.

*****

Unfortunately, the story, a dark shadow of which fell on the Peierls family, did not end in the 1960s.

In 1978 an English journalist and so-called "popular historian," Richard Deacon wrote a book on Russian intelligence in the atomic project. He named only those agents who were dead. He believed, however, that Rudolf Peierls was dead by 1978, and assumed it was safe to saturate the book with slanderous and ludicrous accusations with regards to Peierls. Peierls filed a libel suit which was settled out of court. The settlement invoked "substantial damages" a part of which was donated by Rudolf to the Pugwash Movement.

In 1999, the English magazine *Spectator* published an article by a journalist Nicholas Farrell (a frontman for the military historian Nigel West) with allegedly new evidence of Rudolf and Genia Peierls' involvement with the NKVD-NKGB. The "proof" was based on the fact that among the data obtained through the operation *Venona*, there were names of NKVD-NKGB agents that had remained undecrypted, and one of them, Pers,[10] resembled the name Peierls. It is hardly necessary to comment on the depth of this conclusion. The trouble was

---
[10]Pers in Russian means Persian.

that sensation hunters, like jackals, immediately capitalized on the new allegation.

I dislike journalists. This is an instinct deeply rooted in me after 40 years of Soviet life. With sadness I discovered that in way too many instances the Western media are infected by the same disease: for political reasons or for the sake of sensation, professional ethics is forgotten, as well as integrity. Way too often the journalists distort or omit "undesirable" information. As a physicist, I can immediately recognize those who write about matters of which they know next to nothing.

To give you just a single example, one such "expert" journalist wrote that Genia Peierls was *obviously* an agent of Soviet intelligence, because after the wedding with Rudolf she was allowed to leave the USSR, and "everyone knows that under Stalin, ordinary people were not allowed to leave the country." He did not bother to check that Genia married Rudolf in 1931, two or three years prior to the final closure of the Soviet border.

I was glad to learn (or should I say, happy to learn?) that Rudolf Peierls apparently had a similar opinion in this regard. In his correspondence with family I found the following passage (in the letter of September 23, 1989, page 393):

> A few weeks ago the *Observer* had a profile of Lord Marshall.[11] The first two paragraphs of this were about a conversation between Marshall and myself when he was a student at Birmingham. According to the story he said he was worried how he could ever become a competent theoretical physicist when he was the only Gentile in a department of 42 members. To which I am supposed to have replied: "You are wrong, we are 43; the last man started today — he is also Jewish." The facts are that at the time we were not 43 but 20, and as far as I knew four were of Jewish origin (possibly a few more, as we did not check the ancestry of our students). I tried to contact Marshall, who was abroad, so I wrote a letter to the *Observer*, which has not so far been published. Incidentally, the "profile" also commented on Marshall's strange

---

[11] Walter Charles Marshall, Baron Marshall of Goring (1932–1996) studied mathematical physics at Birmingham University and obtained a PhD there under Rudolf Peierls. He headed the Theoretical Division at Harwell since 1960.

accent, and suggested that part of it was "East-European," which he picked up from me. I did not complain about that one, nor about the statement that Marshall designed the UK nuclear industry. About the last, Lorna Arnold, a science historian working in the Harwell archives, got incensed and wrote to the journalist who had written the piece. He replied and admitted that he was wrong, and that Marshall had not claimed this.

In another letter, narrating the story of a deliberate distortion of one of his interviews, Peierls concludes: "This way or that way, you cannot win over them [journalists]."

PART 3

SIR RUDOLF PEIERLS. HIS DIARY

Chapter 14

Sir Rudolf Peierls by Sabine Lee

SIR RUDOLF ERNST PEIERLS[1]
June 5, 1907–September 19, 1995

Sabine Lee

*Department of Modern History, University of Birmingham,
Edgbaston, Birmingham B15 2TT, UK*

Born into an assimilated Jewish family in Berlin in the early twentieth century, Rudolf Peierls studied theoretical physics with many of the greatest minds within the physics community, including Sommerfeld, Heisenberg, Pauli and Bohr. His Jewish background made a career in Germany all but impossible, and Rudolf Peierls and his Russian-born wife, Genia, settled in the UK, where Peierls took up a professorship in mathematical physics at Birmingham in 1937. Peierls' discovery, together with his Birmingham colleague Otto Frisch, of the theoretical feasibility of an atomic weapon based on a self-sustaining nuclear chain reaction was instrumental in the setting up of the UK government committee studying the possibility of manufacturing nuclear weapons. Peierls continued to contribute to the British and later to the British-American-Canadian effort to produce an atomic bomb, and he became group leader of the implosion group at Los Alamos. After the war Peierls returned to the UK and he built a world-class school of theoretical physics at Birmingham before moving on to Oxford in 1963. Like many of his colleagues who had contributed to the development of nuclear weapons, Peierls devoted much of his time and energy to the control of these weapons, to nuclear disarmament and to the promotion of greater understanding between East and West, most notably through his activities within the framework of the Pugwash Movement.

---

[1] Biogr. Mems Fell. R. Soc., **53**, 265, (2007).

Early Years

The date 5 June 2007 marks the centenary of Rudolf Peierls' birth. He was born as the third child of Heinrich and Elisabeth Peierls (née Weigert) into a non-religious Jewish family in Berlin. Brought up in a materially comfortable position, Rudolf grew up in an environment not atypical of assimilated Jews in early twentieth-century Berlin. His father (1867–1945), who had joined the Allgemeine Elektrizitätsgesellschaft (AEG) in 1888, had worked his way up to become director of the factory in Berlin-Oberschöneweide. Eventually Heinrich Peierls became a member of the managing board (1908) and later a member of the supervisory board (1929). Heinrich's first wife, Elisabeth, died of Hodgkin's disease in 1921, and Heinrich soon remarried. His non-Jewish second wife, Else (née Hermann) was the daughter of a famous actor and the sister-in-law of the playwright Ludwig Fulda, which added a stronger cultural dimension to the Peierls household.

From a young age Rudolf was interested in science and engineering. He was a bright child, who found school work easy and was keen to probe further into areas that interested him most: the sciences. His childhood friends remember many an occasion when he would leave their play in order to 'think,' only to return once he had solved whatever problem puzzled him at the time. Rudolf's original idea had been to follow an engineering career, but his family, doubting his practical abilities, persuaded Rudolf to settle for physics instead. Bowing to parental pressure, he enrolled for a course in experimental physics at Berlin University. He was soon to discover that first-year students were prevented from taking any practical courses because of overcrowding, and thus he became a theoretician almost by default. This accidental choice was to be a decisive career move, as Rudolf firmly established himself in the theoretical field in subsequent years.

Berlin, Munich and Zurich, 1926–32

While a student at Berlin, Rudolf Peierls encountered some of the leading figures in scientific research — Max Planck, Walther Nernst, and Walther Bothe to name but a few — but the real inspiration came with his move to Munich in the autumn of 1926, when he became a student at Arnold Sommerfeld's institute. Unlike Planck at Berlin, whose research genius did not find expression in his teaching, Sommerfeld deservedly had the reputation of being a superb communicator and a great teacher. His lectures were a model of clarity. It is no coincidence that the list of his students and assistants includes virtually everybody who made his name in (quantum) physics in years to come: Pauli, Heisenberg, Bethe, Peierls, von Laue, Kossel, Ewald, Wilhelm Lenz, Herzfeld, Wentzel, Heitler, Houston, Eckart, Rubinowicz, Pauling, Laporte, Brillouin, Condon, Fröhlich, London, Landé and many others.

Being introduced to quantum mechanics in such an inspiring manner and being confronted by Sommerfeld with the topical question of the electron theory of metals proved to be important for Peierls' short-term, medium-term and long-term physics career. As significant was his acquaintance with Hans Bethe, a fellow student one year his senior. They shared an interest in and a passion for physics that resulted in a lifelong friendship that went far beyond the research-related acquaintance.

Peierls' years at Munich, the autumn of 1926 until the spring of 1928, were a time of rapid personal and scientific development. Basic ideas of quantum mechanics had already been worked out by de Broglie, Schrödinger, Heisenberg, Pauli and Dirac, but new formalism had not been tested widely on the problems that had defeated the old quantum theory of Bohr and Sommerfeld. This was the context of Peierls' first seminar paper at Munich. In early 1927, P. A. M. Dirac and P. Jordan, independently but concurrently [1], had proposed a theory for the description of measurements in quantum mechanics, which was to become known as transformation theory. Peierls was to report on the papers of Dirac and Jordan to Sommerfeld's seminar. This was a difficult first assignment, but one that provided a useful learning experience.

When Arnold Sommerfeld went on a world tour in 1928, Peierls continued his studies in Leipzig to work with one of Sommerfeld's most promising former students, Werner Heisenberg, who had been appointed to the chair there in 1927. The move to Leipzig allowed Peierls to witness a completely different style of mathematical physics. Sommerfeld's approach to theoretical physics was best summarized in his own words: "If you want to be a physicist, you must do three things — first, study mathematics, second, study more mathematics, and third, do the same" [2]. In contrast, Werner Heisenberg relied more heavily on his brilliant intuition.

Prompted by Heisenberg, in the summer of 1928 Peierls began the research project that would lead to his first published paper, an examination of the theory of galvanomagnetic effects, a study of the anomalous or positive Hall effect. As long ago as 1879, Edwin H. Hall had tried to determine whether the force experienced by a current-carrying wire in a magnetic field was exerted on the whole wire or whether it was exerted only on what would later be called the moving electrons in the wire [3]. Hall suspected the latter.

Figure 14.1 Leipzig University, 1929. Left to right in the front row are Rudolf Peierls and Werner Heisenberg, with Georg Placzek standing between them. Courtesy of Professor Gerald Wiemers, Leipzig.

The phenomenon of the Hall effect is largely analogous to the deflection of cathode rays in a magnetic field. In some metals, however, it produces a positive sign as though the current were carried by positive carriers. An explanation of this paradox was impossible as long as electrons were visualized as moving freely in the metal. Bloch's theory of conductivity explained that conductivity was caused by jumps of the electrons from atom to atom where their energy could be less than the maximum potential barrier between atoms. There was no classical analogue for the process and, as pointed out in Peierls' paper, with the new understanding it no longer represented force-free motion. Peierls showed that electrons could give an anomalous sign of the Hall coefficient in a regular lattice, when he explained the positive Hall effect in terms of a concept of holes.

In the spring of 1929, Peierls once again moved to another university as a result of a mentor's absence. Heisenberg had accepted an invitation to lecture in the USA, Japan, China and India, and Peierls decided to move on to Zurich to work with yet another Sommerfeld pupil, Wolfgang Pauli. He quickly settled down to the work that would, in the summer of 1929, earn him a PhD at Leipzig: a study of thermal conductivity in crystals with its recognition of the importance of the so-called *Umklapp-process*[2] at low temperatures.

In the autumn of 1929, Peierls took up Pauli's offer to become his assistant in succession to Felix Bloch. At about the same time, the young Russian theoretician Lev Landau visited Pauli's institute. Landau had come on a Soviet government scholarship. Despite the brevity of his initial visit, Landau and Peierls initiated a deep and lasting friendship as well as an intense working relationship that was to result in several important (and controversial) publications in the following years. Peierls was immediately impressed with the depth of Landau's knowledge and with his striking intuition. Half a year younger than Peierls, Landau already had what his German friend judged to be a "very mature understanding of physics." During Landau's first stay at Zurich, in 1929, the most vigorously debated subject was quantum electrodynamics. Before his European tour, Landau had completed some work on the diamagnetism of metals by using quantum mechanics. In Zurich, however, he moved on to a collaborative study with Rudolf Peierls. The two investigated the limitations imposed on the measurability of physical quantities in the relativistic quantum region. Landau and Peierls looked at light quanta (photons) in space, and wrote a wave equation for photons not unlike Schrödinger's equation for electrons. From this they derived sequences of equations for different numbers of photons; however, as the two would recognize later, the results were not only complicated but physically nonsensical.

The main attraction of the summer of 1930 for Rudolf Peierls and a number of younger as well as more established Western and Soviet physicists was the 7th All Union Conference at Odessa. For many

---

[2]Umklappen (Germ.) Fold down.

Westerners it was the first exposure to the Soviet Union, and for many Soviet scientists it was a rare opportunity to encounter non-Soviet scientists.

At the conference, and during a subsequent boat trip across the Black Sea, Rudolf Peierls met many of Lev Landau's Leningrad colleagues, a close-knit community of exceptionally gifted young men (and one young woman) not unlike the community that Rudolf had known in Sommerfeld's institute. They were known as the "Jazz Band," a group formed around George Gamow (Jonny), Dmitri Ivanenko (Dimus) and Landau (Dau), also called the "three musketeers"; Genia Kannegiser and Matvei Bronshtein (Abbot) also played an active role. For Rudolf Peierls by far the most important new acquaintance of the summer of 1930 was the only female member of the Jazz Band: Eugenia (Genia) Nikolaevna Kannegiser. As Peierls recalled later, "she seemed to know everybody, and was known to everybody" (see [4], p. 63), and throughout the meeting and during the travels thereafter, Rudi and Genia got to know each other better. With their German and Russian backgrounds, neither of them could converse in the other's mother tongue and the only common language spoken sufficiently well by both to communicate reasonably comfortably was English. After six months of intense correspondence by letter,[3] Rudolf Peierls travelled to Leningrad again in March 1931, and during his brief stay — much to the dismay of his surprised family — married Genia.

At first sight the two were an unlikely couple. Among the reminiscences compiled by friends for Genia's 70th birthday (see pages 208 and 208) was one contribution from Denys and Helen Wilkinson [5] that contemplated defining "a genia" as a unit of "loudness, big-heartedness, self-confidence, loving concern, bossiness, generosity, stubbornness, unbreakable English, compassion, bad verse, fantasy, hypnotic gaze and irresistible kindness," and it concluded that whatever the final verdict, Genia stood for something "larger than life: the milli-genia should be perfectly adequate for normal purposes." In contrast, Rudolf was unassuming and modest, quiet, and often even shy. However, he was no less determined than his wife, and the two complemented each other in many important aspects. The marriage lasted for 55 years until Genia's death in 1986. The intense affection of their early long-distance relationship remained strong throughout their married life. Genia was not "merely" the professor's wife, she herself was the initiator of many of

---

[3] Sabine Lee, Vol. 1, Chapter 2.

the not strictly speaking scientific aspects of departmental life, ranging from entertainment to housing, from student counseling to health and safety advice.

One of the topics hotly debated at Odessa had been the unresolved question of infinite self-energy of electrons, a topic that Peierls and Landau (who continued his European tour after the Odessa conference) decided to revisit. They felt that Heisenberg's uncertainty relations for non-relativistic quantum mechanics needed extension in the relativistic field. In particular, they discussed the issue of measurability of momenta and accuracy of measurements of the intensity of electric and magnetic fields. Much of the Landau-Peierls work was done during various stays in Copenhagen, most notably their visit in February and March 1931. Niels Bohr, however, did not agree with their key conclusions, in particular the conclusion that it was impossible to measure electromagnetic fields accurately. Apparently, Bohr was so upset about the publication of the paper that he did not want to be acknowledged in it [6].

Despite Pauli's well-recorded aversion to solid state physics, Peierls' next research topic again was within this field. For his habilitation — the qualification necessary to teach independently at tertiary level in Germany and Switzerland — he investigated the question of electrical resistance at small temperatures. He completed this research speedily and gained his *venia legendi*, his tertiary teaching qualification in October of that year.

Rome, Cambridge, Manchester and Birmingham, 1932–39

Peierls spent the academic year 1932–33 as a Rockefeller fellow in Rome and Cambridge. In Rome he renewed his acquaintance with Enrico Fermi and found himself deeply impressed with the latter's abilities as a researcher and academic tutor. Peierls' main research interest still concerned the theory of electrons in metals, and in particular diamagnetic metals. Landau had developed a theory of diamagnetism of free electrons [7], but it was not clear how this theory needed to be modified in view of the presence of atoms in metals. While in Rome, Peierls completed two papers that discussed the general state of the electron band, deriving a general expression for weak fields and considering strong fields and low temperatures. With this work, Peierls could explain the mysterious magnetic properties of bismuth, which showed much greater diamagnetism than any other substance.

Despite the difficult job situation for young scientists in Europe, Peierls turned down an offer from Hamburg to take up the assistantship at Otto Stern's institute,[4] a position that Pauli had held some years previously. Living under Mussolini's regime in Italy had given Rudolf and Genia a taste of fascism, and of course Genia had already had experience of living under Stalin's totalitarian regime, albeit at a time when its worst excesses had not become apparent. Hence, both were skeptical about the wisdom of returning to Germany at a time when National Socialism was becoming an ever stronger force. They were therefore relieved when Peierls was offered a post at the institute of W.L. (later Sir Lawrence) Bragg in Manchester, an offer with the added bonus that the Peierls couple were, once again, united with their close friend Hans Bethe. Bethe and Peierls would often refer to the Manchester year as one of the most enjoyable and productive in their respective careers. They would recall with some satisfaction the anecdote of one of their great collaborative feats, their attempt to develop a theory of the deuteron photo-effect in 1933. In a conversation with James (later Sir James) Chadwick in Cambridge, the latter had challenged the two friends to develop a theory of this phenomenon. On the train back from Cambridge to Manchester, a journey of about four hours, doubtless after intense conversations and with the help of numerous backs of envelopes, they succeeded in developing a consistent theoretical approach.

The subjective impression of productivity is amply supported by the written evidence in the form of publications. The time in Manchester was among the most prolific period of Peierls' career. He still worked on aspects of the electron theory of metals, publishing a significant number of papers. However, in addition to this, he continued his work on Dirac's hole theory.

The Manchester period also marked the beginning of Peierls' more intense interest in nuclear physics. It was a time of fruitful collaboration between Bethe and Peierls, and in February 1934, together with Hans Bethe, Rudolf Peierls published his first paper dealing with something other than solid state or quantum dynamics [8], quickly followed by several other joint papers on questions of nuclear physics.

In 1935 Peierls accepted a position at the Mond Laboratory, the laboratory for magnetism and low-temperature physics that had been built for P. Kapitza. After Kapitza's detention in Russia in 1934, which

---

[4]Instituts für physikalische Chemie der Universität Hamburg.

prevented him from returning to Cambridge, the Royal Society was persuaded by Rutherford to use the earmarked unclaimed salary for the establishment of two fellowships, one of which was offered to Peierls.

The two years at Cambridge were, again, a productive period for Peierls, partly inspired by old and new colleagues at Cambridge, partly continuing earlier collaborations. Among the papers based on his Cambridge contacts was one on superconductors, which showed traces of Peierls' collaboration with David Shoenberg [9], and a statistical mechanics paper [10], which was inspired by work of Ralph (later Sir Ralph) Fowler [11].

Another paper originating at Cambridge was his now famous paper on the Ising model [12]. The advent of quantum theory had sparked renewed interest in Ising's model [13] for ferromagnetism. Heisenberg had replaced Ising's model with one based on exchange forces [14], and Bloch had extended the theory [15]. Ising's model had solved the problem only for the one-dimensional case, and Peierls expanded the model to two dimensions by giving an elementary proof that in two dimensions the Ising model showed ferromagnetism; he concluded that the same held *a fortiori* also for the three-dimensional model.

In 1937 Peierls was offered his first permanent position, a professorship in mathematical physics at Birmingham University. This appointment was the first move to a university that could not be considered as being "at the heart of theoretical physics." Berlin, Munich, Leipzig, Zurich, Rome and Cambridge, and also Manchester with Rutherford's legacy and Bragg's — and, later, Blackett's - presence, could be regarded as such. Birmingham could not look back on a strong history of theoretical physics research comparable with any of these places, and in fact the chair of applied mathematics was only being created at the time of Peierls' arrival. Initially he was the only theoretician among the physicists. Few people would have predicted that within just over a decade this department of applied mathematics would become one of the foremost centers of theoretical physics teaching and research, not just in England but in Europe and arguably across the globe. This was largely due to the effort of its first professor: Rudolf Peierls.

Peierls continued to engage in cutting-edge research. His continued interest in nuclear physics was evident from the fact that he paid several visits to Niels Bohr in Copenhagen between 1937 and 1939, and discussion topics at the time were invariably linked to nuclear physics. Again, it was the compound nucleus that occupied the two scientists and their colleague George Placzek, who was working with Bohr in Copenhagen at the time. In the collision of a slow neutron with a nucleus, the resulting compound system has resonance levels that can be narrower than their spacing. At higher neutron energies, the width of the resonance increases and spacing decreases, leading to an overlap. In this region of overlapping resonances, two different ideas of expressing Bohr's compound nucleus formation, the Breit-Wigner formula on the one hand and detailed balancing on the other, gave conflicting answers. Bohr, Peierls and Placzek eventually arrived at an understanding of the problem and its solution. By the summer of 1938 the challenge of solving the physics problem had been superseded by the challenge of writing up the solution. Bohr was notorious for laboring over the formulation of his research results, often leading to unwelcome and unnecessary delays in their publication. In that respect the fate of the Bohr-Placzek-Peierls calculations was not unusual. In contrast with Bohr's intention of presenting a broad qualitative argument, Placzek and Peierls wanted a more mathematically rigorous exposition. When Bohr came to Birmingham to receive an honorary degree in June 1939, a short note to *Nature* was written to present the core arguments [16]. The full paper had not been finalized by the beginning of the war and work had to be abandoned until afterwards. However, as a result of the numerous drafts that had been prepared and circulated to others for comment, the results of the proposed publication were already widely known and were being used in the scientific literature. Not surprisingly, therefore, the paper gained considerable fame as being the most frequently cited unpublished paper! (Several drafts of the paper survive and some more or less complete versions have been published; see [17], p. 49, note 86, and p. 50, notes 87–89.)

Figure 14.2 The 1937 Copenhagen Conference organized by Niels Bohr. Rudolf Peierls is in the second row. Left to right: (first row) N. Bohr, W. Heisenberg, W. Pauli, O. Stern, L. Meitner, R. Ladenburg, J. Jacobsen; (second row) V. Weisskopf, C. Møller, H. Euler, R. Peierls, F. Hund, M. Goldhaber, W. Heitler, E. Segré, S. Hoffer-Jensen; (third row) G. Placzek, C. Weizsäcker, H. Kopfermann, [...]. The famous A auditorium of the NBI. Credit: Photograph by Nordisk Pressefoto, Niels Bohr Institute, courtesy of the AIP Emilio Segré Visual Archives, Fermi Film Collection, and Niels Bohr Archive, Copenhagen.

Figure 14.3 Rudolf Peierls in Copenhagen in 1938. Credit: Niels Bohr Archive, Copenhagen.

While Rudolf Peierls was establishing himself in his first permanent post in the UK in the late 1930s, he and Genia were reminded frequently that their position was a fortunate one indeed. Many of Rudolf's close friends and colleagues, who had been forced to leave Nazi Germany, struggled to find suitable positions. But even more disconcertingly, both Rudolf's and Genia's families in Germany and Russia were facing uncertain futures. By 1938 Rudolf's siblings had left Germany to settle in the USA and England, respectively. However, his father and step-mother found it difficult to face emigration, although the conditions for Jews worsened by the day in Germany. Finally, in early 1939, Heinrich and Else Peierls got the necessary permissions from Germany and the UK and emigrated to England. In 1940 they continued their journey to the USA, where they settled in New Jersey near Rudolf Peierls's sister Annie.

The War Years, 1939–45

When war broke out in 1939, Rudolf Peierls had lived outside Germany for about a decade, and he had spent more than half of this time in England. He felt gratitude towards the country that had provided him with a safe home at the time when his country of birth had failed to do so, and by the late 1930s it was clear that Rudolf and Genia Peierls had no intention of returning to Germany. In May 1938, therefore, Rudolf Peierls applied for naturalization, but he also felt the need actively to end his association with the country that Germany had become under Hitler, rather than terminating it by default once the process of his naturalization had been completed. This proved technically impossible, because some of the documents needed for an application to end his status as member of the German nation were also required for his naturalization, which of course took precedence.

The war turned Rudolf and Genia Peierls into enemy aliens. Soon tribunals were set up to classify this group of residents. Being classified as "category one" meant that the couple in practice had to endure very few restrictions to their everyday life. What concerned Rudolf Peierls more than these relatively insignificant limitations of his private life were those placed on him with regard to civil defense and to his work and research. The former were lifted when his naturalization was approved in February 1940, but engaging in war work was still not as easy as for native British citizens. This was felt most acutely in the work that Peierls did together with his Austrian-born colleague at Birmingham, Otto Frisch. The latter, together with his aunt Lise Meitner, had developed a qualitative theoretical explanation of the nuclear fission process that had been discovered by Otto Hahn and Fritz Strassmann in 1938.

Using Frisch's knowledge about the fission process and Peierls' theoretical understanding of the nucleus, the two physicists turned to some fundamental questions of this process, and in particular they considered the critical mass of Uranium-235, the uranium isotope that was believed to be the most promising candidate for fissionable material that would allow a self-sustaining nuclear chain reaction. Frisch and Peierls calculated that a chain reaction was not only theoretically possible but also practically feasible. In a memorandum they suggested that the amount of fissionable material (^{235}U) needed for an atomic bomb based on these principles of a self-sustaining chain reaction was far less than previously assumed and that a sphere of metallic ^{235}U of a radius of about 2.1 cm could be sufficient to be explosive, an amount that corresponded to less than 1 kg of ^{235}U [18]. Both scientists were immediately aware of the potential implications of their finding, and communicated them to the Head of the Physics Department, Mark (later Sir Mark) Oliphant. The recognition of the theoretical possibility of producing a nuclear weapon led to the creation of a government committee, the so-called MAUD Committee,[5] to investigate further the feasibility of a uranium-based weapon. When the committee met for the first time in April 1940, Peierls and Frisch, as only recently

---

[5] Rudolf Peierls recollects (page 156 in [4]): "The name of the committee was the MAUD Committee, and that name has an amusing origin. At an early meeting the members tried to find a suitable code name that would not disclose the committee's purpose. At the same meeting, the committee was told about a message from Niels Bohr. When the Germans invaded Denmark, Lise Meitner happened to be visiting Copenhagen, but she was able to leave and return to Stockholm. As she left, Bohr asked her to send a telegram to a friend in England, saying that he and his family were all right and had come to no harm in the occupation. The telegram ended: "Tell Cockcroft and Maud Ray Kent." This seemed mysterious. If the recipient was supposed to know Maud Ray, why add "Kent"? If he did not, "Kent" was hardly a sufficient address. Someone on the committee therefore suggested that there was a hidden message here, perhaps an anagram. Someone actually "decoded" this anagram to contain a message about uranium. Frisch and I were skeptical. Knowing Niels Bohr, we felt he was quite unlikely to send a message in this way, and as for the "decoding" of the anagram, we wrote down half a dozen alternative "solutions" as plausible as the first. When we saw Bohr again during the war, we asked about the telegram but he had forgotten the whole matter. After the war we saw Lise Meitner, and she solved the mystery: the telegram had been truncated in transmission; the original form had had a complete address between "Ray" and "Kent," the address of a former governess of the Bohr family, whose address ended with "Kent." In any case, at the meeting this puzzle led to the suggestion of the name "Maud Committee," or MAUD to look more official. It was unfortunate that many people associated with it were convinced the letters stood for "Military Applications of the Uranium Disintegration"!

naturalized and enemy alien respectively, found themselves excluded from the work, a fact that met with misapprehension and caused consternation among the two scientists. Eventually, the folly of "trying to keep the scientists own ideas secret from them" was recognized by the committee. Frisch and Peierls became members of the technical subcommittee and remained deeply involved in the developments leading to the production of nuclear weapons.

After months of research spread across various academic and research institutes, the MAUD Committee, in its final reports in June and July 1941, endorsed the Frisch-Peierls memorandum in concluding that the atomic bomb was feasible, although very costly (MAUD Committee 1941). The continuation of the project required theoretical physicists of the highest caliber, and among the people recruited by Peierls into the project was Klaus Fuchs, a German refugee who had to flee Nazi Germany because of his left-wing views. Peierls encouraged Fuchs to join him in his work at Birmingham, a step with fateful consequences, because Fuchs eventually passed atomic secrets to the Soviet Union.

After his initial preoccupation with the calculation of the critical mass of U^{235}, Peierls became increasingly involved in the complex problem of isotope separation. Many of his papers between 1940 and 1942 discussed the more or less promising avenues on the way to an efficient separation process and numerous papers on the Simon plant, an early diffusion plant concept. By the autumn of 1941, the nuclear program, now codenamed Tube Alloys, under the chairmanship of Wallace (later Sir Wallace) Akers, had received government backing, and the key technical sub-committee including Chadwick, Hans Halban, Peierls and Franz (later Sir Francis) Simon[6] had been set up.

In August 1943, at the Quebec Conference, the official nuclear relationship between Britain and the USA was agreed on for the duration of the war, with the American President Roosevelt trading nuclear cooperation with the British for Churchill's agreement for a cross-channel invasion of Europe during 1944. Soon after the signing of the Quebec Agreement, Chadwick, Simon, Oliphant and Peierls arrived in Washington as part of a fact-finding mission about the British role within an Anglo-American project, and by the end of this visit in the autumn of 1943 it had become clear that the key figures of the British effort would join their American colleagues in the USA.

---

[6]Sir Francis Simon (1893–1956) was a German and later British physical chemist and physicist who devised the gaseous diffusion method and confirmed its feasibility of separating the isotope Uranium-235.

Peierls had significant roles in the Manhattan Project, first by working on the complex issues involved in the ^{235}U isotope separation process, and second, after his move to the Los Alamos plant in 1944, as head of the hydrodynamics group, also referred to as the implosion dynamics section.

Rudolf Peierls was deeply concerned about the consequences of the development and use of nuclear weapons from the moment he realized the feasibility of a fission bomb. His memorandum with Otto Frisch had contained a rather unusual "non-scientific" section in which the use of a potential weapon was considered. Despite his realization of the destructive power of the weapon, Peierls believed that it was necessary for Britain and America to produce it, at first in case Germany should develop a nuclear bomb and later, after Germany's surrender, because he reasoned that its use could shorten the war in the Pacific and thereby save lives. As Peierls would later put it himself, his work on nuclear weapons ceased in 1945, but his concern with the weapon he had helped create did not. He became a keen supporter of nuclear disarmament and devoted much time and energy to the campaign against nuclear weapons, not because he felt guilty about the role he had played in their development but because he was convinced of the danger of an irresponsible nuclear policy.

Birmingham, 1945–63

After the end of the war, Rudolf Peierls and his family returned to the UK, and although he had attractive offers from several universities, including Oxford, Manchester, London and Cambridge, he chose to remain at Birmingham. He had clear ideas of what he regarded as important for a prosperous theoretical physics community in the UK: a balanced flexible system that provided good training and high standards without prejudicing against students outside Oxford and Cambridge. Peierls had come to Birmingham in 1937 as the first professor of Mathematical Physics and had set himself the task of establishing a school devoted to both first-class research and first-class teaching. The war had put the effort on hold, but as soon as Peierls returned to Birmingham he re-engaged in the process and, virtually from scratch, he developed a school of mathematical physics, or theoretical physics as it would be called later, that was arguably the best in the country and could compete with any in Europe and with most others globally.

He succeeded where many others failed because he had a clear vision and a determined devotion to his subject and to the people he engaged with. In the postwar decades he regarded teaching as his main responsibility. Although he enjoyed his research and recognized its importance as a contribution to a discipline that was undergoing exciting developments, increasingly this research was being done in collaboration with research students and younger research staff and thereby became virtually indistinguishable from teaching.

His enthusiasm for teaching and building up a viable team found expression in time and energy devoted to securing funding for young scholars and finding the best possible people to perform the increasingly complex research. Within a few years he had built a reputation for his institute, that of being an ideal training ground — a reputation that helped in achieving both the above aims — but it also enabled Peierls to work within a group of a critical mass that would always be certain of being supplied with the best of talent from within Britain and from abroad.

Peierls' commitment to his students and research fellows did not end with the completion of their stay at Birmingham. Much thought and letter-writing went into the task of securing future positions and exchange opportunities. In this, the prospects of the individual scientists were as important as the future of his own institute at Birmingham. Collaboration with the USA throughout the war and close contact with many friends and colleagues across the Atlantic had sharpened Peierls' awareness of the role reversal that had occurred with regard to academic physics. As early as September 1945, he expressed, in a letter to Raymond Priestley, the Vice Chancellor of Birmingham University, that "American universities [had] matured a great deal and contact with this country [was] now less important to them, and more important to us." The consequence of this, in Peierls' view, had to be regular academic exchanges that would allow the UK to benefit from scientific achievements of colleagues in the USA. And his attempts to put Birmingham firmly on the academic map in theoretical physics meant that he was keen to secure a sizable fraction of the exchange for this institution.

A supplementary ingredient that could not be found in any other institute was what some would later term the "Genia factor." Genia Peierls was an enthusiastic supporter of her husband's endeavors to attract the best young scientists to Birmingham, a place that — with post-war rationing, shortage of housing and generally meagre facilities — was not the most appealing location. Her hands-on efforts, which ranged from provision of short-term and long-term accommodation to general advice, from organizing social gatherings to job advice for spouse and general counseling, had a significant impact on the cohesion of the growing "Peierls school" [5].

Figure 14.4 George Placzek, Jan Blaton, and Rudolf Peierls in Copenhagen in 1947. Credit: Niels Bohr Archive, Copenhagen.

Although Peierls ceased his involvement in weapons production as such, his expertise was enlisted in consultancy work for the Atomic Energy Research Establishment (AERE) at Harwell. Among the friends and colleagues from the Manhattan Project who were on the staff at Harwell was Klaus Fuchs. His arrest in 1950 on charges of passing secret information to the Soviet Union was a severe blow to the British scientific community as a whole, and it was a particularly traumatic experience for Peierls and his family. Fuchs had been a close friend of the Peierlses; he had lodged with them when he first came to Birmingham, and he had collaborated closely with Rudolf, who had not only hired him in his department at Birmingham but had also been instrumental in securing his appointment at Los Alamos.

Rudolf Peierls never shied away from expressing his views in public. He did so regardless of the effect this would have on is own position. He defended civil liberties in the aftermath of the Fuchs affair in his memorandum "Lesson of the Fuchs Case" (see page 304), although his close association with Fuchs had made him a prime target of suspicion. He was never secretive about his friendships with people from Communist countries and of Communist persuasion; he argued for the re-establishment of scientific exchange with the Soviet Union and its satellites. He rejected the idea of oppressing the voices of dissenters by arguing that this totalitarian measure would bring security at the expense of values that any democracy had to fight to retain. In the aftermath of the arrest of Fuchs, Peierls' overt expression of these views led some to question his reliability, especially in view of the fact that he had access to sensitive and secret information in connection with the UK nuclear program. However, at that time, as on many other occasions during the subsequent decades, it was recognized by people in authority that the views may have been uncomfortable at times but at no point did they undermine the security and values of democracy in the UK, and at all times Peierls proved loyal to the national interest of the UK.

Rudolf Peierls regarded international exchange as one of the most significant prerequisites for securing first-class research in the UK. Since his Munich days he had been establishing contacts with colleagues all over the world, and his work at Los Alamos had added more depth and breadth to his international links. First and foremost within the postwar collaborative network in and out of Birmingham was the link to Cornell, where Hans Bethe had settled in the mid-1930s. Perhaps the most influential of the exchanges orchestrated by Bethe and Peierls was based on the recommendation of Hans Bethe to Freeman Dyson, in early 1949, to spend some time at Peierls' institute. Rudolf Peierls and Robert Oppenheimer, at that time director of the Institute of Advanced Studies at Princeton, where Dyson, the rising star of theoretical physics at the time, was based, arranged a flexible fellowship. Dyson was based at Birmingham but it was agreed that he was at liberty to spend time at Princeton regularly as long as it fitted in with departmental requirements at Birmingham. This resulted in Birmingham's being in direct contact with the development of quantum field theory, which at the time was worked on by Julian Schwinger, Sin-Itiro Tomonaga, Richard Feynman and Dyson. The arrangement demonstrated two essential ingredients that promoted the success of the Peierls School at Birmingham: first, Peierls was excellent at spotting talent, and second, he was flexible enough to make Birmingham an attractive option for scholars to choose his institute despite stiff competition from

Cambridge, Oxford, Liverpool, Manchester, Bristol and other universities. Others similarly made the journey across the Atlantic; the exchange went both ways with, among others, Nina Byers, Elliott Lieb, Jim Langer, Gerry Brown, Richard Dalitz, Edwin Salpeter, Claude Bloch and Stanley Mandelstam moving between the USA and Birmingham.

Another example of Peierls' spotting talent and being slightly unconventional in securing it for Birmingham was the recruitment of Gerry Brown, a young American scientist who would spend almost a decade at Peierls' department and made significant contributions to its functioning, to research, teaching and administration. Brown had studied at Wisconsin and Yale, where he obtained an MS and a PhD. A short-lived membership of the Communist Party, from which he was eventually expelled, put his academic career in the USA at risk, despite his outstanding doctoral work with Gregory Breit. Various inquiries to universities in England led to the now famous threepenny folded airmail return from Rudi Peierls saying, "Come ahead" (see [19], p. 6). In February 1950 Gerry Brown arrived as a political refugee from pre-McCarthy anti-Communist America; in 1960 he left to take up his appointment as full Professor of Theoretical Physics at Niels Bohr's Nordic Institute for Theoretical Physics (NORDITA).

Although not many of Peierls' students arrived as refugees in the same way as Gerry Brown did, many left to take up distinguished positions. The Birmingham department itself was seen as an exceptional training ground for young scientists well beyond the UK. Many of those who came to Birmingham as students, graduates or research fellows in the 1950s later filled lectureships and professorships around the globe: Dyson, Dalitz, Samuel (later Sir Samuel) Edwards, Brown, Byers, Brian Flowers, Mandelstam, John Bell, Paul Matthews), Denys (later Sir Denys) Wilkinson, Lieb and Langer, to name but a few.

The rising numbers of staff and students and their exceptionally high standard caused some logistic and administrative problems, too. As accommodation within the Physics Department was notoriously limited, huts had to be employed to overcome the shortage of space, and on one occasion Peierls had to ask for permission to add a trailer to overcome the departmental space crisis.

If Rudolf Peierls felt strongly on an issue, he was prepared to make his views known, irrespective of whether this would cause difficulties for himself. One such example was the political rat race that his friend

Robert Oppenheimer found himself facing in the 1950s. In 1953 Oppenheimer had been suspended from the Atomic Energy Commission, on which he had served as Chairman of its General Advisory Committee between 1947 and 1952. Concerns had been expressed about his loyalty and reliability, and his security clearance had been withdrawn. Oppenheimer appealed against this decision, and between April and June 1954 hearings were held to determine whether his clearance should be restored. The commission decided against a restoration. However, Oppenheimer continued to speak out on nuclear physics issues, and although he was never officially rehabilitated, in 1963 he received the Enrico Fermi Award, a US government presidential award honoring scientists of international stature for their lifetime achievement in the development, use or production of energy. This served as a measure of reconciliation for what many perceived to be a grave injustice done to Oppenheimer.

Peierls spoke out tirelessly in support of Oppenheimer, and his many letters to "Oppie" are evidence of the deeply felt indignation at the attacks launched against his friend. Peierls himself had his share of "security troubles." His contacts with left-wing colleagues, his friendship with people of Communist persuasion, his marriage to a Russian, his close friendship with Klaus Fuchs — all led to his being viewed with a degree of suspicion by many. When he applied for a visa to attend a conference in the USA, his application met with a long delay, as did his paperwork in connection with his sabbatical at Princeton in early 1952.

Figure 14.5 From *Revues of Unclear Physics*, Vol. 1 №1 (Birmingham, June 1957). Dedicated to the 50th birthday of Rudolf Peierls.

In 1957 Peierls, who at the time was acting as a consultant for the AERE at Harwell, had his security clearance revoked at the request of the American authorities. Disappointed with the action of the Harwell authorities over this matter, Peierls resigned from his consultancy. Even before this episode, Peierls had been challenged by William (later Lord) Penney, then on the board of the UK Atomic Energy Authority, about his contact with Russian colleagues and in particular his intention of participating in a conference in Moscow. In a letter in 1956

Peierls expressed his conviction of the sanctity of the "freedom of scientific enquiry, the freedom of exchange of scientific information, and of objective discussion with any scientist, regardless of person, nationality, or position as long as these do not interfere with his approach to scientific fact or argument," a principle that was subject only to "the overriding requirement of national security."

The more liberal flow of information from Russia brought the West into contact with Landau's work, and his views on renormalized quantum electrodynamics were discussed widely among Peierls and some of his colleagues. The contacts with Landau facilitated the first English edition of Landau and Lifshitz's seminal *Course of Theoretical Physics*, a set of textbooks that had previously been available only in Russian and was to become one of the standard works of teaching and reference for generations of physicists to come [20]. Peierls clearly valued the fact that the restrictions to scientific exchange with Russian colleagues and friends were slowly lifted, and he attempted to encourage an understanding in the West of the work done by Russian physicists. He asked Niels Bohr to use his reputation and standing in Russia to help Lev Landau travel to the West,[7] and it was doubtless on his recommendation that Birmingham University invited him to accept an honorary degree in 1958. By the 1960s, Peierls' impact through teaching and collaboration with younger colleagues outweighed the contributions he made independently. In addition, his focus was shifting to political work and publications in the area of arms control. Having been offered the Wykeham Chair of Physics in 1961, Rudolf Peierls, after lengthy negotiations, decided to accept in early 1962 and took up his appointment in the autumn of 1963. When asked about the reasons for his decision to move from Birmingham to Oxford, he would later refer to the need for change after a quarter of a century at the same university. However, it was more than simply the desire for change: Peierls liked the challenge. After successfully building a school of theoretical physics at Birmingham, he wanted to achieve something similar in Oxford.

---

[7] After 1934 this was never allowed by Soviet authorities. Bohr's interference did not work.

Oxford, 1963–74

If Peierls' role as a senior academic in the UK had already undergone some changes towards the end of his time at Birmingham, this change became even more pronounced during the last decade of his university career, at Oxford. He became more concerned with university administration, teaching reform and, increasingly frequently, work for nuclear disarmament.

Evidently, Peierls' research had undergone a gradual change, which had already been visible in his final years at Birmingham and accelerated during his time at Oxford. He had been among the outstanding figures of the last generation of universalists in physics, and unlike many of his colleagues of his generation he refused to choose one narrow field as a focal point of his attention and instead tried to keep his interests broad. The increasingly rapid pace of developments in subject areas such as particle physics made it difficult to keep up with the trends in the discipline for anybody keen on dividing his attention between different specializations. In addition, Peierls felt that his age was beginning to make itself shown by the speed with which he was capable of picking up and using other people's ideas and concepts.

A recurring nuisance for the Peierls family, and above all for Rudolf Peierls himself, were the continued attempts of some to link him to Soviet espionage circles. As a German-born Jew with a Russian wife and numerous friends in the Soviet Union and Communist contacts elsewhere, as a close friend of Klaus Fuchs', and as a nuclear scientist with access to classified information relating to atomic weapons, Peierls (and his wife) had been subjects of suspicion throughout the Cold War. He had been under investigation by the Security Service from 1938, when he had re-entered the UK after a visit to Russia, and naturally remained so throughout the war and beyond, until his file was closed in 1953. This exhaustive investigation over 15 years uncovered no evidence of any wrongdoing by Peierls; quite the contrary. In 1948, after espionage suspicion had first fallen on Fuchs, and Peierls was closely scrutinized, an MI5 officer minuted, "not only have we nothing against him, but [that] he is a man of very good sense" (see [21], KV2/1658). Further investigations in the early 1950s in the aftermath of Fuchs' arrest and conviction led to the categorical conclusion that "there is no substantial doubt about the loyalty of Prof. Peierls" (see [21], KV2/1662). The award in 1968 of a knighthood must have brought some satisfaction to

Rudolf Peierls, not least because it was a tangible sign of the official recognition of his loyalty to his adopted home country.

Retirement

Rudolf and Genia Peierls had always enjoyed leading a nomadic existence, and during their Oxford years, both before and during retirement, they continued travelling large parts of the world. Three of their four children had settled in other continents, and this provided extra incentive to travel abroad. In the 12 years between Rudolf Peierls' retirement from his chair at Oxford in 1974 and Genia's death in 1986, the two rejoiced in the opportunities provided by the more flexible work arrangements that Peierls' semi-retirement made possible and the opportunities that the plentiful invitations to far-flung places brought. The travel schedule was truly astounding, with regular visits to the University of Washington, Seattle, where Peierls took up a part-time appointment that resulted in his visiting Seattle between February and May each year until his retirement in 1977, at the age of 70 years. Other places visited between 1974 and 1996 included Sydney, Los Angeles, Vancouver, Princeton, Oregon, Mexico, Pisa, Coimbra, Copenhagen, Finland, Russia, Italy, Stanford, Japan, Virginia, Toronto, Japan, India, Greece and Ljubljana (see Chapter 16).

Rudolf Peierls had always tried to keep in touch with his Russian friends, colleagues and in-laws, and his additional time for travel and leisure facilitated this. Often Peierls provided the "semi-Western" angle on biographical material concerning Russian colleagues, or he liaised between Western and Russian colleagues in other history of science projects or even human rights issues. He and Genia had planned on visiting Moscow and Leningrad in the autumn of 1986, a visit eagerly awaited by both in view of the changes brought about by the advent of Mikhail Gorbachev. However, Genia had been unwell and had undergone surgery to have a benign brain tumor removed in 1985, an operation that had given temporary relief. She spent a comfortable year, and Rudolf and Genia spent a "glorious" holiday in Greece in June 1986. But amid preparations for their Russian trip, Genia's condition deteriorated, and she died on October 26, 1986.

Around the time of Genia's final illness, Peierls' autobiography was published [4]: Bird of Passage was the fitting title of a book that was endearing to many of his numerous friends.

Genia had been the warm-hearted center of much of the social life around Peierls' institute at Birmingham and to some extent also at Oxford. She had made other people's problems her own, and had been keen to contribute to their solutions. Some people may have been irritated by her occasionally unwanted concern or interference, but everybody acknowledged that her heart had been in the right place.

This unsentimental approach to life was also visible in Genia's advice about how to deal with a partner's death:

> In our consciousness there are rings like in a tree. After the death of a partner it is important to develop new rings. At first any recollection of the past is painful, because every experience, every place is always linked to the picture of the partner. One ought to travel, find new occupations, new impressions. Then, after a while one will have recollections which are no longer painful.

Rudi Peierls took Genia's advice. He continued leading a nomadic lifestyle and an active social life, spending time with many old friends but also making new ones. In 1986 Peierls had received the Copley Medal, the highest award from the Royal Society, and the Rutherford Memorial Medal, which is associated with a lecture series to be delivered at selected centers in the British Commonwealth overseas. Initially, Rudolf and Genia had wanted to embark on the lecture tour together, but Genia's illness prevented them from doing so. The trip was postponed until the following year, and in November 1987 Peierls delivered his lectures in India, visiting Moscow and Leningrad en route. In the early years of their marriage, Rudolf, when traveling long distances by himself, would always send Genia detailed travelogues sharing his impressions and reactions to new places. Now, again traveling without a companion, he reverted to his habit of sending travelogues, this time to his children in the form of his "Dear Everybody" circular letters. Peierls' journey to Russia in 1987, although of course filled with meetings of colleagues at the various scientific institutes, had a more personal note than many other trips, because of the emotional ties to the place.

Before his first extensive trip without Genia in the autumn of 1987, in June of the same year, Rudolf Peierls had celebrated his 80th birthday. The Theoretical Physics Department at Oxford marked this occasion with a symposium. The meeting was an impressive display of the breadth of physics tackled by Peierls on his own or by his students in collaboration with "Prof" and it was an indication of the significance of the contribution of Peierls to our understanding of the world [22].

Peierls' life was still a remarkably busy and active one well over a decade after embarking on "retirement." When Freeman Dyson commented that he and his wife were "struck dumb with admiration" for his breathtaking travel schedule, Peierls dryly answered that he regarded this as the soft option as opposed to sitting on one's backside and doing more serious reading or thinking, which I find myself more and more reluctant to undertake "this kind of laziness grows with age."

Whether others would agree with his own assessment that he was prone to laziness is debatable. Not only do his several hundred publications indicate the contrary, but also — and perhaps even more so — his willingness to devote his time and energy to causes he regarded as important.

In the early postwar years, Peierls joined many of his colleagues in political activities aimed at controlling the nuclear weapons they had made possible through wartime research. He was instrumental in setting up the Committee of Atomic Scientists, later called the British Association of Atomic Scientists in the UK, as a forum in which the responsible use of peaceful and nuclear energy was discussed and the control of nuclear weapons was debated. Later, he became increasingly involved in the Pugwash Movement, an initiative triggered by the Russell-Einstein manifesto of 1955, which had called on all scientists to work together to prevent nuclear war. Under its first president, Cecil Powell, and secretary general, Joseph (later Sir Joseph) Rotblat,[8] the movement grew, with increasing numbers of scientists getting involved in the annual conferences and regular meetings and workshops. Peierls took an active part in the Pugwash Movement, serving on its continuing committee from 1963 to 1974 and as its chairman between 1969 and 1974. He had always given high priority to the Pugwash Conferences since attending his first such conference in Moscow in 1960; later, in retirement, he still tried to attend and contribute whenever possible. When he was awarded damages from a libel suit in the early 1970s, part of the money awarded to him was donated to the Pugwash Movement. Even in the last years of his life, when his health was declining

---

[8]Sir Joseph Rotblat (1908–2005) was a Polish physicist who worked on the Manhattan Project during WWII. A signatory of the 1955 Russell-Einstein Manifesto, he was Secretary General of the Pugwash Movement from its foundation in 1973, and shared, with the Pugwash Movement the 1995 Nobel Peace Prize.

and it was becoming increasingly difficult for him to engage in travel and writing, he kept up his determination to contribute to the nuclear debates; his last publication was devoted to these issues [23].

Figure 14.6   Rudolf Peierls at home in Oxford, 1990. Courtesy of Natalia Alexander.

More locally, Rudolf Peierls had also become involved in the FREEZE movement, which had been publicly launched in 1985 as an organization mainly concerned with nuclear disarmament.[9] Peierls became a "Patron" in 1985 and a director in 1986 until his resignation in 1989, and he chaired the local Oxford group, with many of the meetings taking place in his flat. In June 1989 it was decided that the local FREEZE group should not continue independent operations but should instead cooperate with the Oxford Research Group, a registered charity that conducted independent research into decision-making, accountability, intergovernmental mediation and other topics with special reference to nuclear weapons. In 1989 Peierls became a "friend" of the Oxford Research Group.

---

[9]The organization had several changes of name, including Towards a Safer World (1988), Safer World Project (1989) and Saferworld (1991). For most of the time of Peierls' involvement (1985–89) it was known as FREEZE.

During the last years of his life, Rudolf Peierls was troubled by a number of health problems and he suffered a deterioration of his eyesight, which restricted his reading and made correspondence more difficult. Despite all this, he continued to lead an active and independent life well into his eighties. In the summer of 1994, however, after suffering a combination of heart, lung and kidney problems, he decided to move into a residential home close to Oxford. Having been independent since leaving home well over 60 years earlier, Peierls nevertheless settled well into his new environment, one of the few residents at Oakenholt who would word-process circular letters to friends and family and read scientific papers in enlarged script on a computer screen! However, his health deteriorated further throughout 1995 and he died on September 19, 1995.

To adopt his wife Genia's well-rehearsed characterization of Rudolf, he was an intellectual tennis player, not a golfer. He needed partners in his research: in the early days these would be fellow students, then colleagues, or later his own more advanced students. He thrived on bouncing ideas off and receiving the return from others, and many of his important achievements occurred as a result of direct and intense contact with others. This was evident in his early collaboration with Hans Bethe, and then in his work with Placzek and Bohr; it was equally true for the Frisch-Peierls memorandum; and it remained true in his numerous collaborative efforts in cooperation with his graduate students and postdocs. Not all efficient collaborators are enthusiastic or good teachers. Rudolf Peierls most certainly was. His own experiences as a student in the late 1920s, at a time of great excitement, stimulation and achievement in physics, were a key to his own passion for the subject as well as his approach to communicating it. His formative period as a scientist was in an environment with a belief in intellectual exchange as an essential ingredient of scientific progress. All his teachers, Sommerfeld in Munich, Heisenberg in Leipzig, Pauli in Zurich and Bohr in Copenhagen, in their distinct ways, created settings that would provide for a spirit of collaboration and communication.

When looking back at his own experiences as a student, Peierls fondly remembered the warm and friendly atmosphere of the Bohr Institute in Copenhagen. He commented that Bohr's keen interest in people turned the personal relations in his institute into a family-like atmosphere and in fact into an extension of the Bohr family, into which members of the institute were allowed to intrude at any time. The same could be said about the Peierls household in Birmingham and later in Oxford. What impressed his numerous students and junior colleagues about the "Peierls experience" was the way in which departmental affairs were allowed to be extended into the "Peierls family." There was no boundary between home and other portions of Rudolf Peierls' life, and to many students and postdoctoral workers the Peierls family became "their family" for the time of their stay at Prof's institute and sometimes beyond.

Acknowledgments

The author would like to thank Professor Gerry Brown and Mrs. Joanne Hookway (née Peierls) for helpful discussions and for comments on the manuscript.

References

[1] P. A. M. Dirac, "The Quantum Theory of Emission and Absorption of Radiation," *Proc. R. Soc.*, **A112**, 243–265 (1927);
  P. Jordan, "Zur Quantenmechanik der Gasentartung," *Z. Phys.*, **44**, 473–480 (1927).
[2] D. J. Kevles, *The Physicists. The History of a Scientific Community in Modern America*, 4th printing (Cambridge MA, Harvard University Press, 1995).
[3] E. H. Hall, "On a new action of the magnet on electric currents," *Am. J. Math.*, **2**, 287–292 (1879).
[4] R. Peierls, *Bird of Passage*, (Princeton University Press, 1985).
[5] *Reminiscences Collected on the Occasion of Genia Peierls' 70th Birthday*, July 1978. Copy in Peierls Papers, Suppl A.119.
[6] A. Pais, *Niels Bohr's Times: in Physics, Philosophy and Politics*, (Oxford, Clarendon Press, 1991), pages 359–361.
[7] L. Landau, "Diamagnetismus der Metalle," *Z. Phys.*, **64**, 629-,637 (1930).
[8] H. Bethe and R. Peierls, "The neutrino," *Nature*, **133**, 532 (1934).
[9] L. Hoddeson, and P. Hoch, *1981 Interview with Rudolf Peierls, 13 May 1981 and July 1981*, College Park, MD, Niels Bohr Library, American Institute of Physics.
[10] R. Peierls, "Statistical Theory of Adsorption with Interaction between Adsorbed Atoms," *Proc. Camb. Phil. Soc.*, **32**, 471 (1936).
[11] R. H. Fowler, "Adsorption Isotherms: Critical Conditions," *Proc. Camb. Phil. Soc.*, **32**, 144 (1935).
[12] R. Peierls, "On Ising's Model of Ferromagnetism," *Proc. Camb. Phil. Soc.*, **32**, 477 (1936).
[13] E. Ising, "Beitrag zur Theorie des Ferromagnetismus," *Z. Phys.*, **31**, 253–258 (1925).
[14] W. Heisenberg, "Theory of Ferromagnetism," *Z. Phys.*, **49**, 619–636 (1928).

[15] F. Bloch, "Elektronentheorie der Metalle," in *Handbuch der Radiologie*, Ed. E. A. Marx, (Akademische Verlagsgesellschaft, Leizpig, 1933), Vol. 6, parts 1–2, pages 226–275.

[16] N. Bohr, R. Peierls, and G. Placzek, "Nuclear Reactions in the Continuous Energy Region," *Nature*, **144**, 200–201 (1939).

[17] N. Bohr, *Collected Works*, Eds. E. Rüdinger and R. Peierls, Volume 9 (North-Holland Pub., 1986).

[18] *The Frisch-Peierls Memorandum of 1940*, in *Selected Scientific Papers of Sir Rudolf Peierls*, Eds. R.H. Dalitz and R. E. Peierls, (Imperial College Press, London, 1997), pages 277–282.

[19] G. E. Brown, "Flying with Eagles," *Annu. Rev. Particle Sci.*, **51**, 1–22 (2002).

[20] L.D. Landau and E.M. Lifshitz, *Course of Theoretical Physics* in 10 volumes, (Pergamon Press, Oxford, 1976-1981).

[21] MAUD Committee, *1941 MAUD Committee Report*. In The National Archives, Public Record Office, AB 1/594.

[22] R. H. Dalitz and R. B. Stinchcombe, *A Breadth of Physics*, (World Scientific Publishing, Singapore, 1988).

[23] C. R. Hill, R.S. Pease, R. Peierls, and J. Rotblat, *Does Britain Need Nuclear Weapons?*, A Report from the British Pugwash Group, London, 1995.

Chapter 15

Farewell

Rudolph Peierls to his Friends

Flat B, 2 Northmoor Road,
Oxford,
OX2 6UP
November 1986

Dear Friends,

This annual letter is earlier than usual, because its main substance will be a report on Genia's last year, which, in her own words, was "the best year of her life."

As we reported last year, the distressing and rapidly worsening symptoms in October [1985] had been diagnosed as due to a (non-malignant) brain tumor causing excess fluid pressure on the brain, and this was relieved by a "shunt." The tumor was left in place, because it was growing very slowly, and to remove it would have risked serious damage to important functions.

Her recovery was spectacular: within days her spirit and energy were back, and after a short period of recovery from the operation she felt better than for a long time, the only remaining trouble was some lack of balance.

She found it exhilarating to recover abilities that she thought had been lost forever. Soon she was walking all over town on shopping expeditions, or working in the garden.

She did not start driving again, because in August 1985 she had a minor accident in which she hit a cyclist, damaging the bicycle, but fortunately not the rider. This no doubt was due to her reactions being slowed down by the brain trouble, though we did not know this at the time. The accident came to court, and she was not only fined, but also disqualified for six months, and required to take a driving test after that. So we decided to sell our car (I already had given up driving) and settled down quite happily to being carless. Genia never enjoyed driving.

By March it seemed reasonable to make the trip to North America which we had to cancel last October, and we went off to Cambridge, Mass. where we met many old friends, on to Ithaca, where we stayed with the Bethes, and where Ronnie and Julie were spending a month, and where their two sons were students, so it was quite a family occasion. Then New York, where I was a guest at Columbia, and Gaby came over from Princeton, and took us to Montclair to see my sister. The next stop for me was Los Alamos, for some historical discussions, while Genia skipped that stop, and went straight to Seattle, where I joined her a few days later. There we were once again amongst many old friends. Lastly to Vancouver, to stay with Kitty and Chris and their charming children. The whole trip lasted just over a month, and Genia enjoyed it to the full.

This success encouraged us to plan more travels, and in June we joined a two-weeks tour of Northern Greece run by an Oxford travel agency. This, too was a huge success. It involved a fair amount of walking in the hills, and Genia took part in almost all these excursions, including climbing up 142 steps to one of the monasteries on the top of cliffs at *Meteora*. The start of the tour was not auspicious: The plane coming in from Athens had a bomb alert, and our departure from London was delayed seven hours. We reached our first stop, a hotel in Delphi, at 5 a.m., but by 11 we were up and admiring the temples on the sacred site.

For the rest of the summer we stayed in Oxford and Genia again had frequent dinner guests, both local friends and visitors (of whom there is always a good supply in summer in Oxford). We made excursions to Cambridge, to Birmingham to see Jo and Chris, and to Manchester, where we were guests of the Manchester Literary and Philosophical Society, a venerable institution with a long tradition in science. It was also an occasion to see my brother and his wife and daughter.

All this time Genia was full of enthusiasm at having been "born again." We were due to go to India in November, where I was to give the Royal Society's Rutherford Memorial Lecture, and as we also had an invitation to spend two weeks in Moscow, we decided to do so on the way to India. We were to leave on October 19, and Genia was busy getting together clothes for the Indian climate, small presents for our various hosts, reading about the places in India we were going to see, and about Indian food.

Then, on 14th October without warning, she was taken ill. It turned out to be a hemorrhage in and around the tumor. She was taken to hospital, where very charming nurses made her as comfortable as possible, but she had a miserable few hours before she fell into a coma, which lasted for eleven days.

All our children came at once, and between us we watched helplessly as she was slipping away, comforted only by the certainty that she was not suffering.

Genia was cremated on 30th October in a simple and brief ceremony attended only by the family and close friends. Her ashes have joined those of her sister, Nina, who died just four years earlier. As we are trying to settle down to life without her, I recall with gratitude her long and rich life, with its share of sorrow, but much happiness, always aware of the needs and problems of others. What she dreaded above all was being seriously disabled by age, and we must be grateful that she was spared this.

I am now being overwhelmed by letters and messages from friends all over the world, and it will be a long time before I can reply to them all. They give a vivid impression of the image Genia's friends retain of her, and the one word recurring again and again is "vitality."

With deep gratitude for all this warmth and best wishes for Christmas,

Yours,
Rudi Peierls

Chapter 16

Rudolf Peierls' Diary

Peierls, as well as other representatives of the old European culture, which I still managed to find in Oxford and Cambridge, radiated some special charm. All these people — Rudolf Peierls, Isaiah Berlin, Nicholas Kurti, who in the early 1990s were already well over eighty, were real intelligentsia. Of course, among the younger people I met were pleasant, honest, smart, bright ones — but never such as these titans. They were inhabitants of Atlantis, who had gone to the bottom of the European culture, destroyed by various "isms" of the 20th century. Alexei Tsvelik

Literally speaking, the contents of this chapter are not quite a diary. Rather, it is composed of fragments of Peierls' circular letters. The first six entries are the Christmas summary "reports" (1979–1985) signed by *Genia and Rudi*, and addressed to their children. These were apparently written jointly.

Even before Genia's death in 1986, Rudolf Peierls during his longer journeys without Genia used to send her detailed travelogues sharing his impressions of new places and new acquaintances. After her death not only had he kept this habit but even extended it. He started sending monthly reports to his children in the form of "Dear Everybody" letters. Sometimes the reports were written on a weekly basis, so, as Rudolf joked, his "diary turns into 'weekery'."

The intensity of his life was remarkable. In 1986 when Rudolf Peierls turned 79, he received the Copley Medal, the highest award from the Royal Society. The schedule of his meeting and talks on physics, history of physics, nuclear non-proliferation and other topics would be hard to maintain even for a younger person. He was deeply engaged in the Pugwash Movement, wrote a number of books, including an autobiography "The Bird of Passage" and published many articles in various journals, in particular, in *New York Review of Books*, of which I will say a few words later (Chapter 17) because of their significance.

Rudolf Peierls led an active social life: he maintained correspondence with dozens of old and new friends, and gave a number of interviews to mass media. His relationship with the media was complicated, as will become clear from the fragments presented below. Among other countries, Rudolf Peierls visited Russia many times because of emotional ties to this country and recollections of his youth. In his circular letters one will find eye witness testimonies of the events in the Soviet Union during the *perestroika* and in Russia after the collapse of the USSR. From the modern perspective they seem very naive — almost none of the expectations of that time came true — but this makes them even more exciting and instructive to read. His "big" journey to Russia in the autumn of 1987, filled with meetings of colleagues and friends in Leningrad and Moscow, had a more personal touch than many other trips. Before this journey, in June of the same year, Rudolf Peierls had celebrated his 80th birthday. For this occasion the Theoretical Physics Department at Oxford organized a special Conference.

This diary mentions many of Peierls' colleagues and old friends whose names grace modern textbooks and are familiar to physicists all over the world.[1] It is captivating to read about them from a

---

[1] I know personally some of the younger physicist friends of Rudolf Peierls, which made my task of preparing his notes for publication even more exciting.

different perspective, human rather than purely scientific. There is a sad note to that too. Almost every letter brings the news of the death of his old friends and even some students. In addition, Rudolf Peierls also reports on a number of conferences important in the history of physics. It seems to me that historians of science of the 20th century will find in this "Diary" so far undiscovered details.

Among other interests, Rudolf Peierls had a hobby: the art of cookery. Being at home, in Oxford, he loved to invite his friends or visiting colleagues for lunch or dinner prepared by himself using the recipes from various ethnic cuisines. The reader will find below many "juicy" examples including the Russian culinary experiences. These lunches and dinners were important in fulfilling Peierls' longing for socializing, a crucial element of his life after Genia's death.

In working on this Chapter, I avoided those parts of the "Diary" which are too personal or might be sensitive for other reasons, leaving only the fragments suitable (and hopefully interesting) for a wider audience. Not only do they exhibit the enormous amount of work carried out by Rudolf Peierls after Genia's death, but they also show his soul, his love and care for friends and students, his wisdom and kindness — and, yes, sometimes, naïveté. A number of issues raised in the "Diary" are also discussed in other chapters because of their general significance.

*****

Christmas 1979

Before Rudi's retirement in 1974 we sent our last Christmas card, signing off since we expected that from now on we would be all over the place, making the dispatch, and even the receipt, of Christmas cards difficult. This was a realistic expectation, as you will see from the list of places where we met successive New Years since then: Hawaii (*en route* from Sydney to Seattle), Leiden, Oregon (*en route* from Seattle to Los Angeles), Orsay, Copenhagen, and now Copenhagen again. But many of you still remembered us, and your cards and greetings eventually do catch up, so we feel we want to give a sign of life and pass on some of our news.

Our nomadic life took us first for three months to Sydney, stopping to stay on the way in Isfahan, Bangkok, Hong Kong and Bali. It was exciting to be in the antipodes not only because the sun is in the North at noon, but because almost any animal or plant you see is unfamiliar. Apart from visits to other towns we treated ourselves to a trip to the Great Barrier Reef, where we could watch the turtles laying their eggs at night. From Sydney we went across the Pacific, stopping in Tahiti and Hawaii, then going on to Seattle, where Rudi had a three-year appointment for six months in the year. We had become very fond of the North West, and Seattle is full of friends, almost a second home.

Figure 16.1   Rudi and Genia in Hong Kong, 1979. Courtesy of Natalia Alexander.

Last summer, just before leaving England, we were involved with the law. A book appeared about the insidious influence of Soviet agents in England and many spies. The book was full of nonsense, but the author was careful enough to name mainly dead people. Unluckily for him, he believed I was dead, too, and said so, after making some pretty libelous remarks. When the publishers discovered that I was still alive, they at once agreed to settle out of court, and it only remained for the lawyers to negotiate the amount of damages. This came out to a nice substantial figure. Surprisingly, the whole affair took only 11 days from my first approach to my solicitor to the final settlement reported in court.

Figure 16.2 Rudolf Peierls and Elevter Andronikashvili.

December 1981 (annual summary)

In mid-March was the date of our golden wedding anniversary, and Los Angeles, or, rather Santa Monica, seemed a good place for a celebration since even in March snow or fog were unlikely to impede the arrival of planes there.

The winter was spent near Paris, where Rudi was consulting for two months at Saclay and two months at Orsay, and we stayed again in one of the houses of the Institute for Advanced Study in Bures-sur-Yvette, Genia's sister Nina, who had visited us in Oxford for the summer, spent the first month with us in Bures, and spent much time, with and without Genia, exploring Paris.

The winter 1979–1980 was spent partly in Ljubljana and partly in Copenhagen where Rudi was working on the Nuclear Physics volume of Niels Bohr's Collected Works. There we had the good luck to get a flat in a house in the old harbor part, which the State Bank converted into a set of flats for distinguished visitors. Beautifully designed and furnished, and in an ideal position. There the film "A Man Called Intrepid"[2] was showing which is full of libelous insinuations about Niels Bohr, and, as nobody else was doing anything about it, Genia demonstrated for three days outside the cinema with an appropriate placard. This, of course attracted a lot of publicity, which was the object. The year before the excitement was a libel suit. Some idiot wrote a book about Russian agents and spies, in which he carefully named only people who were safely dead, but he somehow had the idea that Rudi was dead, so he said some really nasty things about him. So of course Rudi brought a libel suit, which was settled out of court, and the book was withdrawn — which is just as well, because apart from the libel it was full of nonsense. The settlement involved "substantial damages" (at that rate we would not mind being libeled again!) and remarkably it was 13 days from the first approach to a lawyer to the final settlement being announced in the High Court!

The winter of 1978–1979 was also spent in Copenhagen; there too, we had Genia's sister Nina with us for a month.

November 1982 (annual summary)

The year was overshadowed by the death in October of Genia's sister who had been with us since July. She was already unwell when she arrived, with what was diagnosed as spastic colitis, and suffered from loss of appetite, so that she had become very emaciated. With care and medication these troubles improved but she was still very weak and anemic. A stay in hospital for observation did not show the cause, but in October she had to go to hospital for further tests, which showed

---
[2]See page 206.

cancer. She died from an internal hemorrhage, probably caused by the tests, but this was a blessing, as it saved her a period of suffering and misery. We also were relieved that it happened here, where she could get better care and comfort than in Leningrad.

November 1983 (annual summary)

The longer stay at home allowed Rudi to play with his word processor, last year's birthday present (which had already produced last year's Christmas letter). With this new facility he started writing his memoirs, which so far have grown to about 120,000 words, without being near the end. It remains to be seen if anyone will publish them.

From Pisa to Varenna, a charming village on Lake Como, we went for a summer school. Varenna now is more peaceful than it used to be when main road traffic roared through the narrow village street. Now the village is bypassed by a tunnel in the mountain behind it.

We got back in good time for Genia's 75th birthday.

In October, Genia went for a week to Moscow and Leningrad. She had intended to go earlier but all the tours for spring and summer were booked up, and a tour is the only practical way to get there. But now there was the ban on Soviet planes, and we did not know until the last minute whether her plane would go. It did, but not without trouble: there was "industrial action" at Gatwick, the plane was delayed two hours, the passengers had to carry their luggage on to the tarmac, and there was no food on board.

Genia's main object was to look through the papers left by her sister. This proved quite a shattering experience for her, since these papers included all the letters from her mother and stepfather written when they were banished, and when the stepfather spent sometime in prison. He was a wonderful man, and the letters from this period showed this particularly.

November 1984 (annual summary)

After spending April at home we set out for the next transatlantic trip, this time in the East, mostly in three-night stands. The occasion was an invitation for Rudi to give some very official lectures in Toronto. We stayed the first few days with Isabel Jephcott, who had been Ronnie's foster mother during the war, and had a great time with her and her family.

A trip to Italy was to attend a symposium in Pisa marking the retirement of Giancarlo Wick. Pisa was, as usual, very attractive, and the meeting full of warmth. It turned out that of the people there, we had known Giancarlo the longest, except for Amaldi, who met him two months or so before us.

Another coming event is the conferment by the University of London of an honorary degree on Esther Simpson, the charging person who "was" the Academic Assistance council, later Society for the Protection of Science and Learning, and a kind of godmother to all refugees.

On 1 December we go for two months to France. This visit begins with meetings and receptions at the Academy of Science, which recently elected Rudi a Foreign Associate.

November 1985 (annual summary)

1985 has brought us many problems, but it seems they have all been successfully disposed of. To start with the most serious one: In the summer Genia started to have difficulty walking; it seemed her legs would not obey her. She also lost her usual energy. Our doctor could not find the cause, and finally she went to the best consulting physician around, who saw her in mid-October. By then her symptoms had become much worse, she now could not walk even a few steps without help; she was getting very depressed, and began to get confused. The consultant took a lot of samples for analysis, and when these did not provide an answer, he arranged for her to be admitted to hospital, and a few days later she was given a brain scan. This showed a (non-malignant) brain tumor, which had led to a high fluid pressure in the brain. Three days later she had an operation. The tumor is growing so slowly that it could be left in place, but the surgeon put in a "shunt" to drain off the excess liquid. In the last days before the operation her condition had declined spectacularly, but the improvement afterwards was equally spectacular. We are writing 2.5 weeks after the operation, and she has her old *joie de vivre* back, and can walk better than before the summer. She still has to rest much to recover from the operation, but the doctors are very pleased with her progress, and admire her vitality. It was of course very worrying to see her deteriorating without knowing the cause, and we are very relieved by the outcome. Such troubles show you how many good friends you have, who not merely sympathized, but were ready to help with driving, running errands, or anything else that could be helpful.

Life continued uneventful for a while. For our wedding anniversary, which, at 54, had no symbolic significance, we treated ourselves to a night at the "Manoir des Suat" Saisons, a relatively new hotel near

Oxford, which has two stars in the Michelin guide for its food and fully deserves it. The rooms and the service are also an experience, from the four-poster bed and the bathrobes matching the decor of the room to the bowl of exotic fruit and decanter of Madeira provided. The place is owned by a self-educated French chef, who is an absolute waster. He was running a small restaurant in Oxford, for which he earned his two stars, and very unusually Michelin transferred the stars to his new country hotel even before it opened.

Later we celebrated the 60th birthday of Dick Dalitz with a one-day scientific symposium, and a dinner which filled the dining hall of Worcester College to capacity. In April there was a meeting in Cambridge in memory of Paul Dirac, one of the greatest theoretical physicists of our generation, who died last year. This was followed by a dinner in St. John's, his old college. Dick Dalitz and Rudi will have to write the "Biographical Memoir" of Dirac for the Royal Society, which is a substantial project.

In June there was a conference in Finland on the interpretation of quantum mechanics. Rudi was also invited to spend a week or so in Helsinki. Being so close to Leningrad, we decided to visit there for a few days. We had plans to do the trip by car and ferries (except the side trip to Leningrad) and to call on our friends in Sweden on the way, but we got cold feet when there was a widespread strike in Sweden, and flew directly to Finland instead. We enjoyed Helsinki with its 18th century architecture, reminiscent of Leningrad (much of it was built by the same architect), and the conference in a university in a small town in North Karelia was very pleasant, with us trying many local customs, including the sauna.

Three days were spent in Leningrad visiting many friends. We were there as tourists, but Rudi gave a lecture in Russian in the same institute in which he had given his first lecture course in Russian in 1931. June is the right time to see Leningrad, because the city is at its most beautiful during the "white nights." We had beautiful, if cold, weather. The buildings no longer look shabby. Probably their paints are better, and they have industrial pollution under control. This improvement, however, does not extend beyond the house fronts. Look inside any courtyard or staircase, and everything is dirty and neglected.

We stayed in a very pleasant old hotel, though its restaurant was nothing to write home about. We had only one dinner there, otherwise being fed by friends. In the large menu, only a few items had prices written in, so we chose from them, only to be told they were off. When asked what he could offer, the waiter suggested bouillon with *pirozhki*. "But we have no *pirozhki*."

One day it was crowded and another couple sat at our table. Looking at each other, we found they were Sir Richard and Lady Doll. He is a former medical professor in Oxford, the one who established the connection between smoking and lung cancer.

In July came another change: Rudi had to give up driving, as his eyesight no longer satisfies the legal requirement. This is somewhat inconvenient, but since he lived for 30 years without driving, he should be able to manage again.

This year is the centenary of the birth of Niels Bohr, and this has developed into a real industry. Rudi started his involvement with a brief talk at a conference of the European Society for Nuclear Medicine, which wanted to remember Bohr.

March 13, 1987

I arranged a meeting next week at which I and someone else will give a report on the Moscow Forum, and we can also explore local reaction to the formation of a group (and find some people who will do the work!). Also next week, there will be a small meeting in London of a few local Pugwash people, mainly geophysicists, to discuss with government scientists the question of verification of weapons tests. Government policy is to keep saying that tests cannot be verified, and other scientists say this is now easy. So this confrontation will be interesting.

Gino Segré, the nephew of Emilio,[3] is here on sabbatical, and I invited him (and his wife) for dinner. He told very amusing stories about Emilio.

Lady Oliphant died recently. She had been in a coma for more than a year with his nursing her faithfully, so this also can be regarded only as relief.

May 5, 1987

Belousov (husband of Genia's cousin Natasha) appeared at a conference in Abingdon,[4] so I invited him for dinner in college on Monday, with logistic problems about collecting him from Abingdon and getting him back.

---

[3] Gino Claudio Segrè (b. 1938) is a Professor Emeritus of Physics at the University of Pennsylvania. He is the author of several books on the history of science, particularly on atomic physics.

[4] Abingdon also known as Abingdon-on-Thames, is an old town 10 km south of Oxford.

My own plans are as previously reported. I now have the official invitation to come to Coimbra for an honorary doctorate in January, but the dates are not yet fixed.

Last Friday I gave lunch to the Vigarts (the distant cousins who sent me the family tree).

June 11, 1987

There came a call from a television producer asking me to take part in a discussion program starting at midnight on the Friday and going on, live, for three hours. One of the participants, in fact the only one yet definite, besides myself, will be Edward Teller, and the subject Star Wars. I could not resist this temptation, of course, and I found that I can get back in time for it by flying from Turin via Rome! I did not believe that there would be any viewers at this time, but apparently this is a regular series on Channel 4, which usually has some 500,000 viewers!

July 6, 1987

Next day I had dinner at the Griffins (Maulde Halban) — a large and elegant dinner party with many philosophers, some quite amusing, and gallons of champagne.

Saturday was the big symposium, which went off very well. Dick Dalitz had been hard at work planning, and his choice of 40 minutes for each speaker worked very well. One sad gap was that Tony Skyrme, who had promised a talk, was taken to hospital and died a few days before the the symposium. (Apparently he had a duodenal ulcer, which was upset by anti-arthritis medication, so it started to bleed and an embolism developed.)

Then on Sunday I invited the Bethes to lunch. We could have had more people, but I knew that Hans dislikes large parties, and in particular he was anxious to brief me on the impending debate with Teller, which was most useful.

The conference was in a "Center for Scientific Interchange," built originally as a millionaire's villa in a gorgeous position on the top of a mountain overlooking the town. There were some 30–40 people there, and that is probably all they can accommodate. The lectures were interesting, and I learned a lot. The food was good, but of course too much. On the Wednesday evening was the conference dinner, on the terrace. Halfway through the dinner, a thunderstorm started, with a sudden and dramatic cloud burst. But with rapid action by the diners and the waiters (all family of the woman doing the normal catering), the tables and chairs were moved to a dry place, and the meal continued... Next day I gave my talk, and Regge, who is the director of the Center, came to listen and stayed for lunch. Giancarlo Wick also came.

In London I was met by Alfie Fox, the organizer, who took me by car to the Dorchester Hotel, and on the way played me recordings of BBC interviews with Teller and Sergei Kapitza (who had also come to take part) which took place the day before. The Dorchester does not differ much from more modest hotels. The rooms are large and well furnished and in addition to huge towels there is a bathrobe. A traditional-looking cupboard contains the TV and the mini-bar. The only special thing is the service. The receptionist who signs you in also guides you to your room, and your suitcase just appears without your having identified it. The waiters are courteous but not deferential, and their attitude is more or less that to an uncle from the provinces.

Then at 11:15 another car collected me and took me to the studio in Wandsworth, where then also Teller, Sergei Kapitza, Enoch Powell, a woman writer and a catholic nun appeared. The program did not start at midnight as I had thought, but at 12:30, so the debate was "only" $2\frac{1}{2}$ hours. We sat in a circle, (no fuss, no make-up, no volume tests) with a chairman (I forgot his name) who did not interfere much. Teller picked a violent quarrel with him (before the beginning): The chairman had prepared a couple of sentences to introduce each of us, and read these out. About Teller one sentence said that he was sometimes called "father of the hydrogen bomb" but he did not like that name. Teller blew up and said it was discourteous to mention that name when it was known he did not like it. After a violent argument, which almost delayed the program, the chairman gave way and deleted the sentence.

The program itself went off quite well. Of course one always thinks afterwards of points that might have been put better. I had been given to understand that the title was to be Star Wars, but it was "Peace in our time." We did talk about star wars most of the time, but the two women pressed general arguments for peace and against the use of force in any circumstances. They were answered very effectively by Enoch Powell, whom I found very impressive: I disagreed with him only in a few points. Teller is of course very good at taking up time by speaking loudly and slowly (he does not debate, he preaches) and at length, but there was enough time to answer him, and to point out inconsistencies. He made his usual spiel about the Russians being ahead in work on ballistic defense, which gave me the opportunity of saying that his sources were evidently from intelligence, and that people who had access to intelligence reports had a very different impression.

I have been writing a review for *Nature* of Teller's latest collection of essays (which he is not going to like).

September 9, 1987

My trip to Trieste was uneventful. I was met at Trieste airport by a 12-seater minibus as no other car was free. The next day was the formal meeting to celebrate Newton, and also the occasion for the award of the Bogolyubov Prize (which the Center[5] has invented) with Bogolyubov (a distinguished Russian physicist) present. It went to a bright man from Pakistan. By way of introduction there was a long talk about the merits of Bogolyubov, which was given by his son (also a physicist), a somewhat peculiar idea.

Salam also gave out some minor medals of which I got one. In conferring it Salam described me as friend of the Center which is generous, because he knew that I had opposed its foundation. But of course it is true that now that it exists I wish it well.

I also met the second Mrs. Salam who is English, a Somerville tutor in (I think) biochemistry and very charming. She had their two children, aged about 4 and 7 with her. The wife No. 1 was not there.[6]

[I took] two more trains before I got to Gmunden,[7] a pleasant little town on a big lake between mountains, with a modern conference center. The 200 or so members of the Pugwash conference were spread over a dozen hotels, because of the policy of the conference center to let all the hotels share in the profits. This caused of course logistic complications, since walking from most hotels would have taken 45 minutes or so, but the organizers managed to dispatch appropriate buses, except on the last night, see below.

The conference was friendly and useful. The Soviet members on the Council were replaced, and the new ones (new to the council, not to the conferences) are a great improvement. Viki and Ellen Weisskopf were there, but as he still gets tired very quickly, he came only to the first day, and again on the last day when he gave a major address.

---

[5] International Center for Theoretical Physics. Currently, Abdus Salam Center for Theoretical Physics, Triest, Italy.

[6] Mohammad Abdus Salam (1926–1996) was a Pakistani theoretical physicist. He shared the 1979 Nobel Prize in Physics with Sheldon Glashow and Steven Weinberg for creation of the electroweak unification theory. Abdus Salam was an Ahmadi Muslim and had two wives simultaneously.

[7] Gmunden is a town in Upper Austria.

October 9, 1987

Yesterday I had a meeting in London and decided to visit the Soviet Consulate to see whether my visa was ready. This involves queuing for 20 minutes in the street, and another 10 inside, but when I finally got there they said "we can't give you a visa without a telex from Moscow." I am boiling mad, because they could have said that a month ago. Today I tried to ring the Institute in Moscow (now through dialing, which is back, which helps) but there was no reply. I suspect that is their October holidays. So I tried the home number of my host, but he is away and will not get back until tomorrow evening. I still have two weeks and I think with enough hysterics I can still get my visa.

In Trieste I met Kaganov,[8] a Moscow physicist whose daughter has translated the "Surprises."[9] He was disappointed I had not yet received the typescript of the Russian text. The reason for the delay was that their publishing house had the brilliant idea to send it c/o my publishers, so it had to cross the Atlantic twice. With this delay I had to read it urgently. The style seems to be fine, as far as I can judge, but I found two pages of howlers, and if I had read it more carefully, would probably have found more.

October 23, 1987

First news is that I got my visa this morning! After my last letter I got through to Khalatnikov,[10] my Moscow host, and he promised to get a message sent. Indeed on the Tuesday his secretary phoned to say the message had been authorized and all was well. I waited a week and then decided to phone the Consulate. They were engaged all morning except for a few minutes when the phone was ringing, but there was no reply. So I tried in the afternoon and was told the visa section was closed, I should I ring between 10 and 12.30. Then I tried the scientific attaché at the Embassy who was out. Next morning the same situation, but this time I got the secretary of the attaché, who enquired and rang me back to say the visa had been issued — the authorization came in the day before (so it took a week from Moscow to London). Should they mail it? I decided this was asking for trouble and said I would pick it up today, but I would not have time to queue at the Consulate,

---

[8] Moisei Isaacovich Kaganov (b. 1921) is a Soviet theoretical physicist working in condensed matter and plasma physics.
[9] Rudolf Peierls, *Surprises in Theoretical Physics*, (Princeton University Press, 1979).
[10] Isaak Markovich Khalatnikov (b. 1919) is Landau's student who is famous for his work on quantum liquids, superconductivity, quantum electrodynamics, relativistic hydrodynamics, and general relativity.

and she offered to collect it and keep it in her office. And so it was. I do not know what people do who have no strings to pull.

Also on the Tuesday another problem was solved: The B[ritish] A[irways] plane to Moscow leaves very early in the morning, so I preferred the Aeroflot one, for which I had been on the waiting list for four months. (These planes are not really fully booked, but they like to keep the places up their sleeve.) When I was in London a fortnight ago I called in the Aeroflot ticket office and talked to "Betty" there. She remembered getting me the ticket to go to the Forum in February, and when I said I was invited by an Academy institute she said I would get a place, but this would be confirmed only on the 20th. Which it was.

Next I phoned Khalatnikov to give my flight. He said they would meet me but there was a complication: the Academy guest house was full because of a conference, so when I arrive they will give me dinner at their house and then put me on the night train to Leningrad, where I shall spend the first week.

All this week there has been fog in Moscow, and the airport was closed from Monday until today (Friday). You probably heard that George Schultz had to arrive by train. Today there were flights, but the weather is still unstable. If my plane is cancelled I shall add a note to this letter. I have been given the home phone numbers of three senior people of the Institute to communicate in that case.

December 9, 1987

This will be a giant travelogue, perhaps it will have to be done in installments. No travel problems, plane left an hour late and arrived an hour late. Met by Paul Wiegmann,[11] a very nice and bright theoretician whom I met already in February. He came armed with a letter from the Institute to the immigration people asking for their assistance in meeting a distinguished foreigner, and to his surprise this produced the result that he could meet me right in the customs hall, and that my suitcase was not examined.

After we had supper at Khalatnikov's, Wiegmann came with me on the night train to Leningrad, which is really taking hospitality rather far. The sleepers are comfortable, two berths side by side, but no plumbing, and the only washrooms are standard train toilets, one at each end of the coach. Met by Viktor Frenkel (son of a distinguished

---

[11] Paul B. Wiegmann (b. 1952), a Russian physicist specializing in theoretical condensed matter physics. Currently he is the Robert W. Reneker Distinguished Service Professor in the Department of Physics at the University of Chicago.

theoretician,[12] mainly working in the history of science) and Anselm[13] (son of a friend of Genia's who had visited us in Oxford, he works in the Nuclear Physics Institute at Gatchina[14]). I stayed in quite a good hotel, at the end of the Nevskii prospect, with excellent lifts, and efficient service. Food not so hot. For breakfast and lunch there was a good buffet, but for dinner they were reluctant to serve me in the main restaurant, which they said was only for tourists, and I would have to wait for ages. Otherwise there was only the Grill Bar, with a choice of two or three dishes, and mainly for drinking with a horrible pop music.

On the first day Wiegmann and I went on a long walk among the attractive parts of town, somewhat spoiled by the bunting and slogans put up for the coming festivities. In the evening Frenkel had got tickets for a nice chamber concert; I was afraid that after the night on the train I would fall asleep, but I lasted. This was on the Monday. Tuesday I was on my own, wandered around town, visited the Hermitage and a huge department store, had lunch by chance in one of the best restaurants, and had dinner in the evening at Anselm's. He explained that they had moved to a smaller flat (I suspect to separate from his father, with whom relations were not too good) but since then their daughter had got a divorce and moved in with her baby, and at the moment his parents were visiting, so it was rather a slum. But the evening and the dinner were pleasant.

Next day an excursion to Pushkin (names are very confusing, the place was Tsarskoe Selo, before the revolution, then Detskoe Selo) and Pavlovsk. Charming palaces lovingly equipped, and even more lovingly restored. They showed everywhere photographs of what the rooms looked like after the Germans left. So until Thursday I did not set foot in any Institute. This has to be understood because in their institutes the theoreticians appear only once a week for the seminars, and otherwise work at home. Also they had rather short notice of the dates of my visit.

Thursday I was taken out to Gatchina and gave two talks (in Russian), Friday lectures in the Ioffe Institute. In the afternoon Frenkel accompanied me back to my hotel. I decided to go for a walk with my cameras, though by the time I got to the attractive parts of town it

---

[12] Yakov Frenkel, see page 39.
[13] Alexei Andreevich Anselm (1934–1998) was a Russian theoretical physicist, known for his contributions to the theory of complex angular momenta, works on quark models, spontaneous symmetry breaking, and mechanisms of CP violation.
[14] Andrei Anselm, see page 49. Gatchina is a town located 45 kilometers (28 miles) south of St. Petersburg.

was dark. So I wandered up the Nevskii, amongst dense crowds (it was the afternoon rush hour) when someone tapped me on the shoulder. It was Frenkel, who had meanwhile been in a bookshop. We went to the cinema together, but it turned out to be a rather melodramatic adventure film. Still some interesting reflections of life. Saturday I had reserved for the relatives: Masha [Verblovskaya],[15] and her husband Yura. We joined a friend of theirs who is a Dostoyevsky specialist, and who took us around the parts of Leningrad where Dostoyevsky lived, and where all his stories took place. In fact, people claim to know the precise houses in which various episodes happened, but our guide says these are all legends, the writer did not have any specific houses in mind. But the atmosphere comes across beautifully. Dinner in the flat of Masha's daughter, Natasha, who is some kind of biologist, and then to a modern ballet with the parents.

Sunday was difficult, because both the relatives and the Anselms wanted to entertain me. So we started in the morning with the Hermitage, where the Anselms had a friend who could show us the Scythian gold and other treasures not easily shown, and then after a "light" lunch, for which they also had invited a famous local poet, to Masha's house and dinner. After this another concert with Frenkel.

Monday was a ceremonial session in the Ioffe Institute, with speeches by the director, and various representatives from local authorities (my presence was commented upon) and then a review talk by Gorkov[16] from Moscow, a good man, about the new high-T superconductors.

Lunch in Frenkel's flat, but without his wife who was unwell, one of the many casualties encountered.

In the evening I left by the night train being ceremoniously escorted by Frenkel, Anselm, and the relatives.

Met in Moscow by my faithful Wiegmann. Now I had a reservation in the Academy Hotel, but getting there was not so simple. The service office was one floor up, and the lift did not stop there, so Wiegmann manhandled my suitcase up the stairs. They explained I had a room in the second (newer) building, but the exterminators were there, so I would for a time get a room on the fifth floor of this building. We scrambled up another lot of stairs to get the lift, and then pressed the button for the fifth floor, not noticing that the thing took us to floor 8. The woman on the floor was horrified, as the room was not ready, so

---

[15] See page 111.
[16] Lev Petrovich Gorkov (1929–2016) was a Soviet-Russian theoretical physicist famous for his pioneering work in the field of superconductivity.

she asked us to wait while the room was hurriedly dealt with. But then she spotted the error in the floor and we descended to floor 5. Here the floor lady said she had just let the room to someone else, and he had gone out. As I clearly had priority, they stated to remove the other man's junk from the room, when luckily he turned up.

After this I decided I might as well stay where I was, until I was told that now the exterminators would come here!

In the afternoon I gave a talk in the Institute of Experimental and Theoretical Physics,[17] one place where one can find theoreticians in their offices. Dinner with Abrikosov,[18] which was interesting because he eloped with the wife of Nozières.[19] I had heard the story fully from the other end. By now she has left him, and he has a younger and very attractive wife.

Wednesday I talked to various people about various things and then gave a talk in the Kapitza institute. This place has kept Kapitza's spirit, for example you can just walk in at any time, without any passes or other pieces of paper, which is quite unique. The talk was my "reminiscences," which went down well. After a ceremonial tea I was taken to have supper with Styrikovich,[20] the engineer friend of Landau, with whom we walked through the Caucasus in 1934. He is about 85, but definitely all there, very cheerful and lively.

On the following day I was asked to give a talk on the arms race. This was arranged to be in the Space Studies Laboratory (perhaps to keep the size of the audience down, because permission to enter is rather complicated: but many came anyway). Sakharov was there, and when questions were invited, he made a statement almost as long as my talk, but nobody stopped him. He disagreed with me on several points, above all he saw no need to ban nuclear weapons tests. After that a dinner party at Gorkov's and then a TV interview. This was for a program on Landau. On the Friday morning there was an interview with a reporter from the Moscow News. This weekly is the most outspoken paper in Moscow; copies are sold out at 6 a.m. on the day of publication, so that

---

[17] For 20 years this was "my" institute, see e.g. my Foreword in *Lectures on Particle Physics and Field Theory*, (World Scientific, 1999), Vol. 1, page v.

[18] Alexei Alexeyevich Abrikosov (1928–2017) was a Soviet and Russian theoretical physicist famous for his contributions in the field of condensed matter physics. He was awarded the Nobel Prize in Physics in 2003.

[19] Philippe Nozières (b. 1932) is a French physicist whose major contributions describe the behavior of electrons in metals. Winner of the Wolf Prize in Physics in 1985. Worked at Institut Laue-Langevin in Grenoble, France.

[20] Mikhail Adolfovich Styrikovich was a Soviet thermal power engineer and scientist specializing in high-temperature physics.

all my friends subscribe to the English language edition. I haven't yet seen whether the interview got printed.

Then a call on Kapitza, and the rest of the day with the Belousovs[21] (Natasha and husband) who took me on a drive to an old monastery and to a market. The latter was very interesting, being the intermediate level between the government shops and the free-enterprise stalls. It is run by collective farms, with government encouragement. There is fruit and vegetables of all kinds in plenty, but quite expensive.

The Saturday was the great parade, which Khalatnikov had invited me to watch on TV from their flat. There was a problem: he lives in an area close to the Kremlin, which on such feast days is closed to all people. Wiegmann fetched me and we walked towards the place (no public transport) until we were stopped by police. Wiegmann started arguing with the officer-in-charge, who said there was absolutely no possibility of our passing. To me the situation looked hopeless, but Wiegmann kept talking and waving my passport, until a higher officer appeared, and dispatched a policeman to accompany us. The parade itself was disappointing. Dinner at Khalatnikov's daughter's "because she is a very good cook," which she is.

Sunday was Wiegmann's birthday, and he invited me to a large family party. There was some confusion caused by my carelessness: I had not noticed that my plane was an hour earlier than I had been told at first, and that came out only when I casually looked at my ticket. As Wiegmann insisted on driving me to the airport, this made the party somewhat rushed.

I'll try to sum up my impressions. Talking with many people (but all intellectuals), I got a wide range of reactions to Gorbachev: One extreme was "True, there is some more freedom of information, but this does not represent a deep change, Gorbachev is just out to increase his popularity." At the other extreme there is confidence that he will succeed, that things "cannot possibly go back." But all are unanimous that, if Gorbachev fails, there will be a terrible backlash.

January 9, 1988

I was expecting to go to Coimbra for an honorary degree ceremony on 17th January, but before Christmas there was an apologetic phone call from João da Providência[22] saying that the University administration

---
[21] See footnote on page 219.
[22] João da Providência was a professor at the Centro de Física, Departamento de Física of the University of Coimbra. See page 217.

had slipped up by not reserving the hall, which needs two months' notice, and by not ordering the special suit and gown I have to wear for the ceremony. This also involves an "insignia" which requires a good deal of embroidery — the whole thing takes three months. They had not even asked for my measurements. This request came later, and when Providência phoned to check that this was all right, I asked whether they did not also need my hat size. (The Brinks[23] had watched such a ceremony and reported that a hat was also involved.) We agreed I would also add that information. A few hours later Providência had a call from the university: Please don't forget to ask for the hat size!

It was sad that Zeldovich died just after I had been in Moscow. I saw him there, and he gave me a present, a calendar with paintings of Moscow. He was an amusing man, and a first-rate scientist.

April 10, 1988

Last week I fed dinner to the Griffins and Els Placzek, who was visiting. Els is as ever charmingly naive and self-centered, though the charm is beginning to wear off a little.

My two days in Stanford were very interesting. They were spent in their Center for International Security and Arms Control. My host was Sidney Drell who now spends practically all his time there. He had just been in Moscow with a party of senators, who took him and several other people as advisors. They had a two-hour session with Gorbachev and a similar one with Shevardnadze, as well as with many physicists. He was most impressed with both Gorbachev and Shevardnadze. Surprisingly he was also very impressed with some of the senators.

On my second day there was an all-day meeting discussing the verification of any limit on sea-launched cruise missiles (SLCM, pronounced "slickoms," it took me a while to figure out what slickoms were). Such a limit is likely to be part of the next agreement, and the difficult problem of verification of SLCM also came up in the Moscow talks.

On the second day I had an interesting lunch with the solid-state people, including Ted Geballe,[24] and a big dinner party in the evening hosted by Sidney Drell. After giving my talk and listening to Drell's report on Moscow I was taken to San Francisco airport by a girl student, who had also met me there.

---

[23]David Maurice Brink (b. 1930) is Australian-British nuclear physicist. Recipient of the Rutherford medal, 1982 and Lise Meitner prize, 2006.

[24]Theodore "Ted" Geballe (b. 1920) is an American physicist famous for his contributions to the field of superconducting materials. He is Professor Emeritus of applied physics at Stanford University. Recipient of the Oliver Buckley Condensed Matter Prize in 1970.

May 22, 1988

Figure 16.3 Coimbra: Award of honorary doctorate, 1988.

This will be quite some report. I shall start at the end, with the trip to Coimbra. Compared to the degree ceremony there, our *Encaenia*[25] is just a little routine exercise. I had sent my measurements beforehand,

---

[25] Encaenia is the ceremony at which the University of Oxford awards honorary degrees to distinguished persons.

and had two fittings at the tailors, so my outfit did fit. It consisted of a pair of black trousers, a long black coat, buttoned to the neck Indian style, a black cape held by a cord around the neck. Then the "insignia," a heavily embroidered silk stole, covering chest and shoulders, in the faculty color, which for science is light blue. All this I was given before the ceremony. A page carried in the procession a tray with the things I would be given in the process of conferment; the hat, also light blue, shape of a tea cozy with a spike like a German helmet (the tea cozy analogy is obvious, the hat is locally called "intelligence cozy"), a gold ring with a light blue stone, and a book. The procession marches from the old and beautiful library to the assembly hall, or rather it should have, but it was raining so the procession was re-marshaled under cover. In the procession there are all the doctors present, most in their ceremonial garb, also guards with halberds, and a brass band, which as a marching tune plays a Christmas carol.

The precise order in the procession is laid down in a screed of some five pages, which also says that at some point I have to make a "brief and elegant speech," fortunately not necessarily in Portuguese. There were two orators, the senior one praising the candidate, the other praising the presenter, who was Providência. The Registrar, a dashing youngish man with a black cape and a sword, was going around guiding each performer to his proper position for the next operation. He took me to face the Vice-Rector, who presided, and who asked me in Latin for my wishes, to which I had to reply, also in Latin "the degree of doctor in the distinguished science faculty." Then he pronounced the magic formula and the Dean (Urbano, my Oxford DPhil) handed me the book, a leather-bound edition of the statutes of the university of 1599 (a recent edition), put the hat on my head and the ring on my finger. After this I had to embrace the Vice Rector, the Dean, and then all the assembled doctors. Then I had to pronounce another Latin formula, thanking everybody for the benefits received, and here I stammered a bit, because I had learned the phrase, but the Registrar gave me a card with the wording on it, and trying to remember and to read it ended in confusion. But I did get it out. We marched out as we had come in, everybody signed a book, and was given a diploma with an enormous seal. Apart from posing for innumerable photographs, that was the end of the proceedings. We unrobed, and I was given metal boxes in which to carry my insignia, the hat, and the diploma. I could not of course take photographs of the ceremony, but Providência's daughter, Clara, volunteered to do so with my camera, and she did very well. One of her brothers used another camera. Prints will be sent when available.

I learned one interesting fact: The reason that the Spanish, and not the Portuguese, who were then better seafarers, discovered America, was that the Portuguese had not evicted the Jews.[26] They had good Jewish scientists, who had made an accurate estimate of the radius of the Earth, which showed that India was too far to be reached by the boats of that time; the provisions would not have lasted. The Spanish relied on estimates of Arab scientists, who had a low figure for the Earth and thus thought that India was within reach. Goes to show that knowledge can be harmful.

Traudl left on Sunday morning, and then arrived the Migdals.[27] He is an old and well-known Russian theoretical physicist, whom I first met in 1956. (He still remembers a joke I told then!) As I heard they were paying their own expenses, I had invited them also to stay. His (second) wife, a scholar of French literature, is a nice person. She speaks no English. I took them on the usual tour of Oxford on the Monday, gave them a pub lunch, and in the evening they went to dinner at the Brinks. Verena Brink is Swiss and therefore fluent, in French, so there was no problem. Tuesday he spent, in the department, and she was invited to lunch by Fennell, the retired Russian Professor at New College, which worked out well. In the evening there happened to be a guest night in College, with port in the common room by candle light, and the Migdals were thrilled by this. They left, very satisfied, on the Wednesday morning.

July 16, 1988

Since I last wrote, I have been travelling practically the whole of June. It started on the third with a conference at Ditchley Park. This is a beautiful country house near Oxford run by an American (or Anglo-American?) foundation, which organizes high-powered conferences. This one was about disarmament, or rather on the question of whether a complete abolition of nuclear weapons was possible and how the world would be run after that. This was a bit like "have you stopped beating your wife?" but we quickly managed to get away from that question. I was pleased by the fact that nearly everybody present, and that included establishment figures like retired admirals and generals and

---

[26]Spain expelled its Jewish population in 1492, while Portugal did it five years later.
[27]Arkady Beynusovich (Benediktovich) Migdal (1911–1991) was a Soviet theoretical physicist. Among other achievements he coathored the Landau-Pomeranchuk-Migdal effect formula.

ambassadors, seemed to take it for granted that a reduction to a "minimum deterrent" was entirely reasonable, and were thinking of sensible estimates about what this minimum would be.

On the Sunday morning I departed for Tel-Aviv, to go to the Landau Memorial conference. (This year would have been Landau's 80th birthday.) The conference was well attended, including four senior Russians, one with wife. They worked us fairly hard — meetings from 9 a.m. to 9.30 p.m. (Of course with breaks for meals.) The weather was quite reasonably cool, except for the last day. In Tel Aviv one does not see anything of the tensions, and I avoided discussing the political situation with our host Ne'eman,[28] who is a good physicist and quite a nice person but a rabid hawk.

Sunday I went back to London and on to Copenhagen. This was also a Landau conference. (The Russians felt aggrieved about the Israel one, as they feared none of them would get there, so a second conference was arranged in Copenhagen.) There were about 25 Russians there. The meetings were much more relaxed than in Israel, and there was plenty of time. Gerry Brown was there, and invited Traudl to all the social functions. One evening we had dinner in Traudl's flat, with Gerry, Leonardo and Nicky, who seems to be surviving. Yvonne Rosenfeld[29] had died before I got there.

[On the way back] I had to stay the night [in London] because of an early meeting the next day — the fund-raising committee for the British Pugwash Group. We are supposed to have a conference in England in 1990, for which we need £150,000. We have to decide by September whether we can raise this money, otherwise one has to find another country to hold the conference. The same day there was also a meeting at the Royal Society to discuss disarmament problems. This was the first time we managed to persuade the Society to get involved with these problems, but there was a good attendance, and all but one expressed the wish to have further meetings, so this will probably happen.

August 15, 1988

I had better write a brief family letter before I get overwhelmed by the preparations for my next trip. This, as I think I indicated, starts on

---

[28]Yuval Ne'eman (1925–2006) was an Israeli theoretical physicist, military scientist, and politician. His greatest achievement in physics was the 1961 discovery of the classification of hadrons through the SU(3) flavor symmetry, now named the *Eightfold Way*, which was also proposed independently by Murray Gell-Mann. Recipient of the Albert Einstein Award in 1970; the Wigner Medal in 1984.
[29]Yvonne Rosenfeld was Léon Rosenfeld's wife. Léon Rosenfeld (1904–1974) was a Belgian physicist from Bohr's circle.

27th August when I will go to a Pugwash conference in a place called Dagomys, near Sochi,[30] on the Black Sea, which I am told is a rather vulgar resort, some kind of Blackpool. After that I stay a week in Moscow and Leningrad and return here on 12th September. Then on the 19th I depart for CERN to help celebrate Viki Weisskopf's 80th birthday, and on the 21st I go from Geneva to Paris, where I shall stay for two months, working (if "working" is the right word to use) one month at Saclay and one at Orsay.

September 16, 1988

The weather was very tactful: it had been a heat wave in Moscow, but on the evening we arrived it cooled off, so that our London clothes were appropriate, and next morning it was hot again, so we could wear summer clothing ready for the subtropical Sochi. The Moscow hotel was familiar from the first "Forum" in 1986. It has reasonable service, in particular fast food service, but the interior decoration is unbelievably ugly, dark and depressing. The Sochi hotel in the "Dagomys complex" is a huge and formidable structure of Yugoslav design. Twenty two floors above lobby level and about four below. Huge expanses of landings with parquet floors. Outside all concrete terraces and stairways. Rooms have a balcony from which nobody else is visible. The beach is depressing, because it is a narrow strip between a railway line and the sea, filled with concrete for changing rooms, showers, snack-bars, etc.

Entertainments: There was a reception and a farewell banquet, both with lots of good food, and plenty of drinks, including vodka.

Yura Verblovsky, the husband of Genia's relative Masha in Leningrad, was due to start a holiday in Sochi the day we left, so he came a couple of days earlier, and found me in the hotel. It so happened that we had a free afternoon, so I went for a walk and a swim with him.

I had some conversations with Sakharov, who is very pessimistic. He says there is so much opposition to Gorbachev that he has to keep making compromises, and he is likely to make so many that there will be nothing left of *perestroika*. He also said this in public recently.

The return to Moscow was wearing, though it is less than two hours by air. It was raining cats and dogs, and the buses which took us on the one-hour drive to the airport drove us right to the plane, but we then sat in the plane for a longish time.

When we finally arrived in Moscow it took an hour to sort out the

---

[30]Sochi is a resort town in Russia located on the Black Sea coast.

baggage. We finally got to the hotel about 5 p.m. to find that all our room allocations are mixed up in a huge pile, and unhappy girls shuffled through the pile trying to find one's papers. The system seemed specially invented to make life difficult. "There is madness in this system" was one comment. So when after an hour of this we could go and have dinner, this was welcome, as we had nothing since breakfast except a glass of mineral water (or Coca-Cola if preferred) on the plane.

I did not know what program had been made for me so I decided to go that evening to the Beloousovs, who live more or less on the other side of town, but on a direct Metro line. Their son, who lives near the hotel came to accompany me. On the way to the Metro we were stopped by the wife of Wiegmann, a very bright young theoretician, who was my chaperone last time. (What a small town is Moscow for us to meet in the street!) She told me her husband, whom I had hoped to see, had left for New York that evening. This was Sunday, and on Monday morning I was picked up by people from the Landau Institute, whose guest I would now be, and taken to the Academy hotel. I found I was going to Leningrad on Wednesday night, and there were no plans, other than dinner at Khalatnikov's house on Wednesday evening. So I got the phone numbers of a few physicists and made dates with them, which worked out quite well. In the Academy hotel I ran into Walter Kohn[31] who was spending a few weeks there. He told me he was still working on an invitation for me to California for next year. Also he was talking with Abrikosov (one of the Landau collaborators), who wanted to talk to me, so that made another encounter.

At the conference I had seen a lot of Nelly Artsimovich, the widow of the inventor of the *Tokamak* principle for thermonuclear power,[32] and a very charming man. She was a great friend of Genia's and is also a friend of Lydia Cassin. She complained that I had been a few times in Moscow without phoning her. It turns out she lives in the same (academic) apartment house as Migdal, and when Migdal took me home for dinner on Monday we ran into her outside the house. Migdal was surprised I knew her, and invited her, too. Next evening when I had dinner with her, she invited the Migdals. So I had two dinners in almost the same company.

On Wednesday, dinner at Khalatnikov's. His wife was still on

---

[31] Walter Kohn (1923–2016) was a theoretical physicist/chemist. The Nobel Prize in chemistry in 1998 was awarded to him for understandings of the electronic properties of materials.

[32] Peierls means Academician Lev Artsimovich (1909–1973) widely known as "the father of the Tokamak."

holiday, so his secretary was the hostess, and as far as I could see, did all the cooking. Her husband was there too, as well as a Japanese couple. When I left to go to the station, Khalatnikov said I would always be welcome whenever I was willing to come — they regarded me as an old friend of Landau's. It is nice to have one's position spelt out so clearly.

On the train I shared a two-berth compartment with a pretty girl — this is normal there. I was in bed when she arrived and she asked me to turn the other way. After a while I said "You will tell me when you have finished?" She had finished already. "So we can have some *glasnost* again," I said.

At the station in Leningrad there was the usual mile-long queue for taxis but the young man meeting me (incidentally, the son of a good friend of Genia's) spotted a private car willing to drive passengers, so we got to the hotel without suffering. I was taken for the day to the Nuclear Physics Institute in a suburb where also Amati from CERN[33] was visiting and gave a talk. In the afternoon was my talk, and it was the only official activity during a week's visit. I asked Frenkel, my contact in the Ioffe Institute, about going there next day, but he said everybody was still on vacation, so there would be no point, but invited me to dinner; we arranged to meet at the Metro station near his home. So, I spent the day with Masha [Verblovskaya] walking in the park in Pushkin, which was not far from my hotel.

As I walked towards the hotel, on a perfectly flat piece of pavement, I fell and cut my lip. (Without the assistance of a lady cyclist.) Inspecting the damage in my room, I decided I had better have some medical help. The hotel had no doctor, but after some discussion they called their first-aid person (intended for the hotel staff rather than the guests), who dabbed at my lip with some liquid and said I should go to the hospital. The staff offered to call an ambulance, but it was really not urgent enough for that, so they called a taxi, which took half an hour to come (if I had realized this I might have accepted the ambulance). This gave me time to phone a message for Frenkel not to meet me at the Metro station until I phoned again. When the taxi finally came, he charged me 5 rubles (£5) for a 1.50 journey, but I did not feel like arguing. After a 15 minute wait (shorter than it would have been in Oxford), a gruff-sounding, but kind and efficient doctor did the necessary embroidery on my lip, and another taxi was called to take

---

[33]Daniele Amati (b. 1931 in Rome) is a famous theoretical physicist who headed the Theory Division at CERN (Conseil Européen pour la Recherche Nucléaire) for a number of years. He is known for major contributions in the Regge theory.

me back, which took 45 minutes to come. (Funny that this was Friday afternoon, when in Oxford it would have been impossible to get a taxi, but in Leningrad the day and time do not matter.) I thus had time to ring Frenkel again, to find that on getting my message he had phoned the hotel manager and heard that I had been taken to hospital, and then phoned the hospital to find I was just being done. We agreed I would not come to dinner, but he came out to visit me.

Next day Masha's daughter and son-in-law (and Masha) came with their car to take me for an excursion to Petergof, a set of palaces modeled on Versailles. That evening I was expected for dinner by the family of my "guide." However, I did not feel up to it and cancelled, with regrets, I had started a slight temperature before the fall, which I took to be my usual bout, but it made me feel much worse than usual, and perhaps it was a flu virus or something. On the last day, Sunday, Masha had invited me to lunch, but I defaulted on that, too, and instead she came to visit me. I was seen off ceremoniously on the night train by Masha and Frenkel. There had been fun with my ticket. I wanted to know what carriage I was in for the benefit of the Moscow people who were to meet me. The Intourist girl told me my car and berth number and said I was lucky to have a compartment to myself. When the ticket came it turned out that two tickets had been issued for the same non-existing berth, and this was rectified only at the last minute.

November 1988 (annual summary)

This year the travel began at the end of January. I spent two weeks in Los Angeles, as guest both of the physics department and the Centre for International and Strategic Affairs of UCLA. I enjoyed both contacts. A talk I gave concerned the history of nuclear weapons, and this came to grief in the student newspaper.

The student reporter completely misunderstood what I was saying, and reported accordingly. Another student, who had not been present, wrote a letter attacking me violently for what, from the report, she thought I had said. I was comforted when another student came to see me, who had realized the misquotation and volunteered to write another letter to clear things up.

Amongst other visitors there was Joe Sucher from Maryland, with his wife Dorothy, who had just published her, first book, a thriller set in a theoretical-physics institute, "Dead men give no seminars."[34] There was a party at which she signed copies.

---

[34] Dorothy Sucher, *Dead Men Don't Give Seminars*, (St. Martin's Press, New York, 1988). Joseph "Joe" Sucher is Professor Emeritus at the University of Maryland.

Then I had six weeks at home, resuming social contacts, with many dinner guests, and some house guests, including the Migdals from Moscow. (He is a famous theoretician and an old friend.) A weekend trip to see Jo and Chris in Birmingham, and my brother Alfred in Manchester.

In May to Coimbra, where I received an honorary doctorate in a more elaborate and more impressive ceremony than I have ever witnessed. I was given a special outfit consisting of a black suit, black cape and a stole covering chest and shoulders of pale blue silk, heavily embroidered, and a hat, also of embroidered blue silk, like a tea cozy, with a spike on top, and a gold ring with a blue stone. Blue is the color of the science faculty. My host and presenter at the ceremony was João da Providência; I stayed with him and his family and spent a few days in his department, talking physics. In June came a conference in Tel-Aviv, to mark 80 years since the birth of Landau, the famous Russian theoretician. It had a tight schedule, with meetings from 9 a.m. until about 10.30 p.m. (though with breaks for meals). The weather was very mild, except on the last day. Then another conference, also in honor of Landau, but this time in Copenhagen.

This was a little more leisurely, but had many excellent talks. It was a pleasure to be in Copenhagen again.

Returning to Moscow and later at home I felt extremely weak and wobbly. The doctors could not find the cause of this — it probably was just the after-effect of some virus infection. It all passed after a few weeks, but it made me delay my departure for Paris, where I had planned to be for two months, dividing the time between the nuclear physics institute of Orsay and the theory section of Saclay. So because of the late start I had only two weeks in each place. I stayed in a flat belonging to the Institute for Advanced Study in Bures-sur-Yvette. Although the time was short and did not fit in too well with local programs, it was a pleasant and useful visit. One annoyance was a prolonged mail strike in France. All mail forwarded from here arrived two days before I left. I brought it back and added it to a huge pile which was waiting here, and I am still struggling with it.

January 1, 1989

Last night, which was New Year's Eve, I had a party for 12, which was very cheerful. I was in a cheerful mood because I just had a telegram announcing my election as Foreign Member of the Soviet Academy of Science. But this time I did end up with substantial leftovers, both of bits of food and of mulled wine.

On the one day between the conferences, I went to a dinner party given by the new Master of St. Caths in honor of the Russian mathematician [Vladimir] Arnold,[35] a distant relative of Genia's, who is by now very famous. Amongst the very interesting company was Judith May, who had been my copy editor for the "Surprises."[36] Her husband, a former theoretical physicist turned zoologist, has now a job in Oxford, and she has become an editor for the Oxford Press. A small world.

January 29, 1989

That week I attended the usual three-day conference at the Rutherford Laboratory, though I could go only for the first and the last day, as there was a meeting of Pugwash committees on the second. The first day included a talk by Stephen Hawking who copes very impressively with his voice generator. However, when he is asked a question it takes several minutes for him to compose a reply. But of course the audience waits patiently. A remarkable person. I did not believe a word of his talk.

June 16, 1989

Then back to Oxford: and to a mountain of mail, although the secretary who had been looking after my mail had done so very sensibly. Two days after my return I had agreed to be interviewed by an East German film group. They had seen from my book that I grew up in Oberschömeweide, and brought photos of the house in which I was born, of my school and pre-WWI photos showing my father on various official occasions. Apparently, the local AEG factory has a keen archivist, who was thrilled to know they were going to meet me.

The Copenhagen meeting was still following Bohr's ideas about an open world which are now less utopian than they seemed at the time. Sakharov and Sagdeev could not make it because they had to prepare for the meeting of the parliament. Instead there was the poet Voznesensky,[37] who spoke very well, and another man by the same name: though not a relative, who is adviser to the Council of Ministers. He spoke with remarkable frankness about their present domestic policies and the difficulties and failures that are holding them up. There was

---

[35] Vladimir Arnold (1937–2010) was a famous Soviet/Russian mathematician best known for the Kolmogorov-Arnold-Moser theorem. See page 139.
[36] See footnote on page 372.
[37] Andrei Voznesensky (1933–2010) was a famous Soviet/Russian poet and writer, one of the "Children of the '60s." Robert Lowell characterized ham as "one of the greatest living poets in any language."

also a Polish mathematician, a supporter of Solidarity, who had now taken part in the Round Table conference; he gave us copies of the report from that conference, which essentially legalized Solidarity and initiated a democratic constitution.

I was planning to leave for Rome on Tuesday, but as some papers had not reached me, I phoned Amaldi, the organizer of the meeting, and found that on the evening of the 5th, Sagdeev would give a talk about the current political situation in Moscow. His talk had been planned for later, but he could stay only two days, as he could not get leave of absence from the parliament for longer. This, together with the fact that leaving the next morning would involve catching an 8 a.m. plane, i.e. a 6 a.m. bus to the airport, made me leave on the Monday. To make it easier my phone was out of order on the Monday morning when I wanted to phone for a taxi at 7. I did not want to disturb the neighbors at that time, so I walked a few blocks to a public phone, which is not the best way of celebrating a birthday.

The Rome meeting was by invitation of the Italian Accademia dei Lincei to all East and West European Academies, plus the USSR and USA academies to send people to discuss arms control. This is to some extent the usual circus as far as USA and USSR are concerned. Sagdeev was there for two days as already mentioned, and Panofsky's group was strongly represented. There was also M. May of Livermore, who more or less presented the official American view: which is that it is not possible to reduce numbers of strategic missiles much further, because they are needed for "extensive deterrence" which means threatening their use in the event of a conventional attack in Europe, and in that case attacking military targets, such as command and transport centers, troop concentration, and so on. There are so many targets of this kind that you cannot do with fewer missiles. He is no doubt right that on the present US/NATO doctrine this is so, but quite obviously, this doctrine makes no sense.

July 9, 1989

It is hard to believe that it is three weeks since my last letter, and I have not been travelling anywhere. But there were other excitements. There was the annual garden party at the Maison Française, and this is the one occasion when I can wear my Coimbra outfit. It caused a sensation; everybody said that this was the most impressive academic dress. The M.F. librarian came up very apologetically to say would I mind telling her what kind of outfit this was, everybody was asking her. One man addressed me in Portuguese; I thought he had recognized the clothes, but he had heard that this was a Portuguese degree, and assumed I was Portuguese myself. This party used to be famous for its food, but this time the food was unbelievably lousy. Masses of little cocktail sausages, and some revolting pastry bits. There were of course the obligatory strawberries, but in microscopic helpings.

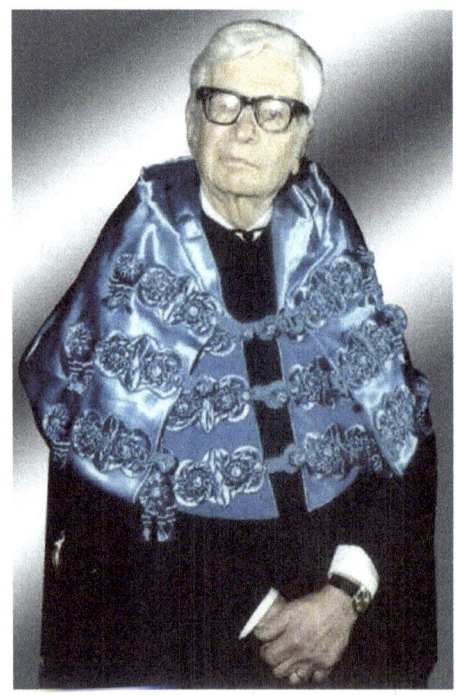

Figure 16.4   Rudolf Peierls' Coimbra Doctor Honoris Causa outfit. Courtesy of João da Providência.

The following week was *Encaenia*, and the honorary graduands included Sakharov and the queen of Spain. Sakharov and his wife stayed with the Berlins,[38] who gave a dinner party the night before the ceremony, to which I was also invited. The guests of honor did not come to dinner, however, because he was busy in London lecturing and giving press conferences. So they came after we all had had dinner, but there was quite a lively discussion, mostly in Russian as he speaks little English. All but one of the men present and many wives spoke

---

[38]Sir Isaiah Berlin (1909–1997) was an influential British social and political theorist. He was born in Riga, Latvia, while most of his professional career was associated with Oxford University. Although averse to writing, his improvised lectures and talks were recorded and transcribed, with his spoken word being converted by his secretaries into his published essays and books.

excellent Russian. The one exception was Rolf Dahrendorf, who is now the Warden of St. Antony's.

Sakharov is extremely pessimistic, and, as you have probably seen in the papers, now very anti-Gorbachev. He says Gorbachev has too much power, which he obtained by undemocratic means. *Glasnost* is fine, but the perestroika is so far only words. When he was asked what other form of government he would prefer, he said "give all the power to the soviets." This means the locally elected councils, but includes also the new national assembly.

Next day I went to the ceremony, where of course Sakharov and the Spanish queen got most of the applause. For the usual official luncheon at All Souls, she had expressed the wish to sit next to Sakharov, which caused havoc with their carefully designed seating plan. Next day Sakharov talked to a small meeting in Balliol, where he was supposed to say "a few words" before inviting questions — he talked for 40 minutes. The news reports on his London lecture had quoted him as saying the West should not give any financial assistance to Gorbachev — if the economic position got worse, he would be forced to do something. But now he said only that if the West did give credits there should be strings attached, to press for a reform of the economy.

The same day was the annual *Conversazione* of the Royal Society, a formal occasion with the rules now relaxed to "white or black tie; orders and decorations." I took Linda Davidson as my guest. She was anxious to see the portrait of Dirac, painted by her grandmother, which her mother gave to the Royal Society, and which was brought back from Ithaca by Dick Dalitz.

We found it hanging, duly framed, to balance a portrait of Isaac Newton. I met, amongst others, Walter Marshall,[39] who reminded me that some years ago he had asked for my reaction to having now two former pupils in the House of Lords (Marshall and Flowers), and I answered: "I had about 100 students; it does not matter if two of them went the wrong way." In principle these parties are occasions to view exhibits of current research, but so much time is taken in working your way through dense crowds to meet acquaintances and to collect food and drink, that you don't study much of the exhibits. The only exception was to see an excellent film about Chaos.

---

[39]Walter Charles Marshall (1932–1996) studied mathematical physics at Birmingham University and gained a PhD there under Rudolf Peierls. He was appointed Chairman of the United Kingdom Atomic Energy Authority in 1981 and Knighted in 1982.

Trieste starts with a symposium for Singwi's 70th birthday,[40] at which I have been asked to give a talk. I decided to look at what happened to a paper which Singwi and I wrote with a student in 1952. This was about the possibility of the universe having started from cold matter, heat developing later from irreversible processes. We found that this was a possible scenario, but would give too many heavy elements, so we dropped the matter. I now find that other people have picked up the idea, and found it a good way of explaining the non-uniformity in the universe; if cold matter expands, it will boil or shatter, according to whether it is liquid or solid. For the difficulty with the heavy elements Zeldovich found an elegant remedy. So I shall give a talk about cosmology!

August 21, 1989
Then off to Boston.

Gaby joined me for some of the open meetings, and a couple of receptions; she was busy on the clam-bake evening. The meeting was interesting and constructive, though we have no real controversy because the American (and other) hawks are not represented. Sakharov was there for the opening session, and sounded less pessimistic than here. He talked mainly about the crimes of the Stalin period and all the fraud and injustice of the time. He also made some critical remarks about China, whereupon the Chinese participants walked out. At a later meeting someone proposed a resolution condemning the events in China. This was out of order because the meetings usually don't pass resolutions, but the Chinese did not walk out; instead one spoke very angrily saying this was an interference in China's internal affairs, and anyway what the media had said was all lies.

On returning and retrieving my mail, with which I am still struggling, I found in particular the letter from Japan, from my friend Kubo,[41] the chairman of the Nishina Memorial Foundation, which had invited us last time. Invitation to a three-day meeting in December 1990 to mark the centenary of Nishina, a great physicist and a great figure in Japan. This comes with all expenses paid, including first-class air fare ("if you bring your wife this can be converted into two business class fares") plus an honorarium of $1000! Here goes my resolution to limit travel next year!

---

[40] Kundan Singwi, Peierls' former postdoc in Birmingham, died in 1990 at the age of 70.
[41] Ryogo Kubo (1920–1995) was a Japanese mathematical physicist, best known for his works in statistical physics and non-equilibrium statistical mechanics.

There was also a message, via the Royal Society, from the USSR Academy asking whether I would like to come and be officially received (or whatever is the right term) into the Academy in Moscow, or whether this should be done at the Embassy in London. I feel I ought to go some time, but I am not willing to go this year.

September 23, 1989

A few weeks ago the *Observer* had a profile of Lord Marshall. The first two paragraphs of this were about a conversation between Marshall and myself when he was a student at Birmingham. According to the story he said he was worried how he could ever become a competent theoretical physicist when he was the only Gentile in a department of 42 members. To which I am supposed to have replied "You are wrong, we are 43; the last man started today — he is also Jewish." The facts are that at the time we were not 43 but 20, and as far as I knew four were of Jewish origin (possibly a few more, as we did not check the ancestry of our students). I tried to contact Marshall, who was abroad, so I wrote a letter to the *Observer*, which has not so far been published. Incidentally, the "profile" also commented on Marshall's strange accent, and suggested that part of it was "East European," which he picked up from me. I did not complain about that one, nor about the statement that Marshall designed the UK nuclear industry. About the last Lorna Arnold, a science historian working in the Harwell archives, got incensed and wrote to the journalist who had written the piece. He replied and admitted that he was wrong, and that Marshall had not claimed this.

October 25, 1989

The next day I had a dinner for eight (counting myself). One lady dropped out at a few hours notice as she had a stomach complaint. It all went very well; the only surprise was that out of the seven people, three wanted teas, so I had a huge pot of coffee left, which however by now has been used up.

I spent a few days translating some Russian papers. These were papers which Goldansky,[42] the senior Russian Pugwash man, had given to Rotblat at some meeting. They contain a long letter by eight intellectuals to Gorbachev, complaining about the sudden rise of vicious anti-Semitism, and asking him to stop this before it gets out of hand. With it were leaflets of a vicious and primitive kind, worse than Goebbels,

---

[42] Vitaly Goldansky (1923–2001) was an outstanding Soviet physical chemist and the Chairman of the Russian Pugwash Committee in 1987–2001.

worse than McCarthy. They accuse the Jews, or as they call them, Yid-masons of everything from the "ritual murder" of the Tsar to Chernobyl and the riots in Georgia and in Armenia. Everybody they don't like is Jewish, including Sakharov, various modern poets, and Bukharin. They want to root out mixed race people to the tenth generation of Jewish blood. Goldansky was anxious that all this should be publicized in the West, hence the need for a translation.

Commentary -MS

I would like to make Peierls' remark more explicit by quoting Soviet press of that time. For instance, on May 16, 1990, *Moscow News* (MN) quoted one A. Kulakov who said: "Liberalism, humanism, communism, democracy — all of them were the weapons with which Judaism intended to destroy and desecrate the Aryan world. The secret power of international Yid-masons conquers the entire world, starting from such vampire as Lenin and ending with the United Nations...". On June 17, 1990, according to MN, the same speaker addressing a meeting proclaimed: "We demand from the Soviet Government to ban the Jewish emigration to Israel and establish People's Tribunal to investigate their collective guilt to convict them." On June 30, 1990, the government outlet *Izvestia* summarized an interview of the famous Russian writer Valentin Rasputin,

> I think, the Jews should feel their responsibility for the sins of the 1917 revolution and the consequences it led to... They should feel responsibility for the terror which was inherent to this revolution and then acquired even a larger scale. They played an important role, and their guilt is great — the guilt for the assassination of God and for the revolution and terror.

On February 22, 1990, *New York Times* ran an excerpt from the Soviet *Literaturnaya Gazeta*:

> It is necessary to prohibit Jews and their ilk from defending dissertations, getting academic titles and degrees or joining the Communist Party; they must be prevented from being elected to government councils or leading Party, government and other posts.

And it goes on and on...

This outburst of anti-Semitism was a rather unexpected byproduct of liberalization in the Soviet society. This was, of course, a bleak precursor of today's activities of neo-Nazi groups in Russia. It is true though, that the problem largely solved itself: in the years that elapsed since 1990 most of the ethnic Jews have left Russia. The remaining few are well assimilated.

Figure 16.5 A rally in Moscow in 1997 under the slogans "Yids and bourgeois out of Russia!" and "Lenin, Stalin, Fatherland, Socialism!" ©Natta Konysheva, from M&M Shifman Collection.

***

November 26, 1989

Last week I had a lot of champagne at a reception given by the German ambassador to confer a high decoration (the Order of Merit, which in German used to be, and perhaps still is, called Pour le Mérite). I had a chance to talk with the Ambassador, which was useful, because I am agitating that the recordings made of the German scientists interned in Cambridgeshire in 1945 be declassified. The government are reluctant to do so for fear of offending the Germans. (One of the internees was von Weizsäcker, the brother of the German President.) The Ambassador saw my point, and will raise it when the publication of post-war records about Germany is discussed shortly.[43]

---

[43] See Chapter 17.

One evening I had another party for eight, including the Dolls[44] (he started the connection between smoking and lung cancer). I find these larger parties attractive because of more varied conversation, and the preparations are not too bad if you select a menu that does not require too much piecework. Another such party next Tuesday, and on the Wednesday I have to feed a Russian couple (Abrikosov), I am hoping there will be enough left from Tuesday, but have a second defense planned.

Cultural life included some concerts, and a student performance of "When I was a girl I used to scream and shout," which is attractive for student groups because it has only four parts and little scenery, but is challenging, because each character changes between different ages, and they did not quite manage.

November 1989 (annual summary)

Then a few days at UCLA, and from there to Irvine, a relatively new campus south of Los Angeles. There I was housed in a beautiful and well-equipped apartment where the kitchen equipment ranged from a dishwasher to a microwave oven. I felt it would be a pity not to use this to entertain. The biggest dinner I had was for five. The three weeks there were enlivened not only by many interesting talks, but also by an earthquake (no damage on campus) and by news of the starting "cold confusion." We had many discussions trying to find an almost believable scenario to account for this, but evidently one should have listened to the old rule "Never rush into making a theory to explain a surprising experiment until you are confident the result is right." I also spent a day and gave a talk at Livermore, invited by the people I had met at Davis, and again I found a very reasonable attitude (but probably not typical of the general view of the laboratory).

December 31, 1989

Christmas dinner — again with venison — was only for the four of us. The day before I had run into Els Placzek, who was visiting the Griffins for a few days before taking them to Egypt. So we asked the three of them over for drinks.

---

[44]Sir William Richard Doll (1912–2005) was a pioneer in research linking smoking to health problems.

February 4, 1990

The sad event during this period was, as you all know, Alfred's death on January 9. The immediate cause appears to have been lack of adequate care in the hospital. The previous day they gave at his lunch a piece of sausage, which they put in his mouth without giving him his dentures, and he choked. Vera was visiting and managed to extract the offending piece and, as Alfred was quite out of breath, called a doctor, who called a physiotherapist, and things got straightened out though Alfred was exhausted. Next day more or less the same happened, but there were no visitors, and he did not recover. One can understand that hospitals are short-handed as they are being kept short of funds, but this goes a bit far. However, he was not getting any pleasure out of life — he was pretty miserable.

I decided that the developments in East Germany and my correspondence with Oberschöneweide made a visit worthwhile. I had written in October to the Leopoldina Academy that I was now willing to give a lecture there, but had no answer, so I wrote to the East Berlin film crew who had offered to look after me. That reply also took a long time to come — it seems they have changed their address — but then they phoned and offered to book a hotel for me (maybe, even pay for it!), meet me at the West Berlin airport, etc. So, I agreed to go on 17th of March for four days. It turns out that the elections are on the 18th, so it will be fun. The next day came a letter from the President of Leopoldina. They had to work out their program for the year, and there were so many changes taking place in East Germany. They would be delighted to have my lecture, and proposed dates in May and June. I can't manage the May one, and the June one is rather inconvenient, so I asked for more dates, but one way or the other it looks as if I am now going to East Germany twice!

Last week I had two Pugwash committee meetings on successive days in London, so I stayed the night with the Flowers.

March 10, 1990

There was a private showing in London of a new film about Los Alamos, "The Shadow Makers."[45] The showing was in connection with Pugwash, and there was a panel discussion following it, in which I took part. The film is a dramatization, to which I am always allergic, and it certainly did not get the atmosphere right. So talking on the panel needed some

---

[45] Released in the US as "Fat Man and Little Boy", this 1989 UK film reenacts the Manhattan Project. Directed by Roland Joffé.

diplomacy, but the discussion was not so much about the film as about the ethical problems it raises. The next day it was shown in Oxford, and I again was there for a question-and-answer session, the panel being just the producer and myself. Here the showing was for the benefit of the Pitt-Rivers Museum, tickets at £10, so I had a select audience. I was rewarded by tea in the museum, where I have not been for ages.

I went to Birmingham for a meeting celebrating 50 years since the invention of the cavity magnetron in the Birmingham department. Oliphant had come over from Australia for the occasion. He is in excellent form, for his 89 years. It was a pleasant occasion, seeing many old acquaintances. I did not realize that the cavity magnetron is now the power source in every microwave oven.

I at last called on the Bishop of Oxford, to talk to him about the neo-Nazi movement in Russia. He promised to write a letter to the head of the orthodox church, or perhaps to get the Archbishop of Canterbury to write, which seems a good thing. By now the story of the "Pamyat" organization[46] is in all the papers.

March 18, 1990

Then for the rest of the time to the [UCLA] campus guest house, which is very civilized, except for the continental breakfast that comes to the room. I wonder what continent this is for. I shared an office with a nice young Frenchman, who asked me a question about diamagnetism on the Monday I arrived. I answered it on Tuesday. By Wednesday we were drafting a paper about it, on Thursday we found it was wrong. Apart from a colloquium, I gave a lecture in Intrilligator's undergraduate course. He was away that week, but it so happened that the topic that week was nuclear weapons, so I could easily fill in, and it was a bright class who asked many good questions. I also gave a lecture in Nina's [Byers] course in quantum mechanics, which was a little less successful.

Nina is in good spirits, because the department is in good shape. They have a new particle theorist, [Roberto] Peccei (if I have got the name right) who was, I believe, brought in partly on Nina's initiative. He proved so popular that after 1.5 years he was made chairman, and everybody seems happy.

April 1, 1990 (no joke!)

I arrived in West Berlin on a Saturday night, and was met by a young

---

[46] A neo-Nazi association active in Russia in the late 1980s and early 1990s.

man of the film people, who took me to the hotel in East Berlin, which they had booked for me. This meant driving by using a detour, because foreigners still are allowed to cross to East Berlin only at Checkpoint Charlie [...]

Then back to my hotel to meet the film producer, who explained to me the rest of the program, and then I watched the election results and comments on both East and West German TV. The results came as a surprise, one had expected the Social Democrats and the Christian Democrats to come about equal (and indeed that was the way the vote went in the North, but the South was strongly conservative). Most of the people I talked to were not only surprised, but disappointed.

Next morning I was picked up by the film people's chief photographer, who took me out to Oberschöneweide. We stopped at the cemetery and found my mother's grave, which is in much better state than I expected. I understand that the row of graves of which this is part, is about to be declared historical monuments, so it will presumably be looked after.

Dinner in the hotel, quite good, very rich and very generous. In the morning I was shown the film, in which I featured (marginally). It was a documentary about Klaus Fuchs, quite well made and fairly objective. What was interesting for me was interviews with Kuczynski, who make the first contact between Fuchs and the Soviet authorities, and with "Sonya," Fuchs's courier.[47] One had heard rumors about them, but here they were, describing their activities not without pride. Also in one interview, Fuchs described a conversation with Herbert Skinner, who said "There are suspicions around about you, but if you assure me they are unfounded, I shall fight for you." This conflict between personal loyalty and political belief was too much, and he gave himself away.

July 22, 1990

On 5th–6th of July there was here a symposium to mark Dick Dalitz's retirement. It was well organized, and there were only invited lectures, and they were about an hour each. It was a nice occasion to see old friends. These included Nina Byers, who stayed five days in my spare room. I have now acquired a new sofa-bed, which is much more comfortable than the old one, so if any of you contemplate a trip to England, bear it in mind.

---

[47]See Chapter 9.

My book ("More Surprises..."[48]) has been accepted by Princeton, and my copy editor will again be Alice Calaprice, who did *Bird of Passage*. But they did not like the typescript; they wanted bigger margins and more legible equations, so I spent the last week or so reprinting it and checking, but it went off again last Friday.

August 25, 1990

A visit from a charming Russian woman, a friend of the daughter of Masha [Verblovskaya] in Leningrad, and more or less raised in Masha's flat. She works in the graphics department of the city library, which had exhibited in Manchester, and she had come to Manchester to recover their property. A Manchester librarian was driving her to Oxford. She phoned one evening to say she would get here at about 10 the next morning, then she phoned to say it would be 2 p.m.; she finally arrived at five! Being young and cheerful, she did not sound depressed about the economic conditions in Russia. Another Russian visitor was a physicist who had been here before, and now came to the Dalitz symposium and stayed on for a while. His name is Nikolai Nikolayevich Nikolayev[49] (Genia called him Kolya-cubed).

I am now working on plans for my winter journey. I am starting out on the 18th of November to spend two weeks in Moscow and Leningrad, then fly to Tokyo for the three-day conference, which is the motivation for the trip (and the source of my first-class ticket!) and then stay another week in Osaka. Then it turns out it is just as cheap to return via the US, so I shall fly across the Pacific to San Francisco, and probably will stop for a few days at Berkeley, where I have a standing invitation. Then I head for Vancouver; possibly stopping in Seattle, if time fits in, and inflict myself on Kitty for Christmas. From there I shall go home directly, missing the East Coast, since I am due again in April to go to McMaster, so I can then visit the East Coast branch of the family. So far that is all my travel plans, except the Pugwash conference at Royal Holloway College (15th–20th September) and a few days in Zagreb around 11th of October. Being in Zagreb, I thought of visiting Ljubljana, but [...]

November 5, 1990

In going to Yugoslavia, I left in the afternoon and so had ample time for packing, which resulted in my forgetting my pajamas. After

---
[48] R. Peierls, *More Surprises in Theoretical Physics*, (Princeton University Press, 1991).
[49] Nikolai Nikolayev is a modern Russian nuclear physicist.

improvising the first night, I set out next morning, when I was free, in search of a suitable shop. My instinct led me to a street with department stores, but in the biggest one I could not find the men's department, and nobody could speak any language. I found the appropriate department in another, and at first was offered a nice-looking pair for £100 (they turned out to be pure silk) but they also had what I wanted. I was looked after by Tadić[50] who had been a postdoc in Birmingham, and his nice wife. I was confused whether I had met her before or not; I knew I had met his first wife, who had left him. (I later discovered that the present one is the third!) I dropped a brick when I asked her whether she had been to England; they had been for a term, in Sussex, and visited Oxford, where I met them, though it was a busy time and I saw them only briefly.

The old part of Zagreb is quite attractive, and I was taken up a hill overlooking the town, with lovely views. The meetings were painless, and my talk appreciated. Then by car to Ljubljana.

February 26, 1991

Just read that Migdal in Moscow died at 80.[51] He was a great charmer. I wrote a letter to his wife, which had to be in Russian, because she does not know much English. I do not know her first name and patronymic, so after some unsuccessful research I had to start the letter (with apology) with the equivalent of "Dear Mrs. Migdal," which in Russian is very bad form.

I am now very likely to be invited to this year's Pugwash conference in September in Beijing! If I go I want to stay a little longer and visit some institutes.

I have just finished a little piece of research on spontaneously broken symmetries, which was prompted by a controversy with Phil Anderson, a Nobel laureate with a very abrasive personality. I enjoyed doing that, and I now have to search the literature to see whether the results are already known — if not I can publish it.

March 2, 1991

Saturday, 2nd March 1991. I am going to Beijing. The invitation came on Thursday, together with a message from Joe Rotblat that this was the last day for an especially cheap fare he had discovered, from the 1st March it would cost more. So I had to decide quickly,

---
[50]Dubravko Tadić (1934–2003) was a theoretical physicist at Zagreb University whose prime interest was the interplay between weak and strong interactions.
[51]In fact, Arkady Migdal died on February 9, 1991, in Princeton, NJ.

and went ahead with the booking. I then contacted the woman in the Royal Society who looks after exchanges with China (whom I know well because she is also the secretary of the group discussing international security). She assured me there was no problem in getting an invitation under the exchange scheme with the Chinese Academy to visit some institutes after the conference. I have now decided not to go to Leipzig in December for a conference to mark Heisenberg's 90th birthday — enough is enough.

Tomorrow I give a lunch for four ladies. Lady Simon,[52] Margaret Gowing, Linda Davidson, and Lorna Arnold. The last is a retired historian from Harwell, collaborator of Margaret Gowing, and a vegetarian. So the menu has to be vegetarian. I have put together a collection of Russian dishes, which I hope will suit [the occasion].

March 23, 1991

Time is passing and my "diary" seems to be becoming a "weekary." But there is not much to report. Last night dinner for seven, when I stuck my neck out by serving things I had never tried before, and it came out quite well. There was a little trouble with the meat course, an interesting microwave recipe for spinach-stuffed saddle of lamb. The recipe said to sew the edges of the meat together to hold in the stuffing, but I did not have a large enough needle, so I tried to tie string around, but this did not work, and I finally used small safety pins. They did not hold too well either, so the stuffing was all over the place, and the meat slices were not very elegant. In addition I was worried that I might not get all the pins out and some guests would eat them. But the taste was excellent.

Actually we were to be eight, but Linda confused the dates and turned up on Thursday evening. She had to attend a meeting on Friday.

More invitations to turn down: A meeting in Mexico in August to honor Mondragón.[53] I like Mexico and I like Mondragón, but enough is enough.

April 7, 1991

Masha in Leningrad sounds very cheerful. She had written from the US, where she was visiting her daughter and son-in-law, very worried about the deterioration of life in Russia, as seen from the Western newspapers.

---

[52] Charlotte Simon, Francis Simon's wife, see page 339.

[53] Alfonso Mondragón-Ballesteros (b. 1932), Mexican physicist and educator, recipient of the Ignacio Manuel Altamirano Prize, Mexico, 1957; Distinguished Service Medal, University Nacional Autonoma, 1979; and Jose Antonio Alzate Prize, Mexico, 1984.

On returning, she found it was not nearly as bad as she had expected, and life was quite reasonable.

My itinerary for May is now pretty well fixed. I leave here on the 12th, stop in Boston with Gaby, and on the 16th we both go to Hamilton (McMaster). On the 18th we visit Mrs. Jephcott in Toronto. On the 21st I go to Ithaca to visit the Bethes, and from there to Port Jefferson on the 24th. Then on the 28th I go to Princeton, to give a seminar next day.

July 11, 1991

For the past three days I was in Cambridge for the annual "Amaldi conference." Amaldi had started conferences of European and US and Soviet academies on international security and arms control. After three such meetings in Rome it was decided to move around, and this year's conference was in England, sponsored by the Royal Society. The meetings were held in the Master's Lodge of Trinity (Michael Atiyah is Master of Trinity and president of the Royal Society). We stayed in Churchill College, which is very well built (trust Cockcroft!) except for a crazily designed shower room (at least on my floor) where you can't close the outer door without stepping into the shower stall, and there is no towel rail!

The meetings were good; we had only about 40 participants so we could sit in a big circle, which facilitates discussion and we had a limited agenda, so that things could be discussed in depth. Talk was about proliferation of nuclear and other weapons, about the forthcoming convention about chemical warfare, and the problems of destroying nuclear and chemical warheads without damage to the environment, which turn out to be very tough problems. I went to and from Cambridge by bus, which is a rather slow, but painless operation.

On my return I found a letter from Helga Singwi. I wrote already that he died last year, and of course I wrote to his wife, and she now replied, explaining that he died of an overdose of Novocain (or some such substance) administered by mistake by his dentist!

July 16, 1991

There had been in the *New York Review of Books*, an article by Bethe, Gottfried (a Cornell theoretician) and McNamara proposing that the US and the USSR should immediately reduce their nuclear weapons to 1,000 each (from about 15,000) and gave good reasons for this. An excellent proposal, except I would like to see fewer than 1,000, but let's get down to 1,000 first! However, they claimed that this would also make a major contribution to the proliferation problem, and that is just wishful thinking. So I wrote a letter to the *New York Review of Books* disagreeing about this, which I faxed a week or so ago.

August 17, 1991

Thursday, I went to London to apply for my Chinese visa and to do some paperwork at the Royal Society.

Received a letter from Rome announcing that I have been elected a Foreign Member of the Accademia dei Lincei. It seems they have their opening meeting for the new academic year in November, which I should attend. Radicati[54] has already written to suggest I might then also spend, some time in Pisa. So this will be my next trip after China. I hope the meeting will not be too early in November, because I have invited Genia's relations from Leningrad, Masha and Yura [Verblovsky], to visit, and they plan to come about 20th of October.

August 22, 1991

The last few days we watched — as no doubt did you — the dramatic events in the Soviet Union, ending with the miserable collapse of the coup. One casualty was Jo and Chris's holiday trip. They were booked to go on a river boat from Moscow to Leningrad, partly by canals, and stopping at some charming old Russian towns. They were supposed to fly out today. Yesterday morning the tour operators cancelled, as it still looked unpleasant, and the Foreign Office had advised against non-essential travel. If the starting date had been one day later they would have made it. They would have tried to contact Genia's relations in Moscow and Leningrad, so I phoned this morning to say they were not coming. To my surprise I got through at once. Natasha in Moscow was still dubious and said only the position might be a little better; Masha in Leningrad was buoyant about the coup having failed, and also thought that it might have improved the prospects.

Sunday I fed lunch to Lady Simon, who is sad because a good friend of hers in Oxford had just died, and also her nephew had died from a sudden heart attack. Supper again at Linda's (not walking this time, but by taxi) with another pair of friends and her daughter and son-in-law. Yesterday she was without house guests, so I fed her a quick supper; today she went off to France where her son has a rather primitive house in the wilds.

---

[54] Luigi Arialdo Radicati di Bròzolo (b. 1919) is an Italian theoretical physicist who contributed to the theory of strong interactions. His most well-known result is the so-called Cabibbo-Radicati sum rule.

August 26, 1991

For the rest of the week I was still glued to radio and TV, with more and more incredible surprises from Moscow. The end of communism is a great pleasure, but of course there are many problems left. I feel very tempted to arrange another visit there.

October 20, 1991

Invitation to talk at a solid-state meeting organized by the American Physical Society in March in Indianapolis, in a session on the history of solid-state physics, coupled with an invitation to give a colloquium talk at Illinois. So I will organize a trans-Atlantic trip around these dates.

October 27, 1991

The BBC Horizon team came to interview me — they are making a program about the German atomic bomb work during the war, and wanted some discussion about the work in the West as background. They expected it would take an hour, but after 1.5 hours of filming they discovered moisture on the camera lens (they previously dropped it in the water but thought they had dried it out), so after getting another camera from London we had to do it all over again, so this made a big hole in my day.

In mid-November I shall go to Rome. I had heard from Radicati that I was invited to a meeting of the Lincei, to be admitted, but there was no direct invitation. So I rang Radicati, who assured me that I was invited, and that the Academy had written to me. That is the Italian mail! So I go on the 15th for a morning meeting the next day, and then for a few days to Pisa, which will be pleasant.

November 14, 1991

Saturday I saw my eye surgeon, Mr. Cheng, who agreed that a cataract contributed to my problem, but not enough to make an operation worth while. He will look at it again in six months' time. I was not as disappointed as I might have been, because I more or less expected this. So I now reorganized my reading, with some success. By covering the book with black paper, except for a slit for one line, I can cut out much of the scattered light, and this makes reading much easier and faster. This is useful because I have agreed to review a Heisenberg biography by a man called Cassidy[55] for the *New York Review of Books*.

---

[55] David C. Cassidy, *Uncertainty: The Life and Science of Werner Heisenberg*, (W.H. Freeman & Company, 1991).

I agreed, even though the book has 650 pages, because I like writing for the *New York Review of Books*, and because there has been much controversy about Heisenberg.

My expected Russian guests (Masha and Yura Verblovsky) have now got all their papers, but are now working on a ticket, which is expected to take a month. So they will be here in mid-December and probably stay for Christmas.

November 23, 1991

I am now back from Italy, and while the trip was in part enjoyable, there were also disasters. The first, minor, disaster happened in Rome airport. There is now a train from the airport into town, and when I had found the station, they were announcing that the train to Rome was leaving, so I rushed to buy a ticket, which was 5,000 lire (£2.50). In the hurry I gave the man a note for 50,000 lire by mistake, and he never gave any change. (The numbers have no commas, and counting the zeros can deceptive.) That was only money, but annoying. After that, a happy coincidence: On the train next to me was Francesco Calogero,[56] a theoretical physicist and now the Secretary-General of Pugwash. So it was a pleasant train ride.

December 12, 1991

I can now see the light at the end of the tunnel. I finished reading the book [by Cassidy] by the end of last week, and wrote the review over the weekend, somewhat handicapped by not being able to read some of my notes and had to go back to the book. I am now taking notes by tape recorder. I was able to fax the review to the publishers on the 9th, which was the date they had specified, but that was not essential, as they said they could wait a while. Knowing them the review probably won't be published for six months anyway. Then I concentrated on Christmas presents. By now all transatlantic presents have gone off, perhaps not ideal, but showing good intentions.

December 20, 1991

It turns out my St. Petersburg visitors will arrive on Christmas Eve, which is worrying, because transport is chaotic. They come by train, and are due in London about 7 p.m., unless they can get on the hydrofoil across the Channel — which does not run when it is too windy. The last train to Oxford is at 10.20, then the railway closes for two days.

---

[56] Francesco Calogero (b. 1935) is an Italian mathematical physicist most known for his research in integrable many-body problems.

Jo and Chris are coming over for Christmas Day. I got a little tree and decorated it to give the visitors a real Christmas. But it is not the real thing without the heated arguments with Genia about where the lights should go!

December 29, 1991

You know in general terms what happened the day after my last entry, except for some of the gory details (gory is the word).

In the morning I felt weak and had to lie down at intervals between bathing, shaving, dressing, etc. At one point I was resting comfortably when I realized I was on the kitchen floor, covered and surrounded by blood. After a while I managed to get to the phone and call the doctor and Jo. The duty doctor on the Saturday morning was a man in whom I did not have much confidence, though he is nice and tries to be helpful. He came in 15 minutes, and said the loss of blood was not as bad as it looked, to keep quiet and one would investigate on Monday.

Jo and Chris came immediately; she talked to the doctor, and he agreed it would do me good to be looked after in hospital, and arranged a room in the Acland (private) hospital. By 5 p.m. I was there. When the ambulance people took me in a wheelchair to my room, I passed out again, which set the bells ringing. They put me on blood transfusion, which went on until Sunday, and made me feel better and stronger immediately. Meanwhile Jo and Chris had tackled the Herculean task of cleaning the Augean stable in the kitchen, putting my clothes through the washing machine, and generally restoring the flat to a civilized state, as well as packing up all my necessities for the hospital. They had to go home on Saturday night, amongst other requirements, to feed the cat, but were back on Sunday.

From then on I was spoiled rotten in the hospital, where the accommodation is like a good hotel, and the nurses are very nice and relaxed, not as harassed as in a state hospital.

Fortunately I was able to put off my Russian visitors. They were coming by train, and then leaving on Saturday. I did not know what time. But I was able to catch them. They were of course disappointed, but we shall now arrange a new date. By now tickets can be got only for hard currency, and a phone call to England is 200 rubles a minute!

In hospital I had an unending stream of visitors — Lady Simon being particularly efficient in spreading the word.

December 1991 (annual summary)

My trip round the world started in November. It was done in great style: The Nishina Foundation, whose Tokyo conference was the occasion for the trip, generously paid first-class fare. I started in Moscow, where I was admitted as Foreign Member of the USSR Academy. A week or so was spent visiting institutes and seeing friends. One heard much of their worries for the future and about everyday difficulties and shortages (which were also illustrated by odd and changing shortages in the Academy hotel). There would be no sense in reporting about this in detail because by now everything has changed again.

In the institutes one felt the effect of the brain drain — many leading people, particularly theoreticians, were abroad, some permanently, some so far temporarily. One Sunday I was taken to Zagorsk,[57] an old town with a famous monastery. The ground was covered with hard-packed snow and ice, so walking became a balancing act, but it was worth it.

A few days in Leningrad gave a chance to revisit the city which, even in the gloom and dampness of the season retains its beauty. But a visit to the Hermitage was a failure — the day was dark, and the lighting poor, so the pictures were invisible. One clear interval was used for an excursion to Peterhof, a palace destroyed in the War and lovingly restored.

Back to Moscow for a day, which included visiting a friend of 1934.

I was afraid that a night on the plane to Tokyo after the previous night on the train from Leningrad would leave me in bad shape, but the night in the first class of Japan Airlines was restful. The conference was well-organized and enjoyable. Each speaker had been given a topic (mine was "Will physics come to an end?"[58]) and only Schwinger disobeyed instructions and talked about cold fusion which he thinks may be real.

Then by the bullet train to Osaka, as guest of the several physics departments. The departments and my hotel were on the edge of town, in a newly-built area, and I went into town only one evening for a banquet. But it was interesting to see the life in modern Japan, including one party in a private house.

---

[57]Currently, Sergiev Posad, a town famous for Moscow Theological Academy situated there. Approximately 70 km north-west of Moscow.
[58]See Appendix on page 427.

Then another overnight flight (crossing the date line and arriving disconcertingly in the morning of the same day as your evening departure) by United, whose first class was no worse than JAL. A few days in Berkeley, being looked after hospitably by Dave and Barbara Jackson,[59] and seeing many old friends, including Stanley Mandelstam.[60]

Otherwise not much to report — correspondence, visitors, social life, including giving dinner parties, which is becoming a hobby.

January 22, 1992

Since I last wrote I had about two weeks at home. This included New Year's Eve, when I had turned down an invitation to a party (stand-up party with lots of people and lots of noise), and I evidently could not give a party of my own. But I did not fancy meeting the New Year by myself, so I phoned various neighbors, and found that the Sullivans had been asked to a dinner party, which would probably not last until midnight, so they came around about 11.30; we opened a bottle of champagne, and ate some appropriate tid-bits, so it was a more or less adequate celebration.

January 28, 1992

I went to my GP[61] today (there by taxi, back by bus), and he was pleased with progress. But he said that the specialist had predicted it would take two months for the clots to disappear completely, and he (the GP) thought this was an underestimate! He gave me permission to lead a slightly more normal life, and to get around a little. He did not seem worried about the Russians arriving on Sunday. I decided not to meet them at Heathrow, but to send Mr. Clare. On Saturday I had dinner at the Christians, and on Sunday, Jo came to visit.

February 9, 1992

The Russians [Verblovskys] have now been here for a week. As I am still limited in the amount of walking, they have been out on their own a lot. Today they are on their fourth trip to London. He speaks enough English to get around, though he is still a bit shy in talking. For example, in the car from the airport he managed to ask the driver whether he had to wait a long time for them, but he did not manage

---

[59] John David Jackson (1925–2016) was an American theoretical physicist, a member of the National Academy of Sciences, known for numerous publications in nuclear and particle physics. Most famous is his widely-used graduate textbook *Classical Electrodynamics*.
[60] See page 141 and Fig. 5.6.
[61] British analog of family doctor.

to ask him to turn down the heating, as they were getting hot in the Russian winter coats. One day last week Linda fed all of us. It helped in communications that her lodger, a Welsh girl studying French, was also at the dinner, and knows some Russian. Yesterday Jo and Chris came over and took us on a tour of the Cotswolds. One of the Verblovsky's friends is Mrs. Haskell,[62] the Russian wife of a well-known art historian. She gave them a guided tour of the Ashmolean and the Christ Church gallery. Last night we went after dinner around the corner to meet with a Russian physicist who is working at Culham, and whom I had met, and his wife. It was very interesting for me to listen to Yura's explanation of the current position in Leningrad, sorry St. Petersburg, to people who have been away for a year. It is more or less what we read in the papers, only the price rises are even more spectacular, and the role played by the "Mafia" or just small-time dishonest people is more than I thought. Also nobody wants to work. Crops in the fields are not harvested, and nearby townspeople with cars or bicycles go out with rucksacks and "pick their own" but without pay.

Their return flight is next Sunday morning at 8.30 — unavoidably. So I booked them a hotel at the airport. I have just discovered that there is a two-hour check-in, and the hotel courtesy bus does not run till 7, so they will have to take a taxi.

February 13, 1992

Just got very pleasing news. Many of us have been fighting to get the so-called Farm Hall transcripts declassified.[63] These are the records of conversations between the German scientists, including Heisenberg, Hahn, Laue, and Weizsäcker, who were interned in this country (at "Farm Hall"). The place was bugged and their conversations recorded, including their reaction to the news of the American atom bomb. This was 47 years ago but the transcripts were still secret. Polite letters to the Foreign Office received a very bureaucratic refusal. So in November, Kurti,[64] Margaret Gowing[65] and I organized a petition to the Lord

---

[62] Francis Haskell (1928–2000) was an English art historian, known for his publications on the social history of art. He was Professor of Art History at Oxford from 1967 until his retirement.
[63] See Chapter 16 for more details.
[64] Nicholas Kurti (1908–1998) was a Hungarian-British physicist. During WWII he worked for the Manhattan project. His hobby was cooking, and he was an enthusiastic advocate of applying scientific knowledge to culinary problems.
[65] Margaret Gowing (1921–1998) was an English historian instrumental in the production of several volumes of the officially sponsored *History of the Second World War*. Famous are her books commissioned by the United Kingdom Atomic Energy Authority covering the early history of Britain's nuclear weapons.

Chancellor (who is formally responsible for such matters), signed by about 20 distinguished scientists and historians, headed by the Presidents of the Royal Society and the British Academy. Now we have gotten the answer that the transcripts will be released to the Public Record Office, and the decision will be announced in Parliament tomorrow.

The Russians are again in London, I think for the sixth time, if my count is right. Tonight we are all invited to dinner at the Haskells (see above). So far the Verblovskys have always come back from London trips one to two hours later than planned, but if they are late tonight it happily is not my problem. On Sunday they were in Stratford, and on Tuesday Linda took us out for a ride. We saw some pretty villages, and then wanted to see the White Horse. Linda, who loves small roads, chose to drive along the Ridgeway, which was a mistake. Mud and ruts got deeper and deeper, and at one point the car seemed stuck. We got out to reduce the weight, and followed the skidding and jumping car probably for a mile. The state of our shoes and trousers can be imagined. But the visitors enjoyed the adventure.

February 18, 1992

The Russians were only half an hour late back from London, which was not too bad. They went to London again on Friday so out of 13 days they had seven days in London. Yura even suggested that on Saturday, after taking an afternoon bus to the airport and leaving their luggage at the hotel, they might take the tube into London and still look at some things there. But I talked him out of it. So they left by an 8 p.m. bus. A woman in London wanted them to take some things to St. Petersburg, and was going to hand these over at the airport hotel, but on my suggestion they got her to meet their Oxford bus and assist them in the not entirely trivial transfer to the hotel. I hope that worked.

February 27, 1992

Sunday lunch *en famille* at the Christians — a vegetarian meal, as Tim is a vegetarian. Monday was the TV program on BBC Horizon about the German atomic-energy work, for which I had been interviewed (the occasion of the wet camera). They were taken by surprise by the release of the Farm Hall transcripts, and had only a week to re-vamp their film to take account of these, and as a result my interview got lost. It was quite an interesting program, though it included a lot of dramatization;

actors playing the parts of the scientists, and this was mixed with real interviews of the surviving scientists. I am quite allergic to that approach.

The transcripts are now in the Public Record Office, but I have not yet seen them. Apparently they are poor copies and very hard to read. Also only the English translations are there, not the original German texts, which the Lord Chancellor told us would also be deposited. Now Nicholas Kurti is frantically chasing them, but nobody seems to know where they are.

March 9, 1992

The Farm Hall transcripts are still a problem. Nobody knows what happened to the German originals, and they may be lost forever. The English text is available on microfiche, made from a poor photocopy of a second carbon copy. I was quite unable to read the microfiche. Some people are now producing a paper copy from it, using special tricks to make it more legible. I have had the first third of it, but some parts are still illegible. Even so one could get some quite interesting points from it, and I have asked the *New York Review of Books* if there is time to amplify my review of the Heisenberg biography to take account of this.

I have now agreed to go to Coimbra for a week in May. Providência has been pressing me for some time even offering to bring me by first class, to ease my leg problem (which I should not accept, as they are not so well off for money), and as it is only 2.5 hours by plane. My GP agreed. Yesterday I called on the Griffins, where Els Placzek was again visiting. She is in very good form, as ever.

The other day Norman Stein, a professor of history not much respected by his colleagues for holding forth about subjects of which he knows nothing, had a preposterous article in *The Telegraph* in which he mentioned my name. So I wrote a letter to the paper, which was published and of which I shall enclose a copy.

March 12, 1992

Sagdeev is in town to give the prestigious Tanner lectures. He is a Russian physicist, for many years director of their space research institute and a member of the house of deputies until it was closed. He is married to Eisenhower's granddaughter and now teaches at Maryland, though he still commutes to Moscow. His first lecture was on the second Russian revolution. Today was his second lecture, on the effect of the communist regime on the Russian *intelligentsia*. I had in the morning a meeting in London, at the Royal Society, my first trip out of Oxford since December (not counting Kidlington). To make it easier, the underground had a signals failure, so it took 45 minutes to get from Paddington to Piccadilly. But luckily there was no problem on the return, so I got back in ample time for Sagdeev's second lecture. These lectures are at Brasenose. When the Principal noticed that I knew Sagdeev well, I was quickly asked to a reception after each lecture and to a dinner after the third lecture on Monday. Meanwhile I have invited Sagdeev and his wife to dinner on Saturday, and will have a small party.

A few days ago I was shopping in the covered market, when I passed three students. One of them stopped me and asked me to autograph his copy of "Bird of Passage" which he was carrying in his hand. He had been my host at a meeting with graduate students at Balliol. He evidently had not trailed me to the market, so it was just a coincidence.

March 23, 1992

The dinner for the Sagdeevs went off splendidly. I had also asked Bas and Susan Pease[66] (he is the former director of Culham, the controlled-fusion laboratory, and is now chairman of the British Pugwash Group). It turned out he was a friend of Sagdeev's. I also asked Margaret Gowing who made a lively contribution to the discussion, though her memory is getting very unreliable, and she is inclined to make the same argument in the same words, several times. Mrs. Sagdeev is a business woman with quite progressive views. One of her businesses is an outfit for economic risk assessment, and in that (or other) capacity she often goes to Moscow — no doubt that is how they met. Her Eisenhower connection comes out in her referring to herself as an "army brat."

On the Farm Hall transcripts, it was discovered that the top copy of the typescript is in Washington, and we have implored American colleagues to get us a decent photocopy. The German original has still not shown up. Another aspect is that Peter Ganz, who worked on the recordings, feels unable to say anything, because he had been sworn to secrecy, and will not talk unless he is released from this undertaking. But we have no idea whose job it would be to give such a release. Finally, Nicholas Kurti arranged for Ganz to visit the Ministry of Defense tomorrow, and perhaps this will lead to progress!

Invitations are piling up. After the Pugwash Conference in Berlin

---

[66]Sebastian "Bas" Pease (1922–2004) was a British physicist who joined the RAF Bomber Command's Operational Research section during WWII.

in September, the next date is 10th November in Rome (50 years since the first chain reaction) in honor of Fermi, where I am supposed to give a talk summarizing the development of physics from 1926 to 1945, probably in something like 30 minutes. Then 15th–20th December a meeting in Chicago also to celebrate the anniversary of the chain reaction. Calcutta from 20th December to 3rd January to celebrate the centenary of Bose, and at an unspecified date in November or December a very prestigious conference in Madrid. The only one I am firmly committed to is Rome. I have not mentioned various invitations I have already declined.

March 26, 1992

The last few days I was in close communication with the *New York Review of Books*, who can act very fast when they want to. I got proofs of my review on Monday, and faxed back corrections on Tuesday. On Wednesday came the final galleys, faxed to an American Express outfit here, who then delivered them by messenger. Wednesday night the editor phoned to check if I had any further corrections, I did not, but I changed one sentence, which he had queried, so as to avoid misinterpretation. Now he phoned again; they are publishing the photograph in Leipzig showing a group which includes Heisenberg and myself, and he wanted to make sure he had identified me correctly. We talked about the other people in the picture, and there was confusion until I realized he had a different picture of the same group. Anyway the issue will be printed tomorrow and out on Saturday.

May 25, 1992

In March I was supposed to talk at the American Physical Society meeting in Indianapolis about the early days of solid-state physics. Of course I did not go, but sent them the text of what I would have said. This was read to the meeting by David Pines (Urbana)[67] and was very well received. Now he wants to publish it in the Reviews of Modern Physics.

June 14, 1992

On one day, a number of us met at the Royal Society to meet Mrs. Brody, the widow of an army officer who helped look after the Germans at Farm Hall. She had felt very frustrated at not being

---

[67]David Pines (1924–2018) was the founding director of the Institute for Complex Adaptive Matter. In the early 1990s he was a member of the Center for Advanced Study, University of Illinois at Urbana-Champaign.

allowed to talk to anybody about what she knew, and enjoyed now being free to talk. We got worried when she had not appeared half an hour after the appointed time — when she did come she explained she had confused the Royal Society with the Royal Institution.

August 3, 1992

On Sunday, phoned Myriam Mondragón, who had come to take her DPhil in person, with her husband and her mother. They wanted to invite me to lunch, but of course I told them to come here. It was lucky that I had a second meal practically ready; otherwise it would not be easy to invite three people for Monday lunch on Saturday evening. It worked out well, even though I had to go to town this morning for some additional shopping, and to call at the lab to send a fax to the *New York Review of Books* to point out the howlers in Jeremy Bernstein's commentaries on the Farm Hall transcripts. The presence of the (German) husband meant that I don't have too much in the way of leftovers. The young Myriam has a job in Munich, and he has one in Bonn, so they see each other at weekends.

August 23, 1992

I had a visit from Rachel Fermi, the daughter of Giulio [Fermi].[68] She is a photographer, and wants to collect photos from Los Alamos for exhibition in some museum. She also wanted information on life in Los Alamos. She says that her father did not remember a thing. I told her that Gaby and Ronnie probably remember a few things, and she may contact you.

Stritinsky, an Ukrainian theorist with a Russian wife, is visiting here for a few weeks. He is a well-known nuclear theorist, now working for two years in Catania of all places. He had a public row with the Ukrainian authorities over excessive nationalism, such as pressure to make all schools teach in Ukrainian.

September 1, 1992

A month or so ago *The New York Review of Books* published excerpts from the "Farm Hall transcripts" with an introduction and commentary by Jeremy Bernstein.[69] He had a number of serious howlers, so I wrote an irate letter to the journal, as apparently did others. As a result he published in the next issue a humble correction and apology.

---

[68] Giulio Fermi (1936–1997) was a biologist and the son of Enrico Fermi.
[69] The expanded version later appeared as the book, J. Bernstein, *Hitler's Uranium Club: The Secret Recordings at Farm Hall*, (American Institute of Physics, Woodbury, NY, 1995).

September 25, 1992

I flew for Berlin on the morning of the 11th. Our meetings were in the old university (where I was a student for two semesters), and we were in two hotels, one, Unter den Linden, and the other, in which I stayed, near to the Alexanderplatz, both in former East Berlin.

Mine was very modern, some 38 floors, but furnished in American style without a single cupboard, wardrobe or drawer, and only an open hanging space, but with a very un-American absence of communications — I discovered only on the fifth day that there was a second breakfast room on the first floor with more space, so one did not have to queue for a table. But both had a very generous buffet breakfast.

Proceedings started with a reception given by the city in the old "Red Town Hall," a very imposing place. This was the biggest Pugwash conference yet: about 300 participants plus numerous wives, etc. Most meetings were in a place about 15 minutes walk from my hotel; the opening session and the meals were in the main university building some 25 minutes away. Good exercise. But halfway through the lunches moved to a different university canteen quite far away so one had to go in a fleet of buses. The reason was apparently that the main canteen had to close for sanitary reasons.

The meetings were constructive and interesting. I was in a working group on conflict prevention and resolution, chaired by a very intelligent former general from Nigeria. As this was one of the five meetings at which a new council is elected, there was some trouble because a man from Egypt, who had been a thorn in the flesh of council, by interminably arguing about issues on which he was in a minority of one, had not been renominated for election, and tried to fight this, claiming there had been breaches of procedure, but of course got nowhere.

Our group produced a very reasonable suggestion: The difficulty of countries paying their subscription to the UN (both US and Russia are seriously in arrears) could be eased if this was charged to the Defense budget, instead of to the Foreign Ministry! One funny episode: Victor Raninowich, a senior American, for long an official of the National Academy, and now secretary of some Foundation, said that in considering conflicts one must pay attention to the symptoms, and continued a long argument about the symptoms. I said I agreed with him, but thought when he said "symptoms" he really meant "causes." When he saw me next day he said "You were right, I meant causes."

One evening I was taken to the house of the daughter of my friend

Heinz Rudolph, who was also there. The daughter is very charming, as is her daughter of 3.5 years old. Her husband George (originally Mohammed) is a Lebanese. He is a civil engineer in a responsible position. Previously he was a graduate in political science, and worked in one of the Lebanese embassies.

I had brought a toy for the girl who I guessed was two, but the toy was still suitable. Another evening George drove me and his father-in-law around West Berlin. Amongst the sights was a very modern congress building, locally nicknamed the Pregnant Oyster, rebuilt after collapsing due to faulty workmanship, luckily when empty. He then took us to a Kurdish restaurant which was interesting.

The traditional half-day excursion was to Potsdam and the Sanssouci Palace, returning by river boat. A huge boat, which the whole conference did not fill, with dinner and dancing on board. I found I could still dance, and this attracted some attention. The conference banquet was in my hotel, conveniently. The food was quite good and the speeches painless. On the last afternoon I went to Oberschöneweide, where my correspondent, Frau Krause, had arranged a tea (coffee) party for a dozen or so pensioners. Most of them had either worked in the AEG cable factory,[70] or had husbnds or wives who had worked there, and many remembered my father. The party was in the Protestant parish house, and the minister was there part of the time, but had to leave early. I was given more old pictures and a tourist brochure in German and English, of the neighboring suburb of Köpenick, which also describes Oberschöneweide. It seems that when Queen Elizabeth visits Berlin in the near future, she will also be shown the cable factory (now owned by BICC, a British firm). The gate (and presumably other parts) is being repainted for the occasion, which is causing great local excitement.

Since my return I was very busy with assorted correspondence. One problem was an article in the *New Scientist* based on one in the *Bulletin of the Atomic Scientists*, claiming that Heisenberg hid the information about the possibility of an atom bomb from Hitler, and thereby prevented Germany from making a bomb. This, of course, had to be contradicted, but it took some effort to do so briefly.

December 1, 1992

This is Tuesday, and I have been home since lunch time Saturday. By now I have looked through the accumulated pile of letters, and dealt

---

[70]Heinrich Peierls, Rudolf's father, was the director of this factory in pre-Nazi Germany.

with the most urgent ones. Yesterday Tom Geballe[71] turned up, who is part of the British infrared astronomy team in Hawaii, and has to come often to England for scheduling meetings. Tomorrow is the meeting at the Science Museum.

The mail contains the news that I have been made an "honorable" member of the Ioffe Institute in St. Petersburg, and should go there next September for their 75th anniversary.

December 1992 (annual summary)

I last wrote about this time, looking forward to a quiet Christmas, with a visit by Masha Verblovsky, a relation of Genia's, and her husband Yura from St. Petersburg. But instead I was taken to hospital with an internal hemorrhage. My daughter Jo and her husband Chris rushed over from Birmingham to give very welcome assistance. Luckily I was able to catch the Russian visitors, who were about to start their journey — they would have been rather lost with me in hospital. The week in hospital was very comfortable; over Christmas there were three patients and four nurses, and they produced an excellent Christmas dinner, to which Jo and Chris were invited.

I got home after a week and was getting back to normal, when I had to return to hospital for another week with phlebitis. This was a nuisance, because the doctors would not allow me to travel any distance, so I had to stay put until May, and I had to cancel a planned trip to America. But then I had visitors here instead. The Verblovskys, whose visit had to be canceled around Christmas, came in February and saw a lot of Oxford, with the help of Jo, and other kind friends with cars, and made many day trips to London. Ronnie came over for a brief visit in February, and Gaby came in April. Of course Jo came from Birmingham more often, though she is very busy and working hard.

There was some excitement, because the "Farm Hall transcripts," the record of the bugged conversations of German atomic scientists interned after the war, which had been kept secret until now, were released as a result of intense pressure from many of us. One would have thought that this would settle once and for all the controversy about how much the Germans knew about possible nuclear weapons, and what they were trying to do, but if anything the controversy has even intensified.

In May I gently started traveling again, and spent a week in Coimbra,

---

[71]Tom Geballe (b. 1945) is an astronomer whose research interests include the Galactic center, the late stages of stellar evolution, the composition of interstellar dust, the surfaces, atmospheres, and aurorae of planets and moons, and brown dwarfs.

staying with my friend João da Providência; it is always a pleasure to visit the ancient town and lovey university. In June I gave the Dirac Memorial Lecture in Cambridge. My host was John C. Taylor,[72] who had been a reader here, and later became the Lucasian Professor (Dirac's former chair), so I was very well looked after.

In July I was at an "Amaldi meeting" in Heidelberg. This is one of a series of meetings of members of many academies to discuss international security and other common problems. Another lovely old town, but full of tourists. I took the opportunity of spending a weekend in Nurnberg, visiting Heinz Rudolph, my best friend at school, whom I had not seen for 65 years. We talked non-stop for two days to try to make up for the gap.

January 1, 1993

On the following Monday I was invited by the Griffins (Mauldely and Jim) because Els Placzek was staying with them. But then this was postponed because Mauldely had what turned out to be pneumonia.[73]

Monday night Nina [Byers] was out, and I invited Els Placzek to dinner, as she was leaving two days later. That was quite successful.

Now I am content to stay put except for trips to London, but on 15th January I go for three days to Spain. Also I am invited in March to a meeting in Triest in honor of Salam, some time in February or March for three days to Leiden, and in April or May for two or three weeks in Sweden. Some time in the summer the Pugwash conference will be held in Stockholm and in September will be the anniversary of the Ioffe Institute in St. Petersburg. And there are also the family weddings to be thought about. So, these are more or less the prospects for 1993.

March 3, 1993

On one weekend (12th–14th February) I went to a British Pugwash workshop in London on Conversion of Military R.&D. I was invited at short notice, because they had six Russian participants who, it was believed, knew no English. When they arrived, it turned out that three of them spoke English perfectly adequately. Two of the others sat next to a semi-professional interpreter, and next to me was the director of

---

[72] John Clayton Taylor (b. 1930) is a British mathematical physicist. He is an Emeritus Professor of Mathematical Physics at the Department of Applied Mathematics and Theoretical Physics of the University of Cambridge.
[73] It turned out that Maulde Griffin was misdiagnosed; in fact she had cancer. See Rudolf Peierls' letters of March 31 and August 8, 1993, pages 420 and 423.

one of the two Russian weapons laboratories. He assured me he was following the talks, and only asked once for a passage to be translated. When his turn came to talk, he moved over to the other interpreter (who was not too good). So I was really unemployed. The meeting was quite interesting, but depressing. There were many interesting ideas for conversion from most countries, but actual progress is very poor. There is no money, no demand, and no customers for the amount of civilian research that conversion could bring. It is evidently hard to get conversion during a recession.

In the week before the workshop I had one of my bouts, and feared I might not get to London, so I asked Joe Rotblat to think of a backup. He did not find one, but my bout only lasted a day, and I was able to go.

Then on 17th March (a Wednesday) I went to Leiden and came back late on Friday. My host was a biophysicist with a young Russian wife. It sounded rather familiar: They met at a conference in Sverdlovsk (she is from Moscow), then corresponded, got married in Moscow, and waited six months for her exit permit.

I spoke on the Wednesday night at the "Ehrenfest colloquium" which is normally held in a room where there are signatures of all speakers and distinguished visitors. I added my third signature, the first one dating from 1930, but the meeting had to be in a bigger room as there were too many people. Next day I went round the labs to talk to people, and on Friday morning I gave a talk on nuclear weapons to a "science and society" course, which had been the purpose of my visit.

There were two good dinners. Everybody seemed to have read "Bird of Passage," and when in a conversation I said "My first visit to Leiden was in 1930 or 1931" they said "In your book you say 30." Leiden has not changed much except for the shops, and is still very attractive. The house with the Lorentz flat, in which we stayed, has been pulled down.

March 31, 1993

You remember that Mauldely (Halban) Griffin was said first to have the flu and then it was pneumonia; it finally turned out to be lung cancer. Also one corner of the heart was affected. She had a lung and a piece of heart removed about three weeks ago but is beginning to live normally — she sounds very cheerful.

In the middle of the month I finally finished my review of the Powers book.[74] It was very negative, but the editor was very pleased. With

---

[74]Thomas Powers, *Heisenberg's War: The Secret History Of The German Bomb*, (Knopf, 1993).

reviews for the *New York Review of Books* you often don't hear anything for a couple of months, but this time the reaction was different and very impressive: I faxed them the typescript on 12th March. On the 16th arrived an edited typescript, with useful suggestions. I was out of town and saw this only on the 18th, and faxed corrections on the 19th. On the 20th came galley proofs — as this was a Saturday, they had guessed the lab would not function, so they sent it to a communications office in Oxford. The proofs were delivered to me by messenger. I phoned in a few more corrections, and on the 22nd had a second set of galleys. After dealing with these by phone, I had next day a call from the editor clearing some queries. The next evening the editor called again; I had referred to a 1977 letter to me from [Samuel] Goudsmit, which Powers had quoted selectively and thereby unfairly. I did not have this letter any more, but could only quote its sense from memory. The editor had found a copy in the archives of the American Institute of Physics. Would I like to have it? So an hour later it arrived again by messenger, with a suggested form of reference to it in my text. I am sure by now the thing is published, but it has not arrived here yet.

April 29, 1993

The most concrete news is that I have acquired a FAX machine, which operates on my phone line. I have had it for about two weeks, and am getting used to its idiosyncrasies. My NYRB review was published with lightning speed.[75] I had a number of complimentary letters, though some people complained I was too polite to the author. So far he had not reacted. The funny thing is that everybody's subscription copies arrived here two weeks before the issue was distributed in the US. Most other reviews were written by literary experts who took Powers's claims at face value, but there was a very sensible review by Jeremy Bernstein in *Science*.[76]

Next February is the centenary of the birth of Yakov Il'ich Frenkel, a distinguished Russian theoretician, to whom I owe a debt of gratitude, as he was the one who had invited me to the 1930 conference in Odessa, introduced me to Genia, and also arranged for my stay in Leningrad in 1931. When he died in 1952, no English language journal took any notice. His son, who is a historian of science has been urging me for some time to persuade the journals to publish something. So now I suggested to *Physics Today* that they should arrange for an article on

---
[75]See R. Peierls, "The Bomb That Never Was," *New York Review of Books*, 1993, April 22 Issue.
[76]Jeremy Bernstein, "Revelations from Farm Hall," *Science*, 1993, March 26 Issue.

the occasion of the centenary. They accepted, and — as might have been predicted — asked me to write it.[77] Of course, I have to collect information. I have a biography written by his son, and a collection of people's reminiscences of him, all in Russian. Now I read slowly, but much more so in Russian. So there was a problem. But I think it is getting solved because the wife of our very bright Russian in Theoretical Physics seems willing to read the material to me, and that will speed things up.[78]

More history: A former Birmingham undergraduate whom I taught, and who is now in TV wants to make a "drama-documentary" about the British work, and asked me to be a consultant. I am not enthusiastic about the drama part, but decided to accept to keep things on the rails as far as possible. He is now waiting to hear if the project has backing.

My annual visit to the eye specialist was disappointing. I had thought there were signs of the cataract developing, which would mean that an operation could improve my vision, but he told me this was wrong — no appreciable cataract, just grin and bear it.

June 6, 1993

Off to Washington, using a new seat on the 747 (new to me, that is) with extra leg room. Met and generally looked after by Wally Greenberg,[79] a nice theoretician, whom I greatly respect for his suggestion of "color" for quarks.

July 10, 1993

At this time I talked to Kitty and happened to mention I was taking a drug called Tenormine against high blood pressure. She found it in one of her books, where it said that one of its side effect was heart failure! So I got the GP's permission to drop it. (He had recently doubled the dose.) Three days later I saw a cardiologist, who confirmed that the Tenormine was the cause, and said I had already improved considerably in the few days without the drug, and that, in spite of some enlargement, the heart was functioning very well. He noticed, however, that the GP, in sending a blood sample for analysis, had not asked the kidney function to be checked. So he took another, and a few days ago came the answer: I also have a "renal failure." So far this was only a phone message from the GP, who explained that it was not very serious — no doubt as undramatic as the cardiac failure. The only

---

[77] Peierls' article "Yakov Il'ich Frenkel" appeared in *Physics Today*, **47**, 6, 44 (1994).
[78] Elena Tsvelik, the wife of Professor Alexei Tsvelik.
[79] Oscar Wallace "Wally" Greenberg (b. 1932) invented three colors for quarks in 1964.

action taken so far is to double the dose of the diuretic. I shall see the GP in a day or so and no doubt get further instructions.

Meanwhile the symptoms have abated so that I feel pretty normal, and live quite normally. I did go to chair Hans Bethe's lecture at the Royal Society, though at that time I was not too well. He spoke well, if perhaps at somewhat too high a level for some of the audience. But there were good questions afterwards. The lecture had been arranged to follow a two-day discussion meeting on a related subject, so there were many experts present. One slightly comic feature was that, as he is rather deaf, Hans could not hear some of the questions, so I repeated them for him; but I could not see who wanted to speak, so he had to point them out to me!

The lecture was followed by a dinner of the Royal Society dining club, to which Hans, Rose and I were invited. I had previously heard only vague rumors about this club. I found that I should have been invited to join on my election to the RS in 1945, but I was then still in Los Alamos. I found however that two more recently elected fellows also were not told.

At these dinners the President normally presides, but in his absence the chairman was Allibone,[80] an old physicist who worked in industry (Metropolitan Vickers) and had a lot to do with making accelerators. He was sitting next to Rose, who was taken aback when he told her that Teller was absolutely right, and it was right to cancel Oppenheimer's clearance.

August 8, 1993

That weekend, Maulde Griffin (Halban) died after a week of terrible agony. Evidently her cancer had spread to some other organ. Her mother, Els, was devastated. The funeral service was in Keble chapel (her husband is a fellow of Keble), the interment was for close family only. There was a reception in the college, where we made polite conversation waiting for the family to return. Afterwards I ran into Isaiah Berlin, who was furious. He had been told it would be a non-religious service. "The only thing they left out was the Lord's Prayer!"

I don't think I have yet reported that in June I had the long-expected visit from a social worker, who explained a lot of useful things. For example, as a "visually handicapped person" I can get a code with which I can make phone directory enquiries free of charge. I also joined

---

[80]Thomas Allibone (1903–2003) was an English physicist, who was involved (during WWII) in radar research, Tube Alloys and Manhattan Projects.

the "Partially Sighted Society," whose bimonthly magazine has useful information about new aids. Above all I have joined a library of books on tape. From their fairly substantial catalogue you select a list of 50 books. You can then have two such books, and when you return one, the computer selects a new one that is available from your list. The post office carry these back and forth free of charge as "articles for the blind." I am just starting on my third book.

September 5, 1993

I have now taken the plunge and ordered a new computer and printer, guided by Jo's expert advice. The printer has arrived, but is not yet installed, because I don't want to change horses in mid-stream. I have nearly finished the article on Frenkel for *Physics Today*. This must be done, at least in draft, by 20th September, when I go to St. Petersburg, where I can discuss it with Frenkel's son. For your information: I go there on the 20th, for the 75th anniversary of the Ioffe Institute of which I am an "honorable member." On the 26th I go on to Rome, for a meeting at the Accademia dei Lincei. Back on the 30th.

Gave a dinner party, where most of the people I asked had either a wife or a husband abroad, so it was mostly grass widows and grass widowers but it was a cheerful party.

Collected my Russian visa. As I expected complications I decided to go in person rather than through a travel agent. The atmosphere in the consulate has not changed. You have to queue in the street, and when I first went to apply, it was an hour's wait, while for collecting the visa you went straight in. So when I came back to collect, the queue for collecting was a mile long, and I waited 1h 15m, while the queue to apply was quite short. You can't win. At least one of the bus lines from Oxford stops practically in front of the consulate.

November 27, 1993

My family letters seem to have got somewhat out of phase. One of my recent entertainments was a TV interview by a team who are making a program about the Russian atomic-energy work, and wanted to include something about the work in the West by way of background. Two days before the date the producer asked if I could also explain the physics, from Rutherford to bombs and reactors, intelligibly to the layman. He wanted this done in front of a blackboard. This was a challenge. I said I needed at least 10 minutes, which was agreed. I did indeed finish in 9.5 minutes, but then they wanted each section done separately, and requiring a close-up of my writing, with only one camera, so I had to repeat the writing of each piece, without talking. This was done in the Nuclear Physics lab, but then they also wanted the interview, for which there was no suitable room available, so we moved to my flat. The whole thing took from 11 to 5, and at the end I was pretty tired. So far they have not offered me any fee for this, but I am querying that!

A week or two ago, Michael Atiyah, the President of the Royal Society had arranged to go with a few other people, to talk to the scientific adviser in the Ministry of Defense to make out a case for a comprehensive test ban. As the meeting was early, I stayed the night in the Royal Society, and as I arrived I was given a copy of a speech made that day by the Secretary for Defense. It was a long document, and I read it late into the night. It all seemed a rehash of known things, until the last few pages, where he stated that the government was now willing to negotiate a test ban treaty! So our visit seemed unnecessary, but it was not, because it turned out this was a new scientific adviser, who had been in office only six weeks, and is very open-minded. We discussed several problems in the relations of scientists with the ministry, so it was worthwhile.

November 1993 (annual summary)

In January was a weekend in Spain, in a Parador (government hotel) in a 15th-century castle in Extremadura. The meeting was about "Science, Technology and Culture." I was one of three scientists amongst about 20 participants; it took an effort to get our views listened to. The food at these Encounters is spectacular – they bring their own chef from Madrid!

In February was a three-day visit to Leiden, a town of which I have fond memories. Going round the laboratories, I found that a team had done an experiment, which called for a slight extension of one of my papers, and later I worked that out, while they completed the experiment, finding good agreement. Some points about the interpretation of the results caused mathematical problems, about which we are still arguing. I have recently acquired a fax machine, which is a great toy, and the argument between here and Leiden is now going on intensely by fax.

Later in September I spent a week in St. Petersburg for the 75th anniversary of the Ioffe Institute, where I had been elected the previous year as Honorary Member (only they called it "honorable" member!). It was an interesting visit. The economic position is terrible, and while the shops are full of goods at prices somewhat below those in the West, salaries are miserable; unskilled workers or pensioners can just about manage to eat, but cannot dream of buying clothes or shoes. The institute, the leading solid-state laboratory in the country, now can spend only 6% of its budget on research; the rest goes into salaries, building upkeep, etc. They get some research grants from foreign foundations; the government taxes these heavily. It is surprising that they have lost very few people through brain drain.

I was there during the strong action by Yeltsin against the parliament, but after the first shock, people in St. Petersburg did not seem worried. From there to Rome for another Amaldi conference, where I began to feel very weak, and finally collapsed with pneumonia. After a night in hospital I returned to Oxford very dramatically, by ambulance to the airport, and another ambulance from there to my familiar hospital here for another six days. It took me a few weeks to recover fully, but by now I am completely back to normal.

Early January 1994

My GP managed to fix up an appointment with a neurologist at short notice. The neurologist again found nothing wrong, but thought the diuretic I was taking might be responsible, since it is known to be very violent, and sends the blood pressure up and down. This would be the third time a medicine prescribed by my GP got me into trouble. I am now taking steps to change to another one. He is retiring in September anyway, but the sooner the better.

March 20, 1994

This was in the John Radcliffe hospital, where the private ward is on the fifth floor, with a glorious view. The medical care was, I think, better than in the Acland; and then food was quite acceptable. It is the same food as in the health service wards, and I approve of that. I had a problem getting in. I was told to come at 11, but there was no bed free, so I sat around in Admissions until 12, then they told me to have some lunch (in their snack bar) and come back at 2.15. This was quite miserable, and I phoned Linda Davidson, who lives fairly close to the hospital, and she came over to keep me company. At 2.15 they said I should go to the ambulance waiting area, where I would be collected, but we decided it was more comfortable to wait in Linda's house. They finally phoned about 6 that they had a bed for me.

October 1994

Although I can now live more or less normally, it was clear that I could no longer do my own housekeeping, so the choice was between having a live-in help and moving to a retirement home. I chose the latter, and I have been since July in this place, which Jo discovered. It stands in large and attractive grounds. The buildings can be seen in the TV series "Waiting for God,"[81] which is being shown here and in the US (though the interior and the events show no similarity!). I have a pleasant room with bathroom, and all meals, mostly excellent, are provided. The staff are charming and helpful. The place is about five miles from Oxford, and there is a good bus service.

A further complication is that, independently of the other troubles, my eyesight has gone much worse. However, with a "scanner" (which shows an enlarged TV picture of any text), and a software scheme, which gives an enlargement of the computer screen, I read and write almost normally, so I am again literate.

Otherwise there is not much to report. The American Institute of Physics is going to publish a collection of my non-technical essays, and that material has been sorted and sent off. Another publisher will publish a collection of my scientific papers,[82] and this is almost ready to go to the publishers; I had much help from Dick Dalitz, who is the editor. The early German papers are being translated by Gerry Field.

*****

Appendix: Commentary to page 408

Unfortunately, I do not know the contents of Rudolf Peierls' talk on the future of physics. It would be very curious to compare his vision with the prediction given by Richard Feynman[83] in 1965:

> A new idea is extremely difficult to think of. It takes a fantastic imagination.

---

[81] A British sitcom that ran from 1990 to 1994 starring Graham Crowden and Stephanie Cole. This show is about two spirited residents of a retirement home who spend their time running rings around the home's oppressive management and their own families.

[82] *Selected Scientific Papers of Sir Rudolf Peierls*, Eds. R. H. Dalitz and Sir Rudolf Peierls, (World Scientific, Singapore, 1997).

[83] Richard Feynman, *The Character of Physical Law*, (Cox and Wyman LTD, London, 1965), Lecture 7.

What of the future of this adventure [physics]? What will happen ultimately? We are going along guessing the laws; how many laws are we going to have to guess? I do not know. Some of my colleagues say that this fundamental aspect of our science will go on; but I think there will certainly not be perpetual novelty, say for a thousand years. This thing cannot keep on going so that we are always going to discover more and more new laws. If we do, it will become boring that there are so many levels one underneath the other. It seems to me that what can happen in the future is either that all the laws become known — that is, if you had enough laws you could compute consequences and they would always agree with experiment, which would be the end of the line — or it may happen that the experiments get harder and harder to make, more and more expensive, so you get 99.9 per cent of the phenomena, but there is always some phenomenon which has just been discovered, which is very hard to measure, and which disagrees; and as soon as you have the explanation of that one there is always another one, and it gets slower and slower and more and more uninteresting. That is another way it may end. But I think it has to end in one way or another.

We are very lucky to live in an age in which we are still making discoveries. It is like the discovery of America — you only discover it once. The age in which we live is the age in which we are discovering the fundamental laws of nature, and that day will never come again. It is very exciting, it is marvelous, but this excitement will have to go. Of course in the future there will be other interests. There will be the interest of the connection of one level of phenomena to another — phenomena in biology and so on, or, if you are talking about exploration, exploring other planets, but there will not still be the same things that we are doing now.

Chapter 17

Peierls, Heisenberg and Farm Hall Transcripts

Farm Hall Transcripts

The making of the nuclear weapon in the 1940s changed world history forever. That's why historians of science paid so much attention to each and every detail of this fateful development. As I mentioned above more than once, Otto Frisch and Rudolf Peierls were *the* pioneers who initiated the process. Peierls was very sensitive to any distortion of the historical truth.

From the early postwar years most historians of science followed a theory according to which, the team led by Heisenberg had deliberately sabotaged the German nuclear project, accounting for its failure. This myth can be traced back to a letter Heisenberg sent to Robert Jungk, see [1]. Thomas Powers, who wrote [2] the monumental treatise *Heisenberg's War*, was an ardent proponent of this theory. This book was severely criticized by Peierls, as seen from his diary, see e.g. page 420.

Two documents completely changed the prevailing historical discourse. In the early 1990s the so-called "Farm-Hall Transcripts" were declassified and made available to public. As the reader could see from the diary (Chapter 16) it was Rudolf Peierls who was one of the initiators of the declassifying process (pages 410, 412, 413, and so on).

I will return to the "Farm-Hall Transcripts" later, since now I have to mention the second crucial document, the unsent letter of Bohr to Heisenberg which starts with the words: "I have seen a book, *Stærkere end tusind sole* [*Brighter than a thousand suns*] by Robert Jungk [1], recently published in Danish, and I think that I owe it to you to tell you that I am greatly amazed to see how much your memory has deceived you in your letter to the author of the book, excerpts of which are printed in the Danish edition." This letter kept in the Niels Bohr

Archive in Copenhagen was released in 2002, and is currently accessible on the Internet [3].

To my mind, the "Farm-Hall Transcripts" and this letter leave no basis for the myth of sabotage. I hasten to add though that polemics still continue among historians, and will probably continue for a long time. History is not an exact science, which leaves space for "deniers" of all kinds.

To my mind, Heisenberg's team worked in earnest. They failed not because they deliberately sabotaged the project, but because from the very beginning they did not take it as passionately as their Anglo-American or Russian counterparts. Even if they did, they were not qualified to solve the problems that arose in the course of the work.

One of the fatal mistakes was Bothe's misinterpretation of the results of his experiment.[1] Another fatal mistake was Heisenberg's estimate of critical mass. Unlike Frisch and Peierls, whose estimate was in the ballpark of kilograms, Heisenberg's original estimate was in the ballpark of tons,[2] which was absolutely prohibitive. Given the absolute authority of Heisenberg, nobody in the German team bothered to check his calculations. This issue is discussed in great detail in Jeremy Bernstein's paper "Heisenberg and the Critical Mass" [5]. Jeremy Bernstein is mentioned many times in Peierls' diary. Apparently, they were in correspondence with each other. Peierls' assessment of Bernstein's article on "Farm Hall Transcripts" in *The New York Review of Books* is highly positive. Later Bernstein published the book entitled *Hitler's Uranium Club* [6] which I consider as one of the best sources in answering this range of questions.

The Farm Hall was bugged so that conversations of all detainees were recorded and transcribed. On August 6, 1945, the detained German physicists learned that a new weapon had been dropped on Hiroshima. They did not believe that it was nuclear.

When they finally were persuaded that it was, they began trying to

---

[1] The graphite he used in his experiment was not pure enough. In the US, Szilard insisted on further purifying commercially produced graphite. Fermi fully agreed and could then use graphite in his pile.

[2] The 1939 wartime report [written by Heisenberg] is entitled "Die Möglichkeit der technischen Energiegewinnung aus der Uranspaltung" ["The possibility of the technical use of the energy gained from uranium fission"], see [4]. In this report Heisenberg derives a rather crude formula for the critical mass. No numerical estimates were presented. Using the Heisenberg formula one can obtain the critical radius and the corresponding mass, 31 cm and approximately one ton, respectively, i.e. approximately 70 times larger than the actual value of the critical mass. These numbers agree with the first estimate Heisenberg presented to his colleagues at Farm Hall. In 1940 Karl Wirtz heard Heisenberg commenting on this calculation.

Figure 17.1   Farm Hall, circa 1945. Ten German physicists — Erich Bagge, Kurt Diebner, Walther Gerlach, Otto Hahn, Paul Harteck, Werner Heisenberg, Horst Korsching, Max von Laue, Carl Friedrich von Weizsäcker, and Karl Wirtz — were interned here from July 3, 1945, to January 3, 1946.

explain it. That evening, Otto Hahn and Heisenberg had a conversation. Heisenberg gave Hahn an estimate based on the data concerning the Hiroshima explosion published in newspapers. Heisenberg reasoned as follows. He knew that the Hiroshima explosion was about equivalent to 15,000 tons of TNT, and he knew that this amount corresponded to the fission of about 1 kg of uranium. Then he estimated that this would require about 80 generations of fissions assuming that two neutrons are emitted per fission. He then assumed that during this process the neutrons flow out to the boundary in a random walk of 80 steps with a step length equal to the mean free path for fission. This gave him a critical radius of 54 cm and a critical mass of several tons. (The correct estimate would give 15–20 kg.)

On August 14, 1945, Heisenberg gave a lecture on this subject at Farm Hall to the other nine detainees. From his August 14 lecture one can infer that he had finally understood the basics of the problem. This time Heisenberg derived a reasonable critical mass of 15 kg.

Then Bernstein continues:

> There is one especially surreal aspect of this discussion that took place after the second bomb was dropped on Nagasaki. The mass of material for this bomb was given in news reports and it seemed too small. The Germans indulged in all sorts of wild speculations as to why this was so. It never occurred to them that the Nagasaki bomb was made of plutonium, despite the fact that von Weizsäcker, who had introduced the idea of transuranics into the German program, was in the audience.
>
> To have admitted that plutonium was used was to admit that the Allies had a vast reactor development and that everything the German scientists had worked on for so long and so hard had been insignificant. Heisenberg's lecture, which represented the high water mark of the German understanding of nuclear weapons, shows that in the end they understood very little.
>
> Prior to Hiroshima the Germans were absolutely convinced on the basis of their own experience that a nuclear bomb could not be built in the immediate future. Their belief was based on the idea of their superiority: they were absolutely convinced that they were ahead of everyone else in their study of nuclear chain reaction. Because they had not been able to build a nuclear reactor, they were sure that no one else had done so. These points are made very explicitly in the Farm Hall transcripts.

In his Diary, Peierls describes two articles on this issue which he wrote for *The New York Review of Books*, one in 1992 [7] and another in 1993 [8]. In Chapter 16, Peierls narrates the story of his work on these articles, which apparently were close to his heart. He used some data from the "Farm Hall Transcripts." Importantly, he relied on his own recollections of Heisenberg whom he personally knew since 1928–29 when he spent some time at Leipzig University, and his intimate knowledge of Heisenberg's research in nuclear physics. I think, Peierls' characterization of Heisenberg's role in the failed German program is both balanced and revealing.

Below I reproduce excerpts from these articles. Both articles [7] and [8] are precious not only because Peierls was the active participant and eye witness in the development of the first applications of quantum mechanics, but also because his narrative is enlightening.

It is a pity that in his time Rudolf Peierls did not have access to the unsent letter of Bohr to Heisenberg mentioned above (on page 429). I wish I could read his comments on this letter. It is reproduced below in its entirety.

*****

The Uncertain Scientist by Rudolf Peierls[3]

[...] After Hitler came to power in 1933 Heisenberg found the effects of the regime on academic life, on science, and on life generally deplorable, but he decided to remain in Germany because of his patriotism, and because he felt a duty to do what he could to mitigate the regime's evil effects. During the war he worked on the German atomic-energy project, which failed to get any results. His actions during that period have caused intense controversy, which continues [...]

Cassidy [9] is clear and reasonable on the controversies about Heisenberg's behavior during the Hitler period. The first question asked by many is: Why did Heisenberg stay in Germany, when he was opposed to much of what the system was doing, and why did he make so many compromises with the system? Cassidy's answer, with which I agree, is that Heisenberg was strongly patriotic. He also had some sympathy for the idea of a national renewal put forward by Hitler, since the social system did not seem to be working under the Weimar republic. But he is on record as strongly disapproving of the sacking of "non-Aryans" and liberals, and the attacks on "Jewish," as opposed to "German," physics.

He often sought the advice of Max Planck, the elder statesman of physics, who felt it his duty to remain in his post as head of the Kaiser Wilhelm Society, the main body supporting research, to use his influence to prevent the Nazis' worst excesses. In fact there was little he could do; his attitude is well characterized by a remark once made by Robert Oppenheimer, in reference to his own past, "As long as I ride on this train, it will not go to the wrong destination."

Cassidy criticizes Heisenberg for trying to help his own students and others in his immediate circle and not anybody outside, but there were limits to what one could do under the system without great personal

---

[3] *The New York Review of Books*, 1992, April 23 issue. Peierls' review of *Uncertainty: The Life and Science of Werner Heisenberg*.

risk. Once he had decided to remain in Nazi Germany, he could hardly avoid making concessions to the regime. Not everybody could be like Max von Laue, who kept out of sight and made as few concessions as possible (it is said that he never left the house without carrying two packages or a briefcase and a package, so that he would have no hand free to give the Hitler salute). This was not a possible way of life for Heisenberg, who wanted to continue being active as a leader in science.

He never openly opposed the Nazi system, but he was outspoken when it came to the attacks on modern physics by Johannes Stark and other fanatical supporters of "German" physics, even though this exposed him to virulent personal denunciation. This culminated in an attack on him in a Nazi newspaper, and he was, as a result, turned down for the appointment to the chair in Munich, for which he was the favored candidate. He complained to Himmler, whose organization sponsored the paper, asking if this attack represented the official line, in which case he would feel obliged to resign his chair in Leipzig. It took a long time before the reply came that such an attack did not have official approval, and would not be allowed to happen again. Indeed the attacks on him ceased, to his gratification, though this result was not achieved without cost. While he was allowed to teach relativity, he had to promise not to mention the name of Einstein. Cassidy thinks that in this struggle, Heisenberg was concerned solely with his personal honor and standing, but it is hard to tell whether or how far he was also thinking of the standing and reputation of his subject and his profession.

Another important controversy centers on the question why the German scientists did not produce an atom bomb. Robert Jungk[4] and Walter Kaempfert of *The New York Times* have claimed that the German scientists were inhibited by moral scruples. Others, including Sam Goudsmit in his book Alsos[5] (though he later withdrew some of his statements) and Jeremy Bernstein in his April 1991 review in these pages of Victor Weisskopf's autobiography,[6] claim that the failure was owing to incompetence. [...]

[Heisenberg and the other German] scientists overestimated the difficulties, because they never made a careful estimate of the critical size needed for a fast chain reaction, and therefore tended to overestimate the size of a weapon. It is also true that there were many errors

---

[4] Robert Jungk, *Brighter Than a Thousand Suns*, (Harcourt Brace Jovanovich, 1958).
[5] S. Goudsmit, *Alsos*, (Tomash Publishers, 1983).
[6] "The Charms of a Physicist," *The New York Review*, April 16, 1991, pages 47–50.

of judgment in their work. For example, they excluded graphite as a moderator to slow down the neutrons because of a mistaken finding that it absorbed too many neutrons (not, as Cassidy says, that it was ineffective in slowing them down). This failure is often blamed on Walther Bothe, who is alleged to have made "wrong" measurements. In fact his measurements were correct; but the "pure" graphite he was using contained impurities in amounts that were too small to be detected chemically, but were fatal for the absorption of neutrons. In the same situation, Fermi and his collaborators in the US guessed that further purification would improve the results. Another bad judgment was Heisenberg's insistence on using uranium plates, rather than rods or cubes of uranium, in the face of theoretical and experimental demonstration that rods or cubes were more effective.

The opinion of Heisenberg and other German scientists about the feasibility of making a weapon during the war is made clear in the "Farm Hall Transcripts," the recently released records of the conversations between German scientists, including Heisenberg, during their internment in England after VE Day, in a house that was bugged by British intelligence. When the first news came of the atom bomb being dropped on Hiroshima, most of them refused to believe this, claiming that the Allies were bluffing. Even when more detailed statements convinced them that this was an atom bomb, Heisenberg kept saying that he could not understand how the Americans had been able to make a weapon in the time available to them. The transcripts also dispose of the idea that the Germans did not make a bomb for moral reasons: they discussed at great length the reasons for their failure, but they did not claim that moral scruples stopped them from working on the bomb. When Weizsäcker tentatively suggested that such scruples had prevented success, Otto Hahn firmly contradicted him. [...]

*****

The Bomb That Never Was by Rudolf Peierls[7]

The recently released "Farm Hall Transcripts," of recorded conversations of German atomic scientists who were interned in England shortly after the war, firmly ruled out the idea of their moral superiority. The transcripts record the reaction of the scientists to the announcement

---

[7] *The New York Review of Books*, 1993, April 22 issue. (Peierls' review of *Heisenberg's War: The Secret History of the German Bomb* by Thomas Powers.)

that a bomb was dropped on Hiroshima. Apart from a tentative remark by von Weizsäcker, none of them mentioned moral objections as reasons for not having made a bomb, although they argued about the reasons for their own failure and the consequences of the bomb for their own future in Germany. Yet the controversy has continued, and in *Heisenberg's War* Thomas Powers produces a new theory. [...]

In the main thesis of his book, stated explicitly in the last two chapters but present in the background throughout, Powers advances his own, new explanation of the failure of the German project. Heisenberg, he argues, deliberately withheld information from the authorities and from his colleagues in order to stop the construction of a German atom bomb. The most important aspect of this deception was that Heisenberg kept stressing that the development of a bomb was too difficult and involved too great an effort to be completed in wartime Germany. No doubt Heisenberg said this on every possible occasion; the questions are, did he believe it, and was it true? [...]

*****

Draft of Letter from Bohr to Heisenberg[8]

Dear Heisenberg,

I have seen a book, "Stærkere end tusind sole" by Robert Jungk, recently published in Danish, and I think that I owe it to you to tell you that I am greatly amazed to see how much your memory has deceived you in your letter to the author of the book, excerpts of which are printed in the Danish edition.

Personally, I remember every word of our conversations,[9] which took place on a background of extreme sorrow and tension for us here in Denmark. In particular, it made a strong impression both on Margrethe and me, and on everyone at the Institute that the two of you spoke to, that you and Weizsäcker expressed your definite conviction that Germany would win and that it was therefore quite foolish for us to maintain the hope of a different outcome of the war and to be reticent as regards all German offers of cooperation. I also remember quite clearly our conversation in my room at the Institute, where in vague terms you spoke in a manner that could only give me the firm impression that, under your leadership, everything was being done in Germany to

---

[8] Niels Bohr Archive. In the handwriting of Niels Bohr's assistant, Aage Petersen. Circa 1957.

[9] These conversations took place sometime during the week of September 15–21, 1941. Heisenberg attended a German-sponsored conference in Copenhagen. Bohr did not attend the conference as a protest against the Germans, but he agreed to personally see Heisenberg.

develop atomic weapons and that you said that there was no need to talk about details since you were completely familiar with them and had spent the past two years working more or less exclusively on such preparations. I listened to this without speaking since [a] great matter for mankind was at issue in which, despite our personal friendship, we had to be regarded as representatives of two sides engaged in mortal combat. That my silence and gravity, as you write in the letter, could be taken as an expression of shock at your reports that it was possible to make an atomic bomb is a quite peculiar misunderstanding, which must be due to the great tension in your own mind. From the day three years earlier when I realized that slow neutrons could only cause fission in Uranium 235 and not 238, it was of course obvious to me that a bomb with certain effect could be produced by separating the uraniums. In June 1939 I had even given a public lecture in Birmingham about uranium fission, where I talked about the effects of such a bomb but of course added that the technical preparations would be so large that one did not know how soon they could be overcome. If anything in my behavior could be interpreted as shock, it did not derive from such reports but rather from the news, as I had to understand it, that Germany was participating vigorously in a race to be the first with atomic weapons.

Besides, at the time I knew nothing about how far one had already come in England and America, which I learned only the following year when I was able to go to England after being informed that the German occupation force in Denmark had made preparations for my arrest.

All this is of course just a rendition of what I remember clearly from our conversations, which subsequently were naturally the subject of thorough discussions at the Institute and with other trusted friends in Denmark. It is quite another matter that, at that time and ever since, I have always had the definite impression that you and Weizsäcker had arranged the symposium at the German Institute, in which I did not take part myself as a matter of principle, and the visit to us in order to assure yourselves that we suffered no harm and to try in every way to help us in our dangerous situation.

This letter is essentially just between the two of us, but because of the stir the book has already caused in Danish newspapers, I have thought it appropriate to relate the contents of the letter in confidence to the head of the Danish Foreign Office and to Ambassador Duckwitz.[10]

---

[10] Georg Duckwitz (1904–1973) was a German diplomat. During World War II he served as an attaché for Nazi Germany in occupied Denmark. He tipped off the Danes about the Germans' intended deportation of the Jewish population in 1943 and arranged for their reception in Sweden.

References

[1] Robert Jungk, *Heller als Tausend Sonnen [Brighter than a Thousand Suns]. Das Schicksal der Atomforscher,* (Stuttgart, 1956), in German.

[2] Thomas Powers, *Heisenberg's War: The Secret History of the German Bomb,* (Knopf, 1993).

[3] Niels Bohr Archive, Draft of letter from Bohr to Heisenberg, never sent. In the handwriting of Niels Bohr's assistant, Aage Petersen.

[4] W. Heisenberg, *Gesammelte Werke/Collected Works,* Eds. W. Blum, H. P. Dürr, and H. Rechenberg (Springer, Berlin, 1989), Ser. A, Pt. II. page 378.

[5] J. Bernstein, "Heisenberg and the Critical Mass," *American Journal of Physics,* **70**, 911 (2002).

[6] J. Bernstein, *Hitler's Uranium Club,* (Copernicus Books, Second Edition, 2001).

[7] R. Peierls, "The Uncertain Scientist," *The New York Review of Books,* 1992, April 23 issue.

[8] R. Peierls, "The Bomb That Never Was," *The New York Review of Books,* 1993, April 22 Issue.

[9] David C. Cassidy, *Uncertainty: The Life and Science of Werner Heisenberg,* (W.H. Freeman, 1991).

Chapter 18

Rudolf Peierls and Nuclear Responsibility

> *[The] scientist is the custodian of the immense inherited wealth of discovery and advance in science and technology of all the ages. He alone has the key and therefore the access to this treasure. The scientist should take the responsibility of this position in our world as seriously as the physician should take his Hippocratic oath.*
>
> Isidor Isaac Rabi [1]

This is the last chapter of the book intended mostly for those readers who familiarized themselves with Chapter 16. Peierls' Diary shows, with a remarkable clarity, how much time and effort he invested in the arms control movement after his retirement. In fact, this issue became one of his main occupations long before the retirement. The early activities aimed at public awareness of nuclear arms and their impact on the humankind are discussed in detail in [2] where the reader can find an extensive list of references. I will briefly summarize Chapter 6 of this book with a special emphasis on Peierls' role and ideas.

The nuclear arms race, which in the 1940s and 1950s was a competition for supremacy in nuclear warfare between the United States and the USSR, started in earnest with the first test of the Soviet atomic bomb in 1949. However, the debate on the public control over nuclear weapons began earlier — first in the USA. In 1945, even before the end of the WWII, a small group of scientists led by James Franck, Eugene Rabinowitch, and Leo Szilard publicly addressed the question whether the nuclear weapons should be used at all. They issued the so-called Franck Memorandum recommending that a demonstration explosion of a nuclear weapon be conducted in a desert to compel the Emperor of Japan to surrender before the weapon was actually used. Needless to say this plan was unrealistic at the time of fierce combat actions in the Pacific.

After the war ended in 1945, the Franck Committee was replaced in the US by a newly organized Federation of Atomic Scientists (FAS) and by its UK counterpart, the Atomic Scientists Association (ASA) founded by Joseph Rotblat in 1946. After Peierls' return to the United Kingdom, he became a driving force behind the ASA. Like his colleagues from Los Alamos, Peierls was driven by a moral responsibility that transpired from the creation of nuclear arms. FAS set up the Emergency Committee of Atomic Scientists (ECAS) in May 1946. Chaired by Albert Einstein, it included the German-speaking émigrés Hans Bethe, Leo Szilard and Victor Weisskopf. Approximately at the same time, Eugene Rabinowitch and Hyman Goldsmith launched the *Bulletin of the Atomic Scientists* — a nontechnical academic journal, covering global security and the dangers posed by nuclear threats, which soon became their main media outlet.

The major objectives of the ASA formulated largely under the influence of Peierls' ideas were: educating both the public and fellow experts in atomic-energy-related matters; advising political decision-makers about nuclear power, and promoting the agenda of the international control of nuclear power. At first the ASA maintained an association with the *Association of the Scientific Workers*.[1] This association was broken shortly because of the political activism of the *Association of the Scientific Workers* that was engaged in political issues representing the left-wing perspective. At the same time, Rudolf Peierls advocated the ideology of 'unpolitical' science. In a letter to Sir Chadwick sent in February 1946, Peierls expressed serious concern about the ASA's dependence on the left-wing Association of the Scientific Workers because, "such a connection would only antagonize certain people." In his opinion, the Association of the Scientific Workers combined two areas — political and scientific — that "do not mix and should be carried out by separate bodies." Christoph Laucht writes ([2], page 132):

> Peierls not only distrusted political activism from the left but harbored a general suspicion towards scientists' involvement in politics.

He insisted from the outset that "a proper division should be made between statements of scientific facts and opinions held by scientists."

Christoph Laucht continues ([2], page 134):

---

[1] A trade union in the UK founded in 1918. The union largely represented scientific, laboratory, and technical workers in universities, the National Health Service and in chemical and metal manufacturing [2].

At the ASA first general meeting on June 15, 1946, at the University of Birmingham, Rudolf Peierls along with Oliphant, Blackett, Rotblat, Kurti and others, were elected as members of the ASA Council. "Peierls then served as one of the ASA's two executive vice-presidents. At the General Meeting in Oxford in late June 1948, Peierls was elected president of the ASA. He was re-elected twice, and stayed in office until October 1950 when he became one of the organization's several vice-presidents. So significant was Peierls' role as a leading functionary in the Association that it prompted the Central Committee of the Soviet Communist Party to gather and compile information on him."

At its peak, the ASA had about 140 full members and around 500 associate memberships. By that time the ASA had become the major forum in the United Kingdom to educate members of the general public about the benefits and perils of atomic power in what they believed was a politically "objective" way.

In addition to his ASA activities — especially regarding his favorite "child," international control of nuclear energy — Rudolf Peierls was a prolific contributor to the *Bulletin of the Atomic Scientists*, and, in the UK, the Atomic Science section in the *New Scientist*.[2]

*****

Here I would like to make a pause for a digression emphasizing two points which seem important to me.

First, in the 1940s–50s, Peierls' efforts at arms control were not always shared by his American friends, as the reader will see from two letters (from George Placzek, 1947, and from Hans Bethe, 1948) at the end of this chapter. Being sympathetic to the general idea, they expressed doubts as to the possibility of advances at that time. For instance, Bethe wrote to Peierls about his "conviction of the futility of these endeavors at the present time." Moreover, on September 6, 1950, Robert Oppenheimer noted in his letter[3]

> I cannot, of course, hold very high hopes for the outcome of even the most earnest collective effort with regard to the problems of peace and atomic control [...].

---

[2] *New Scientist* is a weekly English-language international science magazine, based in London, founded in 1956. Since 1996 it has also run a website.
[3] Sabine Lee, Volume II, page 246.

Second, nothing even remotely reminiscent of the ASA mission emerged in the USSR. The reason is of course obvious: the social consciousness as well as independent social movements did not and could not exist. The Communist Party and the government were all-pervasive and their control over the society was absolute. In a sense, the awakening of the social consciousness occurred only in 1967. On July 21 of this year Andrei Sakharov, one of the fathers of the Soviet hydrogen bomb, wrote a letter to the Soviet leadership in which he explained the need to "take the Americans at their word" and accept their proposal for a "bilateral rejection by the USA and the Soviet Union of the development of anti-ballistic missile defense," because otherwise an arms race in this new technology would increase the likelihood of nuclear war. He also asked permission to publish in a newspaper, a manuscript he had authored to explain the danger to humankind. The government ignored his letter and refused to let him initiate a public discussion. In May 1968, Sakharov completed the essay "Reflections on Progress, Peaceful Coexistence, and Intellectual Freedom" in which he described the anti-ballistic missile defense as a major threat which could lead to world nuclear war. After this essay was circulated in *samizdat* and then published outside the Soviet Union, Sakharov was banned from conducting any military-related research. Eventually he was exiled to the city of Gorky (currently Nizhny Novgorod) which at that time was closed to foreigners.

*****

At the 1947 ASA Council meeting, the Council members agreed that Rudolf Peierls should draft a statement on instituting a system of international arms control and circulate it to ASA Council members for feedback. After some debates the ASA Council ultimately released it to the press in January 1947, and the *Bulletin of the Atomic Scientists* reprinted it. Even the Soviet media covered it, and Andrei Gromyko made references to it in his speech to the UN Security Council (see Placzek's letter to Rudolf Peierls, page 446).

Unlike some other participants of the disarmament movement whose activities bordered on pure propaganda, Rudolf Peierls addressed extremely difficult technical issues to be solved before any progress could have been achieved in earnest. In his interview on March 5, 1986, [3] Rudolf Peierls stated:[4]

---

[4]In the same interview Peierls defends the use of the atomic bomb against Japan: "[The] air raids

> [E]verybody knows that nuclear weapons are not weapons to fight wars but are good only as a deterrent against other people using such weapons. Once you accept that, it's clear that a very limited amount of weapons is sufficient to have an effective *deterrent*.

The question of the critical (i.e. "very limited," in Peierls' language) amount of nuclear weapons sufficient for deterrence was the issue which concerned Peierls for years, see e.g. his article "Counting Weapons" in [4]. The second problem of importance was the issue of verification which was the main stumbling block for any progress.

As it is seen from Placzek's letter on page 446, at the early stages the non-governmental organizations (NGO) working on arms control in the UK and US put forward the idea of inspectors who could freely travel in "foreign" countries and even fly over them. This was more than naive. From the Soviet point of view this idea was totally unacceptable: as we remember from Chapter 11, at that time every large nuclear facility in the USSR was surrounded by the Gulag camps.

The efforts of Rudolf Peierls and his friends bore some fruits only in the Brezhnev era, toward its end,[5] when the Soviet Union started gradually losing the arms race.[6] During the Brezhnev rule the remnants of the Gulag system were dismantled and, more importantly, reliable satellite methods of verifications were developed.

By mid-1950s it became clear that international arms control could not be achieved and, moreover, an ever-increasing number of countries worldwide sought to acquire atomic power for either military or civilian purposes or both. Nuclear non-proliferation had replaced the implementation of a global control scheme as top issue on the agenda.

At the same time the global anti-nuclear movement became highly political, in sharp contradiction with the stated ASA guidelines of "objective science" so cherished by Peierls. In 1959, the ASA was formally disbanded; in fact its operations ceased in 1956, and in that same year Rudolf Peierls along with many former ASA members, joined the International Pugwash Conferences.

---

and killing of civilians were by this time commonplace. The, in fact the number of people killed and injured in Hiroshima, for example, was no greater, about the same as a typical fire raid on Tokyo. So it was not the scale of the casualties that made a difference. [...] So just as we were used to raids on Japanese and German cities, this was another one. But this was the one which had helped to finish the war."

[5] Leonid Brezhnev, General Secretary of the CPSU, died in 1982.

[6] I believe, the final blow which broke the system was Reagan's Star Wars.

In 1969, in an interview conducted by Charles Weiner for the American Institute of Physics Oral History project, Rudolf Peiels summarized the ASA activities in the late 1940s and 1950s as follows:

> [...] Discussions about international control were very academic in those days, of course, because it was fairly clear that there was not much hope of persuading the Soviet government to agree to any scheme that would have been acceptable to other people. It was also, I think, beginning to be clear that even given that, it might not be so easy to get the United States Congress to agree to a very generous scheme. But that didn't even arise. So one came down to a discussion of what might be features of a possible scheme, but without very much hope of implementing it.

Unlike the ASA, the Pugwash movement has survived to the present day.[7] As we see from the "Diary," Peierls was deeply involved in the organization of the Pugwash Conferences practically until his death, generously investing his time, effort and social skills.

Chapter 16 gives ample (and unique) evidence of Rudolf Peierls' involvement in the Pugwash movement in the 1980s–90s. Since the 1980s, the Pugwash Conferences became the leading world forum for the atomic scientists. The major goal of Pugwash was to bring together scientists from all countries East and West of the iron curtain for a productive dialog.

When I wrote "productive dialog," I realized that this commonplace assertion perplexed me. Can one speak of a dialog between NGOs from the both sides of the iron curtain if on the Eastern side NGOs did not exist as physical objects in the first place? Formally they could have been formed as a camouflage, but in fact they were under total control by the Communist officials.

"Peierls — I thought — should have been aware that on the Soviet side the Pugwash movement was just a game. The Soviet participants of the Pugwash delegation were cherrypicked (I believe) at the level of the Central Committee of the Communist Party; they had rigid instructions as to what to say and what not to say under any circumstances, and

---

[7]To my mind, though, after Peierls, it started deteriorating. The global situation has changed as a number of countries which originally were outside the "nuclear club" have developed or are trying to develop their own nuclear weapons. No adequate political response was suggested, nor does today's Pugwash movement have leaders of the caliber of Joseph Rotblat and Rudolf Peierls.

they could not deviate from the scenario given to them. In this sense, it was not the voice of the Soviet scientists at the Pugwash Conferences, but the voice of Brezhnev and his lieutenants. The Pugwash movement was a front for agents whose task was to weaken Pugwash criticism with regard to the USSR and, instead, focus on blaming exclusively the United States and the West for the absence of progress."

I addressed this question to Jo Hookway, Rudolf Peierls' daughter, and to Sabine Lee. From both I received reassuring answers. They are given below.

*****

Jo Hookway (August 6, 2017):

> I do believe that my father was aware of the limitations under which the Soviet delegation was operating but probably felt that he was able to have off-the-record conversations with many of these people, some of whom were his friends. I am sure he felt it was better to continue to engage in dialogue even under these conditions, rather than having no contact at all.

Sabine Lee (July 14, 2017):

> As to Peierls' ability to judge Pugwash for what it was, I believe that he was not entirely consistent here. He was often guided strongly by his emotional links to and recollections of inter-war Soviet physics. His friendship with colleagues such as the "Jazz band" made it difficult to accept just how all-pervasive the regime was. As such, I suspect that he may well have been slightly naive with regard to the Geneva process and also the Pugwash movement more generally. But in the mid-1950s, this may have been understandable, as there was genuine hope among many that after Stalin's death the mood would change, as indeed it did in a small way.
>
> I am sure that Peierls believed that some of the scientists were genuine in their desire to see Pugwash succeed in making the world a safer place. How realistic he would have been about just how deeply political the

Soviet appointments to the group were is difficult to say with certainty, but I strongly suspect that Genia would have had strong views on this, and she would not have held back expressing those. Also it is really important to judge this within the context of a specific time. I suspect that Peierls (and many others) would have changed their views over time, as they got to know their counterparts better and began to understand the political maneuvering better.

*****

G. Placzek to Rudolf Peierls[8]

Schenectady
March 7, 1947

Dear Peierls,

Many thanks for your detailed letter and the statement of the Council of the British Atomic Scientists, which I have read with great interest. It tends to show that, if even scientists of one nation cannot agree in the fundamentals of the matter we need not be surprised to see the politicians land in the present hopeless mess.

You are probably aware that the statement has been extensively misquoted by Mr. Gromyko in yesterday's session of the Security Council.[9] This is of course not your fault since even if it had been possible to write it from a more unified point of view, this would not have afforded protection against distortions. I hope you can get a verbatim copy of the speech, it was rather long and the papers left out the most interesting passages. I happened to listen to it on the radio and it was rather characteristic to hear Gromyko state in terms which were clearly sincere conviction that the idea of an *inspector* who could freely travel in a "foreign" country and even fly over it, could not have been meant seriously.

---

[8] Sabine Lee, Volume 2, page 74.
[9] At that time Andrei Gromyko (1909–1989) was the Soviet Permanent Representative at the UN Security Council. On March 5, 1947, Gromyko gave a speech in the Security Council denouncing American nuclear policy. See *New York Times* of March 5, 1947, page 16. From 1957 to 1985 Gromyko was the USSR Foreign Affairs Minister.

The Russians, of course, have by no means a monopoly on such a mentality; the recent statement by Senator Taft, warning against the danger that communists might "infiltrate" into an international control agency, is on a very similar level.

Enclosed [is] a pictorial record of yesterday's meeting, taken from today's *New York Times*. You will probably enjoy the subscript.

I am afraid I cannot entirely share the Olympian detachment of your American military friend who hopes that chances of an agreement might be better once stories of successful Russian bomb manufacture begin to circulate. I am rather inclined to believe the opposite. But unfortunately, we shall see.

Clayton[10] (from Chalk River) asked me for a copy of the table of the exponential integral for complex argument, so that he could forward it to you. Unfortunately I have no copy here. I believe I left one at Montreal, but whether or not this is so, they cannot find it. The rest of the copies are in Carlson's hands[11] at Los Alamos and to extricate one from there, under present circumstances, might take longer than to have the whole set of tables recalculated. Therefore I recommended to Clayton to get in touch with Lowan, (Math. Tables Project) who might have a spare copy or could perhaps get one reproduced.

I hope you will keep me informed on the result of your and Skyrme's investigations re Wigner Dispersion formula, and gauge invariance of Weisskopf's logarithmic divergence.

Newspaper reports here inspired, in some credulous souls, the hope that Schrödinger might have discovered the laws of the universe. Mr. W.L. Laurence[12] was speeded into action. Jumping at the op portunity to act as a great patron saint of science, he got hold of Schrödinger's manuscript, had it photostated at the *Times'* expense, and distributed it far and wide throughout the country. Of course, it turned out to be bunk. A lesson for the above-mentioned souls showing that the more traditional channels of scientific information still seem to be fully adequate.

---

[10] Henry H. Clayton (1906–1989), was an English-born physicist who had studied at the University of British Columbia. In 1945 he joined the Montreal Laboratory and moved to Chalk River in 1946. In 1950 he became head of the theoretical physic branch, a position held until his retirement in 1969. - SL

[11] Bengt Carlson (1915–2007), was a Swedish-born physicist who had studied at Stockholm and Yale. He worked in Placzek's group at Montreal from April 1943. -SL

[12] William Leonard Laurence (1888–1977) was a well-known science journalist working for *The New York Times* in the 1940s and 1950s.

With best regards to you and Genia,

Yours sincerely, G. Placzek

❧

Hans Bethe to Rudolf Peierls[13]

Ithaca
April, 1948

Dear Rudy,

Thanks for your letter of March 23. I am rather sad that the conference cannot be moved. The starting of American Universities has always been around September 20. The trouble was only that Uncle Nick has always disregarded university schedules and chooses the dates for his Copenhagen Conference at random.

Some visitors, like Vicky, were too anxious to go to the grand re-opening of the Copenhagen meetings[14] last year and therefore did not return in spite of the semester. (I appreciate that the Bristol meeting was the first to get fixed and I protested immediately to Powell about its date, but without success.)

I have pretty much decided to go to your conference but none of the later ones.

I hope that I will hear everything in Birmingham then and earlier in the summer.

The meeting sponsored by the American Emergency Committee[15] is not likely to serve a useful purpose. It was Szilard's idea to have such a meeting to bring together scientists from all countries east and west of the iron curtain. Those from the east of the curtain have already refused to attend, as was to be expected. In the present situation, it is my opinion, that nothing useful can be done from the side of atomic energy or of scientists in general and that any stress on atomic energy

---

[13] Sabine Lee, Volume II, page 134.
[14] The conference had attracted a large number of physicists who spent some time in Copenhagen during the last two weeks of September 1947. Among them were Kramers, Weisskopf, Pais, Rosenfeld, Peierls, Blackett, Placzek, Wheeler, Rossi, Ferretti, and Klein. See Wolfgang Pauli's letter to Otto Stern, August 19, 1947, in *Wolfgang Pauli. Wissenschaftlicher Briefwechsel mit Bohr, Einstein, Heisenberg u.a.*, Ed. K.V. Meyenn, Vol. 3: 1940–1949, (Heidelberg: Springer, 1993), page 471.
[15] In 1948 the American Emergency Committee of Atomic Scientists was headed by Albert Einstein.

can only deepen the international conflict. I was very much against holding this conference and this opinion is shared by other people such as Oppenheimer and Weisskopf.

By the way, I resigned from the Emergency Committee, not because of the conference but because of my conviction of the futility of these endeavors at the present time. I have also heard some rumors that the Federation of American Scientists is giving up the idea of the International Conference. This seems sensible to me, but of course the Emergency Committee is an independent agency. Anyway, I would much rather talk physics with you than have you go to that conference.

Thanks very much for the parliamentary debates of the House of Lords.[16] I am of course very proud.

Yours sincerely,

Hans

References

[1] Isidor Isaac Rabi, *Science: The Center of Culture*, (World Publishing, New York and Cleveland, 1970), page 114.
[2] Christoph Laucht, *Elemental Germans: Klaus Fuchs, Rudolf Peierls and the Making of British Nuclear Culture, 1939–59*, (Palgrave Macmillan, 2012).
[3] "War and Peace in the Nuclear Age," aired by PBS in 1989, http://openvault.wgbh.org/catalog/V_28F1C920696B486F8796D1E449FF8391
[4] R. Peierls, *London Review of Books*, Vol. 3, No. 4–5, (1981), page 16.

---

[16] Almost certainly, this refers to the parliamentary debate on the perils of atomic warfare launched by the Archbishop of York, February 18, 1948, *Hansard*, HoL, Vol. 153, cols 1178–1213. -Sabine Lee

Index

Abrikosov, Alexei, 376, 384, 396
Akers, Wallace, 339
Akhmatova, Anna, 55, 159
Aldanov, Mark, 167
Alexander, Natalia, 162
Allibone, Thomas, 423
Amaldi, Edoardo, 366, 389, 403
Amati, Daniele, 385
Ambartsumyan, Victor, 58, 71, 93
Anderson, Philip, 401
Andronikashvili, Elephter, 130
Andronikashvili, Elevter, 363
Andronikov, Iraklii, 130
Anselm, Alexei, 374, 375
Anselm, Andrei, 49, 130
Aragon, Louis, 269
Argo, Abram Markovich, 129
Arnold, Henry, 249
Arnold, Katya, 163
Arnold, Lorna, 402
Arnold, Vera Stepanovna, 163
Arnold, Vladimir, 111, 388
Artsimovich, Nelly, 384
Artuzov, Artur, 148
Atiyah, Michael, 403, 425
Attlee, Clement, 238

Bagge, Erich, 431
Balzac, de, Honoré, 49
Barbier, Henri, 115
Barwich, Heinz, 249
Bassis, Henri, 269
Beauvoir, de, Simone, 267
Bell, John, 24, 344
Belousov, Natasha, 218, 219, 368, 377

Beria, Lavrentiy, 271, 272
Berlin, Isaiah, 359, 390, 423
Berman, Morris, 204
Bernal, John Desmond, 204
Bernhardt, Karl, 184
Bernstein, Jeremy, 415, 421, 430, 432, 434
Bethe, Hans, 10, 12, 176, 178, 179, 183,
    184, 196, 200, 201, 203, 214, 234, 327,
    332, 343, 352, 369, 403, 423, 440, 441,
    448
Bethe, Rose, 179, 191, 214, 403, 423
Beurton, Ursula, 232
Blackett, Patrick, 333, 439, 448
Blaton, Jan, 342
Bloch, Claude, 344
Bloch, Felix, 10, 73, 78, 98, 200, 329
Bogdanova-Belsky, Pallada, 168
Bogolyubov, Nikolai, Jr., 371
Bogolyubov, Nikolai, Sr., 371
Bohr, Margrethe, 436
Bohr, Niels, 16, 18, 108, 189, 200, 206,
    303, 325, 327, 333–335, 338, 346, 352,
    364, 429, 436
Born, Max, 54, 230, 233
Bose, Satyendra, 54
Bothe, Walther, 39, 54, 95, 326, 430, 435
Botticelli, Sandro, 82
Bragg, Lawrence, 12, 332, 333
Breit, Gregory, 3, 344
Bretscher, Hanni, 178, 191, 192, 222
Brillouin, Léon, 9, 327
Brink, David, 378, 381
Brink, Verena, 381
Broglie, de, Louis, 327
Bronshtein, Matvei, 49, 51, 53, 58, 60, 64,

67, 71, 75, 77, 82, 93, 95, 109, 130, 330
Brown, Gerry, 24, 201, 210, 212–214, 344, 382
Buber-Neumann, Margarete, 267, 268
Bukharin, Nikolai, 7, 394
Burde, Fritz, 265
Bursian, Victor, 95
Buter, Stuart, 211
Byers, Nina, 24, 344, 398, 399, 419

Calaprice, Alice, 400
Calogero, Francesco, 406
Carlson, Bengt, 447
Carr, Alan, 188
Carter, Peter, 206
Casimir, Hendrik, 200
Cassidy, David C., 405, 406, 433–435
Cassin, Lydia, 384
Cathcart, Brian, 29
Chadwick, James, 11, 20, 234, 332, 339, 439
Cherenkov, Pavel, 39
Churchill, Winston, 271, 339
Clayton, Henry, 447
Cockcroft, John, 19, 224, 338
Compton, Arthur, 21
Condon, Edward, 327
Crowther, James Gerald, 204
Curtade, Pierre, 269
Curtis-Bennett, Derek, 247

Dahrendorf, Rolf, 391
Daix, Pierre, 269
Dalitz, Richard, 24, 213, 344, 367, 369, 391, 399, 427
Davidson, Linda, 391, 402, 426
Deacon, Richard, 320
Delbrück, Max, 153
Diebner, Kurt, 431
Dirac (Wigner), Margit, 178
Dirac, Paul, 11, 200, 223, 327, 367, 391
Doll, Richard, 368, 396
Dorfman, Yakov, 95, 109
Drell, Sidney, 378
Duckwitz, Georg, 437
Dvoretz, Lev, 284
Dyson, Freeman, 24, 201, 209, 212, 213, 343, 350

Eckart, Carl, 327

Edwards, Samuel, 344
Ehrenburg, Ilya, 266
Eichenbaum, Boris, 117, 123, 125, 128
Einstein, Albert, 206, 434, 440, 448
Esenin, Sergei, 167
Euler, Hans, 335
Ewald, Paul, 327

Farrell, Nicholas, 320
Feklisov, Alexander, 227, 232, 238, 246, 251
Fenigstein, Lilly, 71, 73
Fermi, Enrico, 189, 196, 331, 414, 430, 435
Fermi, Laura, 189, 196
Fermi, Nella, 193, 194, 196
Fermi, Rachel, 415
Ferretti, Bruno, 448
Feuchtwanger, Leon, 263
Feynman, Richard, 211, 220, 228, 343, 427
Field, Gerry, 427
Flanders, Donald, 196
Flanders, Jay, 196
Flowers, Brian, 24, 344
Fock, Vladimir, 56, 93, 95, 108
Fowler, Ralph, 333
Fox, Alfie, 370
Fröhlich, Herbert, 201, 327
Franck, James, 439
Frenkel, Sara Isaakovna, 64, 69, 93, 108
Frenkel, Viktor, 373, 375, 385
Frenkel, Yakov, 39–41, 44, 95, 421, 424
Freud, Sigmund, 75
Frisch, Otto, 2, 10, 18, 192, 201, 224, 315, 325, 338, 339, 429, 430
Fuchs, Klaus, 20, 59, 192, 201, 210, 220, 221, 227, 231–239, 241, 244, 246, 249, 251, 252, 255–257, 265, 266, 303, 304, 308, 311, 315, 339, 342, 343, 347, 399
Fulda, Ludwig, 326

Gabeler, Eva, 104
Gamow, George, 7, 9, 40, 42, 49, 51, 52, 55, 57, 71, 72, 83, 90, 93, 95, 130, 200, 233, 330
Ganz, Peter, 413
Geballe, Ted, 378
Geballe, Tom, 418
Geiger, Hans Wilhelm, 100
Gerken, Evgeny, 129
Gerlach, Walther, 431

Gibbs, Josiah Willard, 50
Ginzburg, Vitaly, 1
Goebbels, Joseph, 393
Gold, Harry, 220, 233–235, 246, 247
Goldansky, Vitaly, 393
Goldhaber, Maurice, 335
Goldsmith, Hyman, 440
Gorbachev, Mikhail, 7, 348, 377, 378, 383, 391, 393
Gorelik, Gennady, 51
Gorkov, Lev, 375, 376
Gorlova, Zinaida, 266
Gornfeld, Arkady, 117
Gottfried, Kurt, 403
Goudsmit, Samuel, 421, 434
Gowing, Margaret, 402, 410, 413
Greenberg, Wallace, 422
Grenier, Fernand, 267
Griffin, Maulde, 369, 378, 396, 412, 419, 420, 423
Gromyko, Andrei, 442, 446
Gross, Charlie, 175
Gross, Gaby, 4, 135, 163
Groves, Leslie, 185, 194, 234
Guber, Pyotr, 117
Gul, Roman, 166
Gumilev, Nikolai, 52, 53, 55, 159
Gurvich, Alexander Gavrilovich, 125, 157, 161
Gurvich, Anna, 130
Gurvich, Natalla, 130
Guzenko, Igor, 238

Hahn, Otto, 18, 338, 410, 431
Halban, Els, 423
Halban, Maulde, 369
Hall, Edwin, 328
Hamburger, Ursula, 232
Haskell, Francis, 410
Hawking, Stephen, 388
Heifets, Fanny Aronovna, 284
Heisenberg, Werner, 2, 54, 86, 87, 325, 327, 329, 335, 352, 405, 410, 414, 417, 429–434, 436
Heitler, Walter, 327, 335
Hellman, Hans, 9
Hemingway, Ernest, 81
Herz, Gustav, 69
Herzfeld, Karl, 327
Hildebrandt, Olga, 168

Himmler, Heinrich, 434
Hitler, Adolf, 2, 11, 183, 209, 241, 249, 263, 264, 417, 433
Hoffer-Jensen, S., 335
Hookway, Joanna, 4, 32, 205, 445
Houston, William, 327
Houtermans, Charlotte, 69
Houtermans, Friedrich (Fritz), 25, 39, 41, 69, 79
Hund, Friedrich, 335

Intrilligator, Kenneth, 398
Ioffe, Abram Fyodorovich, 130
Ioffe, B.L., 1, 20, 148, 229, 272
Ioffe, Valentina, 130, 133
Isaacóvich, Mikhail Aleksandrovich, 164
Ising, Ernst, 333
Ivanenko, Dmitri, 49, 51, 58, 60, 69, 81, 88, 89, 93, 130, 233, 330

Jackson, Barbara, 409
Jackson, John David, 409
Jacobsen, Jacob, 335
Jaroš, Otakar, 279
Jephcott, C.M., 184
Jephcott, Isabel, 184, 365, 403
Joffé, Roland, 397
Joliot-Curie, Frédéric, 267, 269
Jordan, Pascual, 93, 327
Jungk, Robert, 429, 434, 436

Kaempfer, Walter, 434
Kaganov, Moisei, 372
Kannegirser, Maria Abramovna, 130, 218
Kannegiser (Saker), Rosa, 168
Kannegiser, Elizabeth, 167
Kannegiser, Genia, 1, 3, 9, 40, 44, 49, 50, 52, 54, 58, 61, 330
Kannegiser, Ioakim Samuilovich, 116, 155, 167, 168
Kannegiser, Leonid, 111, 116, 119, 166–169, 315
Kannegiser, Maria Abramovna, 114, 134, 144, 218
Kannegiser, Nikolai Samuilovich, 111, 112, 114, 168
Kannegiser, Nina, 13, 15, 54, 58, 60, 67, 75, 77, 81, 82, 89, 112, 133, 144, 155, 158, 159, 161–164, 175, 218, 364
Kannegiser, Samuil, 111

Kannegiser, Sergei, 168
Kapitza, Pyotr, 8, 9, 11, 13, 14, 60, 271, 332, 376, 377
Kapitza, Sergei, 370
Kellerman, Bernhard, 49
Khalatnikov, Isaak, 372, 373, 377, 384, 385
Khariton, Yuli, 294
Kharitonova, Anastasia Mikhailovna, 121, 123, 144
Khrushchev, Nikita, 132
Kiebel, Ilya, 71
King, Truby, 12
Kirillov, Zhenya, 288, 289
Kisling, Moïse, 68
Kittel, Charles, 14
Kleiber, Erich, 85
Klein, Oskar, 448
Klemperer, Otto, 85
Kochankov, Leonid, 228, 229, 251
Kohn, Walter, 384
Kolhörster, Werner, 75
Konysheva, Natta, 395
Kopfermann, Hans, 335
Korsching, Horst, 431
Kossel, Walther, 327
Kozintsev, Grigori, 158
Kozyrev, Nikolai, 130
Kramers, Hans, 448
Kravchenko, Victor, 266
Kremer, Simon, 230, 232
Krutkow, Yuri, 95
Kubo, Ryogo, 392
Kuczynski, Jürgen, 230, 231
Kuczynski, Ursula, 231, 232, 399
Kurchatov, Igor, 251, 271, 272
Kurtada, Pierre, 267
Kurti, Nicholas, 359, 410, 412, 413, 439
Kuzin, Boris, 158, 159
Kuzmin, Mikhail, 117, 155, 157
Kvasnikov, Leonid, 228

Ladenburg, Rudolf, 335
Landé, Alfred, 327
Landau, Lev, 1, 9, 13, 14, 49, 51, 58–60, 73, 89, 93, 95, 98–102, 107, 109, 130, 146, 147, 200, 233, 256, 329–331, 346, 376, 387
Langer, James, 24, 213
Langer, Jim, 344

Laporte, Otto, 327
Laucht, Christoph, 439
Laue, von, Max, 327, 410, 431, 434
Laurence, William, 447
Lee, Sabine, 32, 43, 61, 445
Lenin, Vladimir, 163, 232
Lenz, Wilhelm, 10, 327
Levin, Boris Danilovich, 123
Levin, David, 111
Levin, Maria Abramovna, 68, 111, 112
Levinson, Evgeniy Maksimilianovich, 120, 121
Levinson, Natalya Maksimilianovna, 120, 121
Lieb, Elliott, 213, 344
Lifshitz, Benedikt, 117, 130
Lifshitz, Evgeny, 346
Liliencron, von, Detlev, 115
Lindhard, Jens, 211, 212
Loevberg, Maria, 117
London, Fritz, 327
Lorentz, Hendrik, 49

Mandelshtam, Benedikt Yemelyanovich, 112
Mandelshtam, Isai Benediktovich, 13, 49, 50, 54, 112, 113, 123, 125, 127, 128, 131, 144, 159, 167, 218, 219
Mandelshtam, Leonid, 68, 151
Mandelshtam, Maria Abramovna, 119
Mandelshtam, Max, 112
Mandelshtam, Osip, 9, 28, 111, 115, 158, 159
Mandelshtam, Sergei, 68, 72
Mandelstam, Stanley, 24, 213, 344, 409
Manley, John, 191, 234
Manteith, James, 54, 169
Mark, Kathleen, 191
Marshak, Ruth, 191
Marshall, Walter, 391
Marx, Karl, 232
Masaryk, Jan, 257
Masereel, Frans, 68
Mason, Katrina, 193, 194, 196, 241
Matarasso, Leo, 266, 267
Matthews, Paul, 344
Maupassant, de, Guy, 76, 80
Maurois, André, 81
May, Alan Nunn, 306, 308, 311
McCarthy, Joseph, 394

McKibbin, Dorothy, 186
McNamara, Robert, 403
Meitner, Lise, 18, 100, 335, 338
Mercader, Ramón, 265
Meshcheriakov, M.G., 273
Migdal, Arkady, 381, 384, 401
Mironova, Olga Nikitishna, 284
Mises, von, Richard, 39
Molotov, Vyacheslav, 261, 271
Mondragón, Alfonso, 402
Mondragón, Myriam, 415
Moon, Winifred, 191, 192, 222
Moskovskaya, Fania, 62, 97–99, 102
Moss, Norman, 221, 227
Mott, Nevill, 232
Moynet, André-Remy, 267
Mussolini, Benito, 332
Møller, Christian, 335

Nabarro, Frank, 211
Neéman, Yuval, 382
Negin, Evgeny, 294
Nernst, Walther, 326
Neumann, Heinz, 267
Newton, Isaac, 391
Nikolayev Nikolai, 400
Noether, Fritz, 9
Nordmann, Joe, 266
Nosik, Boris, 267
Nozières, Philippe , 376

Odoevtseva, Irina, 159
Oliphant, Mark, 19, 338, 339, 398, 439
Oppenheimer, Robert, 10, 22, 185, 188, 194, 211, 234, 251, 343, 345, 423, 433, 441, 448
Ostrovsky, Alexander, 294

Píka, Heliodor, 281, 282
Pais, Abraham, 448
Panfilov, Alexei, 231
Panofsky, Wolfgang, 389
Papaleksi, Klara Efimovna, 163
Pasternak, Boris, 159
Pauli, Wolfgang, 2, 39–41, 54, 61, 73, 77, 79, 80, 86, 95, 100, 102, 152, 200, 223, 325, 327, 329, 332, 335, 352, 448
Pauling, Linus, 327
Pease, Sebastian "Bas", 413
Pease, Susan, 413

Peccei, Roberto, 398
Peierls, Alfred, 387, 397
Peierls, Elisabeth, 326
Peierls, Else, 326, 337
Peierls, Gaby, 12, 175, 193–195, 197, 203, 241
Peierls, Heinrich, 43, 326, 337
Peierls, Ilse, 150
Peierls, Joanna, 175
Peierls, Kitty, 175
Peierls, Nina, 65, 70
Peierls, Ronald, 165, 203, 214, 241
Penney, William, 224, 345
Pereltsveig, Vladimir, 167
Peters, Bernard, 224
Petrarca, Francesco, 115
Picasso, Pablo, 68, 269
Pines, David, 414
Placzek, Els, 191, 192, 222, 378, 396, 412, 419
Placzek, George, 10, 147, 256, 333–335, 342, 352, 441–443, 446–448
Planck, Max, 326, 433
Pomerants, Grigory, 257
Pontryagin, Lev Semyonovich, 135
Powell, Cecil, 350, 448
Powell, Enoch, 370
Powers, Thomas, 420, 421, 429, 436
Priestley, Raymond, 341
Providência, da, João, 217, 377, 378, 387, 412, 419
Pudovkin, Vsevolod, 158

Rabi, Isidor, 439
Rabinowitch, Eugene, 439, 440
Radicati, Luigi, 404, 405
Raison, Maxwell, 199
Ramsauer, Carl, 39
Raninowich, Victor, 416
Rasputin, Valentin, 394
Ravenhall, Geoff, 213
Rayleigh, John William, 55
Renoir, Pierre-Auguste, 68
Rerikh, Georgy, 291
Riefenstahl, Charlotte, 41
Robeson, Paul, 269
Roosevelt, Franklin, 339
Rosenberg, Alfred, 288
Rosenberg, Ethel, 249
Rosenberg, Julius, 249

Rosenfeld, Léon, 83, 200, 448
Rosenfeld, Yvonne, 382
Rossi, Bruno, 448
Rotblat, Joseph, 205, 350, 420, 439
Rubinowicz, Wojciech, 327
Rudenko, Sergei, 266
Rudolph, Heinz, 417, 419
Ruhemann, Barbara, 65
Rutherford, Ernest, 13, 200, 333

Sack, Heinrich, 69, 73, 77, 81, 82, 85, 95
Sadoul, Georges, 269
Sagdeev, Roald, 388, 389, 412
Sakharov, Andrei, 1, 251, 290, 300, 376, 383, 388, 390, 392, 394, 442
Salam, Abdus, 371, 419
Salpeter, Edwin, 201, 344
Sanderson, A. C., 184
Saypol, Irving, 249
Schlippe, von, Wladimir, 273
Schrödinger, Erwin, 327, 447
Schwinger, Julian, 343
Segré, Emilio, 189, 335
Seligman (DeWitt), Bryce, 224
Serber, Robert, 234
Sharpe, Tom, 220, 221, 227
Shawcross, Hartley, 247, 249, 252
Shevardnadze, Edward, 378
Shiryaeva, Olga, 273, 301
Shoenberg, David, 14, 256, 333
Sikorski, Władysław, 275
Simon, Charlotte, 402, 404, 407
Simon, Franz (Francis), 39, 64, 70, 80, 95, 108, 339, 411
Simpson, Esther, 366
Singwi, Helga, 403
Singwi, Kundan, 392
Skardon, William, 238, 240, 251, 252
Skinner, Herbert, 399
Skriabin, Aleksandr, 85
Skyrme, Tony, 211, 234, 369, 447
Smirnov, Alexander, 118
Smith, Alice, 191
Sokolskaya, Irina, 49, 54, 130
Solzhenitsyn, Alexander, 266
Sommerfeld, Arnold, 39, 325–327, 329, 352
Stalin, Joseph, 2, 9, 60, 218, 241, 249, 256, 257, 260–263, 265, 266, 271, 282, 284, 315, 332, 392, 445

Stark, Johannes, 434
Stein, Norman, 412
Stern, Otto, 332, 335, 448
Stevenson, William, 206
Strassmann, Fritz, 18, 338
Styrikovich, Mikhail, 13, 376
Sucher, Dorothy, 386
Sucher, Joseph, 386
Svoboda, Ludvik, 275, 277, 279–282, 292
Swiatecki, Wladek, 211, 213
Szasz, Ferenc, 190, 192, 221
Szilard, Leo, 430, 439, 440, 448

Tadić, Dubravko, 400
Taft, Robert, Sr., 446
Tagore, Rabindranath, 68
Tamm, Igor, 39–41
Taylor, John, 419
Teller, Edward, 10, 200, 221, 369, 370, 423
Teller, Mici, 221
Tharp, Robert, 316
Thomson, George, 19
Thorner, Hans, 65, 72, 90, 99, 104, 105
Thorner, Ilse, 153
Titterton, Peggy, 191
Tizard, Henry, 19
Tomonaga, Sin-Itiro, 343
Trauberg, Leonid, 158
Trotsky, Leon, 265
Trueman, Harry, 271
Tsvelik, Alexei, 359, 422
Tsvelik, Elena, 422
Tsvetaeva, Marina, 116, 168
Tuck, Elsie, 191
Tuck, James, 191
Turski, Cesar, 288

Ulmann, André, 267
Unna, Issachar, 215
Uritsky, Moisei, 166, 167

Vavilov, Nikolai, 284
Vavilov, Sergei, 284
Verbenski, Bozhena, 278
Verblovskaya, Masha, 17, 111, 112, 124, 130, 159, 203, 207, 219, 383, 385, 386, 400, 402, 404, 406, 409, 411, 418
Verblovsky, Yuri, 383, 404, 406, 409, 411, 418
Vokhmintseva, Lyubov, 7

Volodarskaya, L.I., 112, 218
Voznesensky, Andrei, 388
Vraskaya, Olga, 131
Vraskiy, Stepan Borisovich, 133
Vyshinsky, Andrei, 232, 252

Wallace, Edgar, 83
Wallenberg, Raoul, 257
Waller, Ivar, 86, 109
Walter, Bruno, 85
Warren, Stafford, 194
Weiner, Charles, 4, 83, 145, 203, 204, 443
Weissberg, Alexander, 265
Weisskopf, Victor, 10, 147, 200, 213, 335, 383, 434, 440, 448
Weizsäcker, von, Carl, 335, 395, 410, 431, 432, 435–437
Wells, Herbert, 101
Wentzel, Gregor, 102, 327
Werner, Ruth, 232
West, Nigel, 320
Wheeler, John, 448
Wick, Giancarlo, 365, 370

Wiegmann, Paul, 373–375, 377, 384
Wigner, Eugene, 447
Wildgans, Anton, 81
Wilkinson, Denys, 208, 330, 344
Wilkinson, Helen, 208, 330
Wilson, Jane, 191
Wirtz, Karl, 430, 431
Woodhouse, P.G., 97
Wooster, W.A., 178

Yang, C.N., 226
Yeltsin, Boris, 426
Yurkun, Yuri, 117, 155, 157

Zababakhin, Yevgeny, 294
Zaltsman, Lotte, 159
Zaltsman, Pavel, 157–160
Zarniko, Barbara, 65
Zeldovich, Yakov, 1, 290, 293, 294, 301, 378, 392
Zernov, Pavel, 291, 292
Zimin, Georgy Aleksandrovich, 291

www.ingramcontent.com/pod-product-compliance
Lightning Source LLC
Chambersburg PA
CBHW060453300426
44113CB00016B/2570